Historical Geography and Geosciences

This book series serves as a broad platform for contributions in the field of Historical Geography and related Geoscience areas. The series welcomes proposals on the history and dynamics of place and space and their influence on past, present and future geographies including historical GIS, cartography and mapping, climatology, climate history, meteorology and atmospheric sciences, environmental geography, hydrology, geology, oceanography, water management, instrumentation, geographical traditions, historical geography of urban areas, settlements and landscapes, historical regional studies, history of geography and historic geographers and geoscientists among other topically related areas and other interdisciplinary approaches. Contributions on past (extreme) weather events or natural disasters including regional and global reanalysis studies also fit into the series.

Publishing a broad portfolio of peer-reviewed scientific books Historical Geography and Geosciences contains research monographs, edited volumes, advanced and undergraduate level textbooks, as well as conference proceedings. This series appeals to scientists, practitioners and students in the fields of geography and history as well as related disciplines, with exceptional titles that are attractive to a popular science audience.

If you are interested in contributing to this book series, please contact the Publisher.

More information about this series at http://www.springer.com/series/15611

P. J. Capelotti
Editor

The Coldest Coast

The 1873 Leigh Smith Expedition to Svalbard in the Diaries and Photographs of Herbert Chermside

Springer

Editor
P. J. Capelotti
Division of Social Sciences
Penn State Abington
Abington, PA, USA

ISSN 2520-1379 ISSN 2520-1387 (electronic)
Historical Geography and Geosciences
ISBN 978-3-030-67879-1 ISBN 978-3-030-67880-7 (eBook)
https://doi.org/10.1007/978-3-030-67880-7

Published in association with the Grenna Museum.

This Springer imprint is published by the registered company Springer Nature Switzerland AG
The registered company address is: Gewerbestrasse 11, 6330 Cham, Switzerland

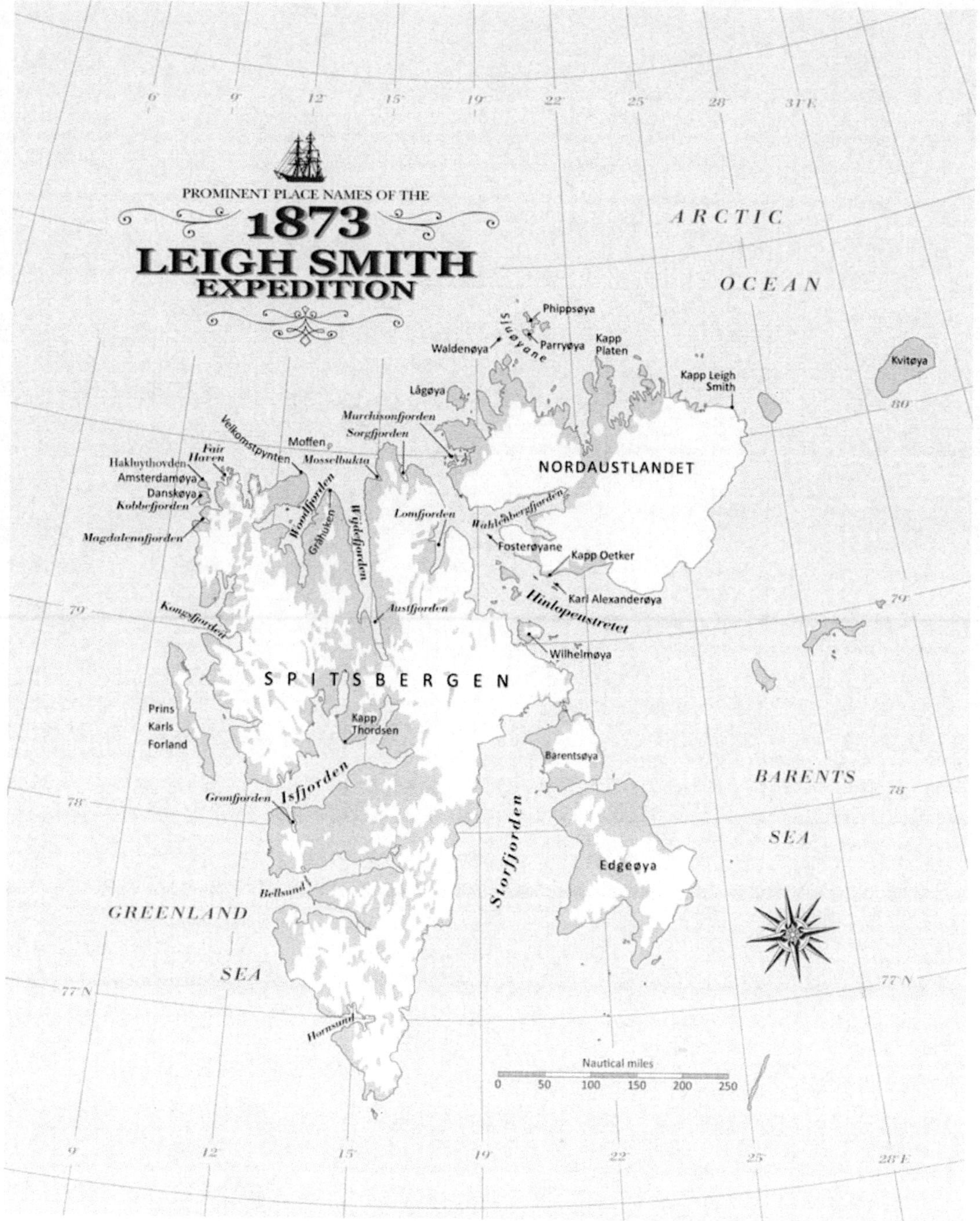

Svalbard/Spitsbergen with prominent place names from the 1873 Leigh Smith Expedition. Map created by As the Crow Flies Cartography, Canada. © P.J. Capelotti

Foreword by Charlotte Moore

"He has personal *influence* that subtle thing that makes a person a power without even uttering a word."

This was how Benjamin Leigh Smith's sister Nannie described her older brother. Her observation is borne out by the facts. Without that "influence," or charisma, Leigh Smith could not have conducted his four successful Arctic expeditions, let alone guided the entire crew to safety when on the fifth and last expedition his ship sank, leaving them stranded in Franz Joseph Land facing the prospect of a long Arctic winter with no means of communicating news of the disaster to friends at home.

I am Benjamin Leigh Smith's great-great-great niece. "Uncle Ben" is both hero and villain in our family. Hero, because, as his oaken grave marker puts it, "five times he sailed the Arctic seas and for his country's good braved countless hardships and perils." Villain, because that powerful "personal influence" was not always put to good use. After the death of his father in 1860, Ben became head of the family. In this role, he exercised excessive control over the professional and personal lives of his nephews and nieces. His possessive affection for his beautiful niece Amy (my great-grandmother) was so strong that he blocked her marriage for more than four years. Eventually, she succeeded in marrying the man she loved, but it was another 20 years before Ben would abandon the feud and re-establish contact with her.

Uncle Ben was a complex character with an unusual background. His father, Benjamin Smith, was a radical member of the British House of Commons, one of the ten children of William Smith MP who had been a close associate of Wilberforce and Clarkson in the campaign against the slave trade. The Smiths were Unitarians. This variant of the Christian faith upholds "Freedom, Reason and Tolerance," and puts emphasis on deeds, not words. Philanthropy, altruism and a belief in social progress are all essential. Unitarians reject belief in the Holy Trinity, and regard Jesus as a man to be followed rather than a God to be worshipped. They have strong family and friendship networks and are involved in each other's lives to an unusual extent. The wealthy and cultured Smith family spread its influence widely in nineteenth-century Britain; Florence Nightingale, pioneer of hospital and army reform, is its best-known member. She was Uncle Ben's first cousin.

Benjamin Smith senior rejected marriage as a restrictive institution within which the wife became little more than one of her husband's chattels. On a visit to his Nightingale cousins in Derbyshire, he fell in love with a working-class girl, a milliner named Anne Longden. When she became pregnant, he set her up in a house in Whatlington, East Sussex (he owned much property in the area) and treated her as his common-law wife. The first child, Barbara, was born there in 1827, followed by Benjamin in 1828. There were three more children–Nannie, Bella and Willy–before Anne Longden died of lung disease at the age of 32.

The bereaved father enlisted the help of Anne's sister, "Aunt Dolly Longden," and his own sister, "Aunt Ju," in looking after the five children, who therefore had the interesting and unusual experience of being reared by two surrogate mothers, one upper class, one working class. The family took up residence in two houses in Pelham Crescent, Hastings, one house for the adults, one for the children. Pelham Crescent was an elegant newly built terrace overlooking the English Channel. I imagine that this proximity to the sea–during storms, waves would crash within feet of the houses–gave young Ben his taste for ocean adventure. An interest in travel was encouraged by his father, who commissioned an enormous horse-drawn coach in which the entire family journeyed through Europe and the British Isles. Mountains were climbed, rivers were swum in, landscapes were sketched, plants were collected. The Leigh Smith children (their father added the family name "Leigh" to distinguish them from the mass of their Smith cousins) had an unusually free and adventurous upbringing. Their schooling was similarly unconventional. Ben attended Bruce Castle, a progressive school in Tottenham founded by the utilitarian philosopher Jeremy Bentham, where science and engineering were taught at a time when the English public school syllabus was dominated by Latin and Greek. There was no corporal punishment. Outdoors activities were encouraged; there was a swimming pool. Ben's headmaster was Rowland Hill, inventor of the postage stamp.

Ben went on to Jesus College, Cambridge, where he studied mathematics. He was already a proficient yachtsman and an excellent shot, and though, after Cambridge, he began reading for the Bar, he was never cut out for an indoor career. His restlessness was solved when the death of his father left him with a considerable inheritance. (The discovery at their father's death of a second illegitimate family, raised in London, was a shock to the Leigh Smiths; nevertheless, they buried old Ben next to their own mother Anne). The 32-year-old inheritor could now abandon his fledgling legal career and take over the management of his Sussex estates; all his life, Ben was deeply interested in farming, especially hop-growing, and in rural pursuits such as shooting and beagling. Family life took up much of his time. Letters and diaries from the family archive show him taking his nephews and nieces to London theatres and trips to Paris, holding parties so they could get to know their multitudinous Smith cousins, providing party frocks for the girls, and sporting equipment for the boys. He took his turn with family difficulties, too, sitting up all night with his sister Bella when she suffered psychotic episodes, bringing toys and invalid foods to the sickbed of a 4-year-old victim of diphtheria, lending houses to invalids in need of country air.

In 1865, another relative bequeathed Ben a handsome fortune. He was now rich enough to launch himself as an explorer. He had already obtained the Board of Trade "ticket" enabling him to command his own ship, and had travelled extensively in Europe, Egypt and the British Isles. By 1870 his thoughts were turning to the Arctic and the lure of unchartered territory. On 28 April 1871 Ben's brother-in-law and friend General John Ludlow noted in his diary, "Ben has bought a big & stout yacht without having seen her. She has been cased in the North Sea & was built for going among the ice. Ben says he is 'going to the North Pole' ... his object is sport." Sport, of course, meant shooting bears, reindeer, seals and seabirds, but if this was the main objective of Ben's first voyage, his scientific and enquiring mind soon realised that there were more interesting things to do in the Polar regions. On each of his subsequent voyages, Ben pushed forward the frontiers of knowledge, as this book amply demonstrates.

The three voyages of the 1870s increased the already high standing of "Uncle Ben" in the eyes of the family. He came home bearing gifts of haunches of reindeer venison, the skins of reindeer, seals and Polar bears, and a wolverine tippet and muff for his little niece Milicent. The children could visit the live bear, Samson, donated to London Zoo, and revel in the tale of how, on the return journey, Samson had made a break for freedom, but Uncle Ben had grabbed him and wrestled him to the deck as he was about to dive into the sea. They pored over Ben's sketches and Lieutenant Chermside's thrilling photographs, quite unlike anything they had seen before. After the 1873 relief of the Swedish expedition, they gloried in their uncle's heroism and fame. The extended family had gathered at General Ludlow's large house and the children were staging some amateur dramatics when, as young Harry Ludlow wrote, "a cab drove up to the door and we found to our joy that it was Uncle Ben." The children quickly incorporated news of his voyage into their play, "as reported in a newspaper brought in by a page. It was very effective, and it amused Ben mightily," General Ludlow recorded. "He was the spirit of our party last evening- and told [the governess] that he was 'game for anything.'"

Family acclaim was one thing, but Ben shrank from public attention, dreading fame "more than ice," as he told the General. His reticence, and his failure to "write up" his travels, plus the fact that he was illegitimate and a non-Conformist, therefore never a member of the British "establishment" and not likely to be awarded a knighthood or peerage, explain why his achievements have been overlooked by polar historians.

Ben's Arctic career now suffered an interruption, partly because of the family crisis that blew up when his favourite niece Amy announced her intention of marrying Norman Moore, an impoverished but talented Irish doctor inaccurately denounced by Ben as a "gold-digger." The romance was fostered by Ben's sister Barbara Bodichon, the remarkable feminist pioneer and artist; Ben's bitter opposition created a painful rift. Amy had been Ben's favourite person in the world. Now, he rebuffed her attempts at explanation with icy clarity: "I am very glad I am not your father. Please do not write again or think of coming here."

Another reason for the seven-year gap between voyages was a cab accident in 1878; Ben's left hand went through the window and an artery was severed. No sooner had he recovered than he fell dangerously ill with typhoid. His cousin Florence Nightingale kept him supplied with flowering plants throughout his convalescence. After an operation to cut the tendons of his wrist in order to restore movement in his hand, he revisited his Polar ambitions, and a year later had the *Eira* built to his own specifications.

Eira's first trip, in 1880, was a resounding success. The family was proud that Ben named previously unchartered territory after them–Cape Ludlow, Mabel Island and more–though as Amy was still in disgrace there was no Amy Island. In the following June, *Eira* was launched on her second, and, as it proved, final voyage. Family anxiety mounted when she failed to return by the end of September. All through the long winter following the wreck of *Eira,* Ben and his crew sheltered in their makeshift "Flora Cottage," subsisting on Polar bear meat augmented by the rations they had managed to salvage from the sinking ship. They survived ten months in territory never before inhabited by humans, due in no small part to Ben's dauntless and disciplined leadership. Back at home, guessing that disaster must have befallen, his siblings and cousins helped organise a rescue expedition to which they contributed generous sums.

The rescue vessel, the *Hope*, set sail in June 1882, bearing family letters to Ben; young Milicent Ludlow kissed the letters for good luck. In August, she learned that the kisses had worked. "What do you think?" she wrote in her diary, "We got a telegram from Uncle Ben! He is all right, he says I think 'Got back safe but *Eira* is awa.'"

Not long after Ben's astonishing return, General Ludlow died suddenly, leaving Ben *in loco parentis* to his three orphaned children. He installed them in a house in Hastings and presided as Lord of Misrule. "The meals are wonderful affairs," wrote Amy's mother, "a very long table a very dirty tablecloth and all kinds of queer dishes–Uncle Ben's idea of bringing up his young family is to go & buy them endless quantities of goodies ... chocolate creams—candied fruits—and all sorts of incongruous things appear." Ben's new responsibilities put plans to return to the Arctic on hold, especially when one niece suffered a mental breakdown and became suicidal. Ben kept her incarcerated in his London house, 64 Gower St, for months, appointing himself as nurse and jailer. "I will take the responsibility," he told the doctor. "You will have to," was the doctor's disapproving reply.

Though now well into his fifties, Ben certainly intended to continue exploring and made enquiries about ships. Plans were delayed when, in 1887, to the astonishment and dismay of the family, he became engaged to Charlotte Sellars, beautiful, penniless, half French, Roman Catholic, and only 19 years old. "Men certainly about your Uncle Ben's age seem to go crazy, and I am sure it ought to be illegal," Nannie wrote to Amy. Uncle Ben, for so long single, childless and rich, was no longer the source of bottomless financial and practical support.

Ben and Charlotte ("Aunt Charley") had two sons, Valentine and Philip. Several factors–another cab accident, advancing years, the needs of his sons, marital difficulties–combined to crush Ben's dream of returning to the Arctic. He contented himself with managing his estates and rebuilding his largest house, Glottenham Manor. But the longing never disappeared. In his eighties, deaf and showing signs of dementia, Ben was interviewed by Dr. Norman Moore (Amy's husband, but the rift had at last been healed) to assess whether he was fit to manage his own affairs. He was confused about dates, money and the identity of those looking after him, but though "he could not tell me when he last went on an Arctic expedition, [he] said if anyone asked him to go he would very likely go again."

Ben died on 4 January 1913, aged 84. In his decline he was cared for by Dr. Neale, the intelligent and resourceful medical officer on board *Eira* whose skill had contributed hugely to the healthy survival of the stranded men. He was buried in the churchyard at Brightling, East Sussex, next to his sister Barbara and their Aunt Dolly Longden. Ben had drifted from the Unitarian creed of his youth, but perhaps a church burial was not inappropriate. In her old age, Aunt Charley described how towards the end of the terrible winter in Franz Joseph Land, bears and seabirds were scarce, and Ben was really worried by the prospect of starvation, he lay on his bunk in Flora Cottage, turned his face to the wall and recited the Lord's Prayer. Then he got up, went out with his gun and at once shot a bear. "So you see," said Aunt Charley, "he was a religious man in his way, my Ben."

Hancox, Whatlington, UK Charlotte Moore

Foreword by Håkan Jorikson

I samband med att Stiftelsen Grenna Museum flyttade in i nya lokaler 2002 fanns det möjlighet att vidga museets samlingsområde. Med utgångspunkt från samlingen efter S. A. Andrées polarexpedition skapades Polarcenter för att omfatta fler expeditioner i polarvärlden utöver Andrées. Alltsedan dess har förvärv gjorts genom i huvudsak gåvor. Samlingen är idag Sveriges största med artefakter och annan dokumentation från polarexpeditioner i Arktis och Antarktis. Det är det breddade insamlingsområdet som har lett fram till föreliggande publikation—The Coldest Coast.

Publikationen The Coldest Coast är tillkommen genom ett samarbete mellan Scott Polar Research Institute (SPRI) i Cambridge, England och Grenna Museum—Andréexpeditionen Polarcenter. På respektive institution finns objekt med den engelska löjtnanten Herbert Charles Chermside som gemensam nämnare, och med en väsentlig anknytning till svensk polarhistoria.

Sommaren 1873 företog den engelska aristokraten och polarforskaren Benjamin Leigh Smith (1828–1913) sin tredje expedition till Spetsbergen. I sällskapet ingick löjtnant Herbert C. Chermside (1850–1929) med uppgift att bland annat dokumentera resan i ord och bild. Hans dagbok från resan finns på SPRI och ett album med hans fotografier finns på Grenna Museum. På historikern Peter J. Capelottis initiativ har nu detta unika material sammanfogats till en bok. Här publiceras för första gången Chermsides dagbok och hans bilder. Utöver Capelotti medverkar ett flertal andra historiker och sakkunniga för att belysa materialet ur olika perspektiv. Det gäller till exempel polarhistoria, djurliv och bilden av platsen idag genom refotografering.

Chermsides dagbok löper från den 10 maj till den 26 september 1873. Men berättelsen här tar sin början sommaren 1872 när Leigh Smith befinner sig på Spetsbergen och sammanträffar med den svenska polarforskaren A. E. Nordenskiöld. Den senare ska nu med sin femte expedition göra en övervintring och ett försök att nå Nordpolen, som kom att misslyckas. De båda vid det här laget luttrade expeditionsledarna utbyter sannolikt erfarenheter av is och möjligheten att ta sig norrut. Innan de skiljs åt, den 30 augusti, lovar Leigh Smith att i samband med nästa expedition kommande sommar uppsöka sin kollega för eventuella nyheter.

För Nordenskiölds del går det som sagt inte som man tänkt sig. På grund av isförhållandena byggs deras övervintringsstation på en något sydligare punkt än planerat, man har alltså längre väg till Nordpolen. Och marschen mot samma mål kommer av sig på grund av samma is men också för att merparten av de renar som ska dra packningen rymmer. Men värre ändå är att de två eskortfartyg som åtföljer Nordenskiölds fartyg Polhem till Spetsbergen fryser fast vid stationen på Mosselbukta och nu ställs inför en ofrivillig övervintring som varken de eller expeditionsledaren har någon beredskap för. Det är här som Leigh Smith och Chermside kryssar sig fram genom isen och in i historien igen.

Som lovat uppsöker Leigh Smith sommaren 1873 Nordenskiölds övervintringsläger. Nordenskiöld är vid tillfället iväg för sitt försök mot Nordpolen men Leigh Smith finner ett övrigt manskap i en prekär situation med många sjukdomsfall och ett dödsfall. Leigh Smiths fartyg Diana får genast svenskbesök och Chermside förundras över hur lite svenskarna äter och menar att det måste bero på ovana att äta. Den proviant som var tänkt för ett fartyg med drygt 20 man har måst räcka till ytterligare två fartyg och totalt 67 man och en kvinnlig kokerska. Leigh Smith låter genast lossa vad som kan undvaras av proviant. Det är en undsättning som kommer att rädda åtskilliga liv.

Om denna undsättning och åtskilligt mera av upplevelser berättar löjtnant Herbert C. Chermside i sin dagbok från Spetsbergen den sommar han fyller 23 år. Det är hans första men troligen också sista resa till Arktis. Men tillsammans med hans fotografier utgör dagboken några av de tidigaste intrycken av Arktis. De följande åren tillbringas i militär tjänst utöver en period som guvernör över Queensland i Australien. Han dör i London 1929 men är alltjämt ihågkommen på Svalbard genom den ö—Chermsideøya—som A. E. Nordenskiöld namngav 1875.

Det ska tilläggas att i A. E. Nordenskiölds sällskap fanns också en fotograf. Det var Axel Enwall (1844–1912) som kombinerade läkarsysslan med att i bild dokumentera resan. Det fotoalbumet finns också i Grenna Museums samlingar. Enwall var vid tidpunkten för expeditionen bataljonsläkare vid Flottans station i Karlskrona och hade dessförinnan seglat på varmare breddgrader. Det var den bakgrunden parad med ett intresse för fotografering som ledde fram till resan norrut. Han kom att göra bra ifrån sig. Så pass att han senare även erbjöds läkarbefattningen i Nordenskiölds Vegaexpedition 1878–1880 men tackade nej. Han blev kvar på landbacken och kom istället att lägga mycket kraft och energi i ledningen för Kristinebergs nya hospital och utvecklingen av den psykiatriska vården. Likt Chermside finns även han kvar på Svalbard genom Enwalløya.

Fotoalbumen med Herbert C. Chermsides respektive Axel Enwalls fotografier skänktes till Grenna Museum år 2014 av framlidne läkaren Dan-Axel Hallbäck. Han hade dessförinnan räddat dem genom att köpa dem i samband med nedläggningen av ett litet museum över mentalsjukhuset Marieberg i Kristinehamn. Om inte för Dan-Axel Hallbäck hade de albumen inte existerat idag.

Grenna, Sweden

Håkan Jorikson
Director of Grenna Museum

Acknowledgements and Editor's Note

The 1873 diaries of Herbert C. Chermside reside in the archives of the Scott Polar Research Institute (SPRI) at Cambridge University. They are brought to light here through the kind assent of Marie-france le Fel, the legatee of the papers of Benjamin Leigh Smith and the Christopher Leigh-Smith family archives. The editor expresses his great appreciation to Ms. le Fel, as well as the staff of the Thomas H. Manning Archives at SPRI, especially Naomi Boneham, and the director of SPRI, Julian Dowdeswell. The forty-seven black and white photographs that compliment Chermside's 1873 diaries appear here through the kind permission of the Grenna Museum's Andréexpeditionen Polarcenter in Gränna, Sweden, and its director, Håkan Jorikson.

The captions for the images of Magdalenefjorden are the work of Tyrone Martinsson and combined with his modern imagery of this iconic Svalbard location. For Chermside's imagery from the rest of the Svalbard journey, these captions and the modern views alongside them are the product of Andreas Umbreit and his many years of field experience to fix these locations. Andreas is a travel author and was the first travel operator registered in Svalbard, first with Spitsbergen Tours and later with terra polaris. As some of the images taken by Chermside are clusters of sceneries in almost similar locations (or situated closely together), some captions refer to others to avoid too much doubling of text.

In order for modern readers to more easily follow the course of the 1873 expedition, the editor has placed modern spellings of Arctic place names in brackets, as standardized in *Place Names of Svalbard*, immediately following Chermside's first use of his 1873 spellings of, for example, 'Spitzbergen,' and other nineteenth-century place names in and around Svalbard. A list of the two variants, modern and Chermside's 1873 usage, is also found in Appendix 1.

Finally, a warm thanks to my colleagues who contributed to this volume and to those, like Magnus Forsberg, whose constant encouragement kept the project moving across many years. They have been a source of knowledge, inspiration, and friendship for many years and I am grateful to each and every one of them.

Introduction: Herbert C. Chermside and his Chronicle of Benjamin Leigh Smith's 1873 Exploration of Svalbard

Introduction

The recent discovery in Sweden of a rare volume of photographs from a British nineteenth-century Arctic expedition is the result of the unusual events that took place on the north coast of Svalbard in 1872–73.[1] The three ships of the Finnish-Swedish Arctic explorer Adolf Erik Nordenskiöld (1832–1901) and his expedition to explore towards the North Pole became beset at Mosselbukta in the fall of 1872, just after a rendezvous with the Englishman Benjamin Leigh Smith (1828–1913) in northwestern Svalbard.[2]

The following summer, Leigh Smith concluded his private oceanographic and geographical explorations in the seas around Svalbard that he had begun in 1871 and continued in 1872. The logistics of the 1873 expedition, however, were far more complicated than those of the first two voyages. Rather than using a single ship as he had done with the sailing vessel *Sampson* during the previous summers, Leigh Smith chartered the wealthy Scotsman James Lamont's Arctic steamer *Diana* and employed *Sampson* as a reserve supply tender. With the added supplies *Sampson* afforded, Leigh Smith planned to round the northeast limits of Svalbard—which he had discovered in 1871—and survey Kong Karls Land.[3]

But it was his relief of the Swedish polar explorer and his expedition that cemented the relationship between Nordenskiöld and Leigh Smith. As Lieutenant P. M. von Krusenstjerna, commander of the expedition vessel *Gladan*, observed, the fresh provisions delivered by the English expedition allowed those suffering from scurvy to recover in little more than a week and so "join in the severe work of sawing a passage through the ice in order to deliver us out of our ice-prison."[4] A crewmember of the main expedition ship, the *Polhem*, went further, writing on June 19 that all of "the sick have rapidly improved and that already in six days. Such a generous and at the same time beneficent gift deserves indeed to be made known."[5]

The expedition's medical doctor, Axel W. Envall, wrote in a report to the Swedish Board of Health that they would "not have got off with less than one or more deaths" if Leigh Smith had not arrived when he did.[6] The relief of Nordenskiöld and his expedition led Oscar II, the King of Sweden, to award Leigh Smith the Nordstjärneorden, the Order of the Polar Star.[7]

The ever-reticent Englishman shunned any public ceremony and so the medal was sent him by post from the Swedish Legation in London.[8]

Leigh Smith himself produced little in the way of written records of his expeditions, but at the conclusion of each would create bound, illustrated memorials that he could give away to friends and colleagues. Since Leigh Smith came to the urgent assistance of Nordenskiöld and his crew, copies of his 1873 memorial, illustrated in the main by Herbert Chermside's photographs, were no doubt presented to Nordenskiöld and other members of the Swedish Arctic expedition of 1872–73. We know that at least one such collection of images from the summer of 1873 found its way to Sweden, for it was eventually given over to the Grenna Museum's Andréexpeditionen Polarcenter in Gränna, Sweden. The complete collection is published here for the first time.

Chermside

Many if not all the photographs in the collection at the Andréexpeditionen Polarcenter were the work of Herbert C. Chermside (Intro-Fig. 1). Chermside was a young officer in the Royal Engineers when he joined Leigh Smith's cruise to the north coast of Svalbard in the summer of 1873. The photographs he took and the logbooks and diaries he kept offer some of the earliest descriptions and imagery of this archipelago in the Norwegian Arctic.

Born in Wilton in southern England in 1850, Chermside graduated at the top of his class and was commissioned into the Royal Engineers in 1870. He would turn 23 during his Arctic summer, apparently the only such visit to the high north during a lifetime of service as military attaché, consul, commander and, from 1901–04, governor of Queensland, Australia. Chermside was knighted (KCMG) in 1897, retired as a Lieutenant General in 1907, and died in London in 1929. A marker placed in the chapel where he worshiped described him as a man who "served his country with integrity, wisdom, and justice, fearless in danger, setting duty above all things."

Chermside's Arctic adventure with Benjamin Leigh Smith produced three oversized ship logbooks that included a detailed personal journal/diary of the expedition. These were all written in Chermside's large, loping hand, and filled with both the expedition's details and a steady stream of cleverly acerbic personal asides on the course of the expedition: the flora, fauna, and landscape of the Arctic and, as interesting if not more so, his companions. And then there are the little special extras, as when he sees a cloud in the shape of a three-corned hat descend over Nelsonøya and imagines the ghost of Trafalgar paying a return visit to Svalbard.

Chermside's entries tend to run on and on with barely a comma to hold his breath. Broken into more digestible bits, as the editor has done in this volume, they can at times seem almost like a form of free verse, as in the following paean, delivered with typical irony, to the expedition naturalist, the Reverend Alfred Edwin Eaton:

Intro-Fig. 1. Herbert C. Chermside. Long after his summer in the Arctic and after a long military career, Chermside was Governor of Queensland from 24 March 1902 to 10 October 1904. He appears here in military dress for this portrait (courtesy State Library of Queensland)

The slopes up here
were really for this land
covered with luxuriant vegetation,
endless cushions
of a little smooth leaved green plant
like a shamrock,
scurvy grass
with its heads of
countless white flowers,
yellow flowers the size of
the common buttercup in flower
& many others,
all of which
with their proper Latin names

(far too long to be inscribed here)
written down,
with all care & precision,
on the lids of the nice
little wooden coffins
prepared for any rarity
vain enough to bloom,
& pining for sepulchre in a museum,
by the indefatigable Eaton.[9]

For this first publication of his diaries, the editor has replaced with periods many of the commas in Chermside's run-on sentences. Commas have been inserted where they improve the logical flow of the writing. Other than these additions and corrections for consistency in spelling and the use of modern toponymy, no other editorial alterations have been made to the diaries.

Chermside participated in the 1873 expedition—apparently his first major field experience—as surveyor, hunter, photographer, and chronicler. A graduate of Eton, he had received the Pollock Prize as the top graduate of the Royal Military Academy at Woolwich in 1870.[10] This earned distinction set him far apart from the typical line officer of his day, most of whom paid for their commissions through the purchase system until it was abolished the year after Chermside's graduation.[11]

Chermside's surveying and photographic skills would have naturally drawn Leigh Smith to him. But it is in his diary entries that he reveals himself as a humorous, adaptable companion, a good shot, and a man curious about the world around him. These same traits apparently served him well throughout his life. Thirty years later, as a colonial governor in Australia, he would be described as someone with a "readiness to share sacrifice, [an] approachable personality, wide range of interests, clear and forthright public speeches and [a] willingness to learn by travel…"[12] The 1873 expedition foreshadowed this later success, as it would end with a major island off the north coast of Svalbard named for Chermside on Nordenskiöld's revised map of the area published in 1875.

Chermside's logbooks and journals from that summer have resided in the archives of the Scott Polar Research Institute (SPRI) at Cambridge University in the U.K. since the 1930s. These previously unpublished diaries, along with a log kept by *Sampson's* captain, William Walker, detail an expedition that, while it failed in its primary objective to round Nordaustlandet, did succeed in relieving Nordenskiöld's Swedish polar expedition beset at Mosselbukta, and maintained an array of contacts with whalers and sealers—for example, the Peterhead whaler David Gray and the Norwegian skipper Frederick Christian Mack—regarding local conditions around Svalbard.

Background to the 1873 Expedition

Benjamin Leigh Smith carried out five private scientific expeditions to Svalbard and Franz Josef Land between 1871 and 1882. His interest in the

Intro-Fig. 2. Benjamin Leigh Smith, from a newspaper illustration around the time of his final expedition to Svalbard (courtesy of Hancox Archive)

Arctic was motivated by a desire to hunt Arctic wildlife, to make discoveries, and to fill with his own funds a long gap in British government-sponsored polar exploration, as his father had attempted when he contributed to the search for Sir John Franklin and his lost Northwest Passage expedition. As Chermside wrote after the 1873 expedition, it had been "nearly fifty years since attempts at reaching the highest latitudes [was] abandoned by British public enterprise and all such exploration [was being] carried on by foreign nations," especially Germany and Sweden.[13]

In the summer of 1871, Leigh Smith launched his first Arctic expedition, a geographic and oceanographic exploration of the north coast of Svalbard from the decks of an eighty-five-ton ice-strengthened ketch named *Sampson*. Captained by a Norwegian sealing skipper and explorer by the name of Erik Andreas Ulve (1833–1896), *Sampson* took advantage of favorable ice conditions to define the northeast corner of Svalbard as far as a point later named for Leigh Smith himself. The 5-month expedition returned to England after gathering an extended series of data on seawater temperatures that made an important contribution to the study of deep-ocean currents in the Arctic. The results of this first voyage to the Arctic appeared in two articles and two charts in *Petermann's Mittheilungen* the following year.[14]

Leigh Smith had *Sampson* fitted out for another voyage to the Arctic in 1872. The expedition left Hull on 13 May and mapped several craters on volcanic Jan Mayen Land before sailing for Svalbard, where ice conditions were considerably worse than a year earlier. A Royal Navy Captain, John C. Wells, continued Leigh Smith's collecting of oceanographic data in the form of soundings, dredging, ocean temperature readings, and measurements of currents. When *Sampson* was damaged and beached for repairs at Widjefjorden in September, Leigh Smith retreated to England, without sailing nearly as far to the north and east as during his first expedition. The much more meager results of the 1872 expedition appeared in rather disjointed fashion in a popular account by John C. Wells.[15]

The 1873 Expedition

For his third voyage, Leigh Smith seems to have chosen Chermside not only for his abilities mentioned above but also so that he could combine the results of all three of Leigh Smith's expeditions into a comprehensive account that Chermside would later deliver before a meeting of the British Association for the Advancement of Science in August of 1874.[16]

The British Association paper is divided into two sections, the first twenty-five pages consisting of a recap of the three expeditions and the final forty pages given over to a general discussion of the oceanographic conditions around Svalbard and its presumed value as a route to the North Pole. For the day-to-day details of the 1873 expedition, one must refer to Chermside's rather massive, three-volume log kept on board *Diana* from May 10 to September 26, 1873.[17] Additional details come from the logbook of *Sampson*, kept by its captain, William Walker, during its voyage that summer from Hull to Svalbard and back.

His first two Arctic experiences—in particular the difficult ice conditions of 1872—led Leigh Smith to reconsider the use of a single sailing vessel in the Arctic. Searching for a more durable vessel for his 1873 expedition, Leigh Smith chartered James Lamont's steamer *Diana*, a vessel especially constructed in 1869 for Arctic cruising and hunting. Lamont's expedition had even left the ship's name behind in the place of Dianabukta, an open bay on the southwest coast of Edgeøya where the vessel briefly anchored that summer.[18] The screw steam yacht was powered by 30 hp compound steam engines and its hull strengthened below the waterline with double layers of Australian bark and the bow clad with iron plates.[19]

With *Diana*, Leigh Smith thought that he would for the first time possess the power to attempt to round Svalbard and survey Kong Karls Land. He was also committed to finding Nordenskiöld's whereabouts, as he met the Swede at Fuglefjorden the previous summer. The addition of *Sampson* as a reserve tender allowed Leigh Smith to carry a larger load of supplies, and many of these would come to Nordenskiöld's aid early in the Arctic summer of 1873.

Departure and Early Days

On May 10, 1873, Chermside came on board *Diana* at Dundee, where he was given, as he writes, "a spacious chamber 5'6" x 3'6" which the steward was pleased to call my 'state room.'"[20] *Diana* had a crew of seventeen, including a harpooner. Chermside was joined by fellow Etonian, seventeen-year-old Richard Potter (1855–1947) and a naturalist, the Reverend Alfred Edwin Eaton (1845–1929) as guests of Leigh Smith. Another thirteen sailed on board *Sampson*, which had already departed Hull on April 30 and arrived at Lerwick on Tuesday, May 7. On May 10, the reserve vessel began laboring her way through heavy seas to Svalbard.

The two vessels, outfitted with a year's worth of provisions, planned to sail for Svalbard and rendezvous on July 1 at Kobbefjorden, a small bay on the west coast of Danskøya. Replenished from the stores on *Sampson*, *Diana* would then sail east along the north coast of Svalbard as far as an Arctic chimera known as Gilles Land, "should such exist."[21] If the mythical Gilles Land did in fact appear, they would course along its western coast, using the land as a barrier to any pack ice flowing from the east, and sail as far north as possible. The extra provisions were also crucial to Leigh Smith's other stated objective: to bring relief to Nordenskiöld's expedition.

Chermside was grateful to Leigh Smith for giving him the chance to take part "in an expedition in these regions, the exploration of which has always had a strange fascination in my mind."[22] He was perhaps less grateful as *Diana* left the Firth of Tay and began to roll in the open sea and the soldier found himself absent from dinner for the first time in his life. Rather than fight a north wind, *Diana* put in to Lerwick in the Shetlands, where Chermside found himself "astonished at the absence of beggars,"[23] which is perhaps more a comment on 1873 London than the Shetlands. Delayed in the islands for a week, Chermside met an old man who recalled to him the arrival of Edward Parry's polar expedition in Lerwick in 1827. It was absence of the British government from the polar field since Parry that had provided much of the impetus for Leigh Smith's private efforts.

The expedition departed the Shetlands under steam on the 18th, and Chermside was re-introduced to the sea. When the winds turned favorable, the ship's steam was run down and the sails hauled up. Just as this was completed, the winds died and the ship wallowed, making Chermside sick once again until steam could be got up again and the ship propelled northwards. Not until the 20th did they get a strong SSW wind that sped *Diana* along towards Svalbard.

Diana met the ice at 72 °N, 1 °6'W on the 23rd, and just as it had with Leigh Smith two years earlier the new experience gave Chermside a chance to exercise his pen. "The ice floats about in white snow-covered detached blocks never much above five feet from the surface & of every shape & form, the pieces being eaten into most fantastic shapes by the action of the water. The sides are a beautiful green & if only a bright sun were upon it, the effect would be lovely. Bump, there we go, as I write, bang against a block. As far as the eye can see the sea is covered with these floating blocks... Bump, bump... We are now in the long wished-for Polar regions..."[24]

They met their first Arctic ships three days later. The first was the Swedish vessel *Vega*, which had been north hunting seals and was "evidently surprised at our flying the 'Blue Ensign' [of the United Kingdom] & at once asks if we are an 'expedition.'"[25] A few hours later, they met a Peterhead whaler called *Active*, which relayed the news that fast ice lay just to the north.

Five days later, a much longer information exchange took place when *Diana* came alongside another Peterhead whaler, the steam-powered *Eclipse*. *Eclipse's* captain, David Gray (1827–1896), was a life-long mariner and a friend and contemporary of Leigh Smith. Leigh Smith had provided *Eclipse* with deep-sea recording gear in 1872 and with it Gray had recorded nine ocean stations on a northeasterly line running from Greenland to Norway. An avocational scientist, Gray was first and foremost a deep-ocean hunter. He shared with Leigh Smith the news that *Eclipse* was returning from its spring hunt with 250 tons of oil from a catch of two whales and 48,000 seals. (This seems an extraordinarily high number of seals. It is possible Chermside mistakenly added an extra zero to the total of seals, and the number should be 4,800. Yet 48,000 is what he writes in his diary.) So much oil was processed that they had been forced to throw some of the steamer's coal overboard to make room for more blubber.

Leigh Smith was carrying papers, letters, and news for Gray, who had not been ashore since March. In the mist, *Eclipse* was rafted alongside *Diana* for a day as the officers and guests shared lunch on board *Diana* and then "a capital dinner ... with a long conversation & numerous yarns."[26] These included Gray explaining how he could use the remains of the blubber extraction process to drive the ship if his coal stocks became low. Using this method, he told his dinner guests that *Eclipse* had "steamed over 100 miles on the tail of a whale..."[27] When Chermside mentioned that he was surprised at the absence of blubber smells on board *Eclipse*, Gray explained that it was cold enough that none of the seal or whale products putrefied at these latitudes. Gray then gave a tour of his cabinet of natural history curiosities, which included parasites collected from whales, narwhals and the stomachs of bearded seals.

By June 6, impatient with the slow progress north, Leigh Smith ordered *Diana* east toward Prins Karls Forland. When the mountains and glaciers came into view the following day, Chermside had his first views of Svalbard and busied himself mixing chemicals for his photographic gear.

After *Diana* anchored near the glacier at the head of Kongsfjorden, Chermside and Potter rowed to a near-shore lake covered with birds. "I think I have never enjoyed an evening more," he wrote as he took in the pyramid-like summits of the Tre Kroner. "There was a glorious Arctic sun still high in the heavens, that even at midnight shone in all its pride above the northern hills, casting on their coverings of the purest snow a sheen of golden light so pure, so dazzling, as to seem unearthly."[28]

The next day, Potter and Chermside ascended a nearby mountain to find a cairn already there. When they dismantled it, they discovered notes in French and Swedish proclaiming that the Swedes had reached the spot in 1861. In the meantime, the much slower *Sampson* on May 22 had fallen in with the ice around 70° North, and the crew began working north along the edge of the ice, sealing as they went. They sighted Prins Karls Forland on June 20 and the next day, as Captain Walker wrote in his log, "came to anchor in Magdalena Bay in 13 fathoms above the small island at the head of the bay; caught 3 seahorses."[29]

The Relief of Nordenskiöld

Departing Kongsfjorden on June 9, *Diana* rounded Hakluyt's Headland in the evening of June 10. After some exploring in this area and leaving word for *Sampson* in Kobbefjorden, they met up with a Norwegian fishing smack early in the morning of June 13. The captain, whom Chermside called "Charlie the Norseman," told Leigh Smith that Nordenskiöld and the rest of the Swedes were lying beset and starving in Mosselbukta.

"This is news indeed," wrote Chermside, and Leigh Smith wasted no time in ordering full speed to the rescue.[30] Charlie the Norseman was rowed smartly back to his smack with a bottle of rum and Leigh Smith's compliments.

Diana reached Mosselbukta, which Chermside described as a mere "unprotected indulation [sic] in the coast on the E side of Wiide Bay" in four hours.[31] (Intro-Fig. 2). There, the three Swedish vessels, the 200-ton, 108' iron steamship *Polhem* along with the steamer *Onkel Adam* and the brig *Gladan*, were frozen into the northeast corner of the bay, in ice that Chermside estimated at three to seven feet thick and with more than three miles separating the ships from open ocean.

All the flags on *Diana* were run up. The ladder was put over forward but the "hardy Swedish sailors" ignored it, swinging themselves on board to be greeted with tinned meat and schnapps. "We smoke & chat & drink each other's health & talk & have quite a spirited party."[32]

Leigh Smith had met Adolf Erik Nordenskiöld and twenty-two other members of his Swedish polar expedition on board *Polhem* the previous August when *Sampson* anchored in Fuglefjorden in late August. They were of similar ages (Leigh Smith was forty-four and the Swede was thirty-nine) and shared social status, education, and wealth. They had exchanged visits on their respective vessels and formed a bond of shared experiences and ambitions in their Arctic research. If there was a difference between them it was that Nordenskiöld was willing to leave his ship and try to reach the pole over the ice, whereas Leigh Smith was the classic gentleman-adventurer who would go ashore for a long hike in search of his fill of game birds but seems never to have seriously contemplated a polar sledge expedition.

Nordenskiöld planned to sail to Sjuøyane and construct a base camp on Parryøya with several prefabricated structures. From this point, Nordenskiöld hoped to lead an expedition to the pole in April 1873, using boats and sledges

Intro-Fig. 3. Nordenskiöld's 1875 chart of Svalbard, showing his route from Sjuøyane southwestwards to Nordauslandet and back to Mosselbukta (from Kjellman 1875). The map also shows the continued fascination with the mythical "Giles Land" thought to exist somewhere to the east-southeast of Svalbard

pulled by reindeer driven by four Sámi. When the two explorers parted from one another on August 30, Leigh Smith promised Nordenskiöld that he would return to Svalbard early the following summer and call in on the Swede for news of his polar expedition.

Soon thereafter, Nordenskiöld sailed north and discovered that he was not able to penetrate further than Mosselbukta. Worse, all three Swedish ships were caught by surprise and frozen in for the winter. The huts and observatories meant for Parryøya were instead erected at Mosselbukta and Nordenskiöld was forced to divide rations amongst sixty-seven people from three ships over a Svalbard winter (Intro-Fig. 3). The following April, the winter hardships notwithstanding, Nordenskiöld made an attempt on the pole with three teams, several sledges, boats and a single reindeer. The conditions were extremely difficult, with broken seas and fog. One crewmember, a båtsman (boatman) called Snabb, who left in search of driftwood, was not seen again.

The expedition managed to reach Sjuøyane where they saw the hopelessness of attempting to cross the vast fields of broken ice to the north. Instead of returning directly, Nordenskiöld retreated via Nordaustlandet in order to explore that area, so when Leigh Smith arrived off Mosselbukta he was still away (Intro-Fig. 4).

By evening on the 13th, Leigh Smith had his crew members moving provisions by sledge from the British to the Swedish ships. It was, Chermside wrote, "a great triumph for the judgment of B.L. as regards ice, current,

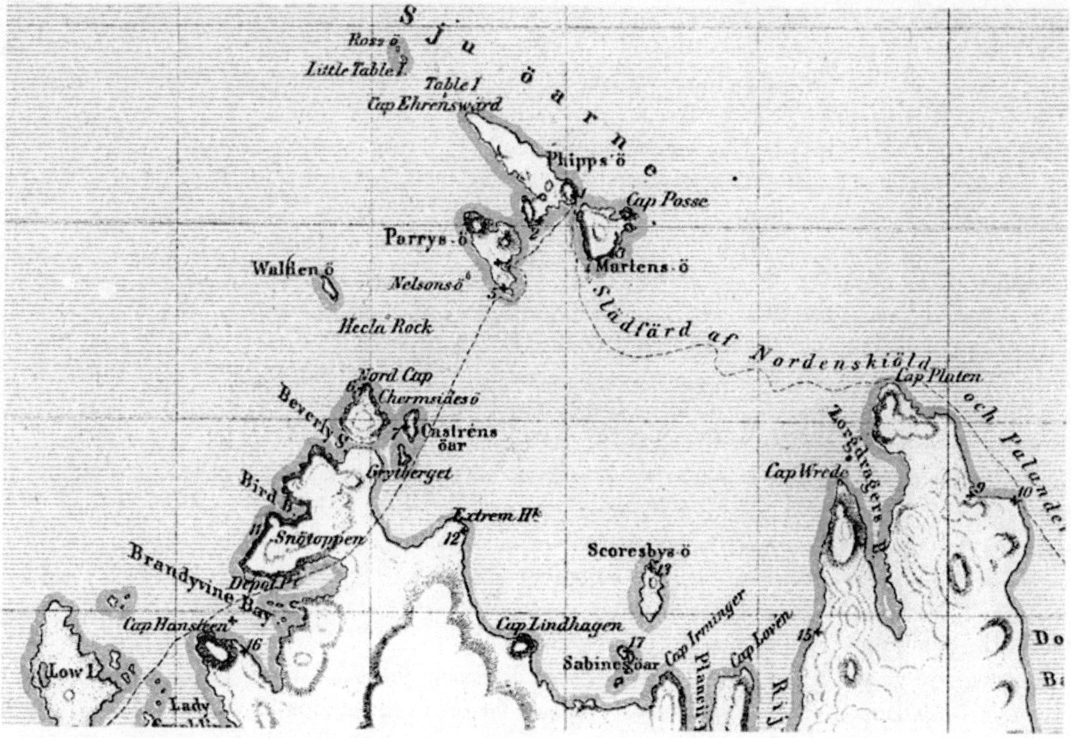

Intro-Fig. 4. Detail from Nordenskiöld's 1875 chart of Svalbard, showing the area of northern Svalbard bounded by Rossøya, Lågøya, and Kapp Platen (from Kjellman 1875). It was in this area where Leigh Smith spent much of his 1873 expedition in search of Nordenskiöld, and where he decided that Beverly Bay was in fact a sound or channel separating Nordaustlandet from the island to the north

winds, etc., as in his letter to *The Times* in the winter, & ever since, in spite of many contrary statements, he had predicted that they would be lying there. This he takes in his usual quiet way."[33]

From the Oxford and Cambridge Club the previous November, Leigh Smith had written a letter to *The Times* in which he spelled out his concerns over Nordenskiöld's fate. He believed, correctly, that the Swedish expedition had put in to some safe harbor, unloaded their cargo but, before the two supply ships had been able to get away, they had been locked in the ice. He felt that the Swedes would survive the winter on the strength of their supplies and their ability to live off the land. In a rare moment of both expression and introspection, Leigh Smith wrote of the last time he saw the Swedish fleet at Fuglefjorden, "rosy with the ray of the evening and the morning sun."[34] As *Sampson* cleared the fjord on its way southwards, Leigh Smith wrote that he felt "something like the shame of desertion"[35] and promised he would search for Nordenskiöld when he went north again in the spring. These are clearly the words of someone who saw Nordenskiöld as a friend and colleague rather than a polar competitor.

The survival of most of Nordenskiöld's crew (another died of scurvy) was in sharp contrast to the deaths by starvation and exposure of nineteen Norwegian whalers trapped in Svalbard by the severe ice conditions the previous summer. By mid-September, 1872, six whaling vessels had been frozen in near Velkomstpynten. Seventeen whalers managed to escape in small boats to a hut at Isfjord but, once there, all perished over the winter. Two other whalers refused to leave their uninsured ship and were later found dead in a small boat, their ship having been crushed.[36]

Despite their own successful overwintering, Nordenskiöld's expedition was in a bad way by early June 1873. The expedition's chief scientist, the physicist August Wijkander, noted that the first vessel to reach the Swedes arrived on June 7, followed the next day by two fishing vessels from Tromsø. They delivered letters and newspapers and a little butter and flour. It was apparently this partial relief on the part of Nordenskiöld's agent in Tromsø that convinced the Swedish government that a large relief ship from Sweden was not required.

They could not have been more incorrect. Wijkander appealed without success to the fishing vessels to return to the mainland for supplies, but the Norwegian skippers were reluctant to abandon their seasonal work before it had even begun. "By this time all but one of the people aboard *Onkel Adam* had come down with scurvy, and half of the crew of *Gladan*, too. Even if all of the ice had magically disappeared from Mossel Bay … there just were not enough [healthy] men available to sail them home."[37]

Just as the Swedes began to sink into renewed despair at the worsening situation, *Diana* and *Sampson* appeared just beyond the ice in Mosselbukta. Wijkander wrote of the almost miraculous appearance of "an English gentleman, Mr. Leigh Smith [who] offered the expedition with great generosity … lime juice, tobacco, rum, fresh potatoes, and preserves for more than two weeks for all men, which we received with great gratitude."[38]

As Nordenskiöld's Sámi, "in their curious hats & long blue frocks … and bright leggings, smoking their pipes," looked on, Chermside acted as quartermaster as provisions were off-loaded from *Diana* for the Swedish survivors. "Half a ton of beef, 40 tins of cabbage, 20 tins of salmon, 1 case of sherry, 1 case of brandy, 10 tins of carrots, 2 cases of lime juice, 59 lbs of tobacco, 5 bags of potatoes & 10 gallons of concentrated rum."[39] Along with Eaton, Leigh Smith took lunch with the Swedes, going over their scientific results for the winter. Chermside, for his part, had perfected his sly style of revealing the quirks of his compatriot's characters: "Eaton was much pleased with their collection of sea-weeds."[40] Eaton had earlier in the year written to Leigh Smith's sister Barbara Bodichon, apparently at her request, that he had compiled lists of all of Svalbard's flora as well as "of the Seals and Whales, Birds and Shells" so that in between his collecting forays he would be able to see at a glance where he had gaps in his efforts and strive to fill them in.[41]

Chermside visited the Swedes' small village of science huts and saw the three ships locked in four feet of ice about a quarter mile from shore. The *Onkel Adam* was in the center, with *Polhem* to the left and *Gladan* to the right. He thought that they could not have picked a worse place to be stranded, as between the ships and the open sea was a screen of grounded ice and several large hummocks. From his few weeks of experience with ice conditions around Svalbard, he thought it unlikely the rotting floe would keep the three vessels trapped for much more than another month.

Diana remained at the edge of the floe near Mosselbukta until the morning of June 15, when steam was gotten up and the ship moved off toward Sjuøyane. Leigh Smith hoped to catch up with Nordenskiöld there, for if the Swede had managed to reach the pole, he was scheduled to return to Parryøya on June 23 and would almost certainly welcome if not require some assistance. When ice blocked *Diana*'s approach to Parryøya, ice anchors were deployed to moor the ship to a floe halfway between Parryøya and Waldenøya. Leigh Smith, along with Chermside and Captain Fairweather, *Diana's* skipper, rowed the three miles to Waldenøya and climbed to its highest point to gain a view to the north.

From this perch, they could see that all the islands of the northwest of the archipelago were linked by ice. Ice blocked the ways to the north and east as well. The three regained the ship and began a long explore along the ice edge, gliding through still, clear waters and making occasional soundings as they went. "A deep sea sounding [on Sunday, June 22] of 800 fathoms gave no bottom, & we got plenty of exercise hauling in the line."[42] They again returned to Sjuøyane, as "the owner is very anxious to get to Parry's Island to see if there is news or trace of the Swedes."[43]

Once again, they anchored to a flow and this time Leigh Smith, the captain, and two crew members trudged across the ice toward Nelsonøya in search of Nordenskiöld. Chermside and Eaton rowed back to Waldenøya to collect some lichen. There Chermside found an old coffin filled with stones and a bone or two. They had, as Chermside thought of them, brought miniature caskets of their own, for Eaton's botanical collecting. They ascended to the highest point of the island, collecting lichens, flowers, and grasses along the way, and then wrote the names of their prizes in Latin "with care and precision, on the lids of the nice little wooden coffins prepared for any rarity vain enough to bloom, & pining for sepulcher in a museum, by the indefatigable Eaton."[44] They reached *Diana* just as Leigh Smith returned to announce that he'd found no sign of Nordenskiöld on Nelsonøya.

On the 29th they made a journey to Tavleøya and, climbing to its height, stood for a long time gazing north. Chermside realized that he would very likely never again set foot on land so far north, so he looked intently "over the ice towards the mystic pole, that defies approach." When it was time to return to the ship, "it is almost with a feeling of awe that one turns away from the outmost (or inmost) threshold of the dread unapproachable tract of frozen ocean…" As they reached the shore, lit their pipes, and waited for the launch, the euphoric feeling began to recede until they were once again "ordinary."[45]

Exploring Hinpolenstretet and Wijdefjorden

Having arrived at Magdalenafjord on June 20, Captain Walker of *Sampson* had sent one of his small boats north to Danskøya and the planned rendezvous point at Kobbefjorden. It returned the following day with the news that they had reached the rendezvous spot and brought back "letters from the *Diana* and found a [separate] letter, on a small island near Kobbe

Bay [Kobbefjorden], from the Danish [likely Swedish] expedition dated 11th Oct. 1872."[46]

On July 1, *Sampson* got underway toward the Sjuøyane and they soon ran into *Diana*, which was maneuvering south. The steamship took *Sampson* in tow and brought her into Sorgfjorden the same evening. As coal, stores, and freshwater were transferred to *Diana* over the next three days, Chermside climbed to the top of what he called Parry's Hill (Heclahuken), where two years earlier Leigh Smith had found a flag staff left by Parry in 1827. The staff was now gone, and Chermside relates a rather extraordinary tale of its disappearance from the 1,505 ft/486m high summit. He writes of a group of Norwegians trapped at Sorgfjorden, presumably during the catastrophic freeze-up the previous fall, who "nipped it as a mast to a piece of drift ice + proceeded on their way."[47]

On July 6, Leigh Smith was on board *Sampson* to talk with Captain Walker and review his charts prior to making his main push north and east. Once the transfer of supplies between *Sampson* and *Diana* was complete, a few of the men went ashore to "collect whalebones, ancient + very bleached."[48]

Chermside in the meantime hiked to Mosselbukta, presumably alongside the low-lying streams and swampy ground that run up from Sorgfjorden and then down to Mossellaguna. There he found Nordenskiöld's huts barred, the windows boarded, and no smoke rising from the chimneys. The Swedes had made their escape at last.

Nordenskiöld himself had journeyed south over the ice from Phippsøya to Kapp Platen. Stopped east of Kapp Platen by open water, Nordenskiöld sought to cross Nordaustlandet to its southeasternmost point at Kapp Mohn. Discovering an impenetrable, almost alien landscape of 'ice canals,' Nordenskiöld and Palander studied the area for a time before giving up any hope of traveling further south.

The impassable glacial icescape forced Nordenskiöld west, toward Wahlenbergfjorden, which he and his men reached in mid-June just as Leigh Smith was relieving the Swedish base camp at Mosselbukta. Nordenskiöld and his party finally returned to Mosselbukta on June 23, less than a week after Leigh Smith had departed. No other expedition of Nordenskiöld's, writes George Kish, was "as beset by trials and bad luck as the expedition of 1872–1873."[49] Only the timely arrival of Leigh Smith had prevented the Swedish expedition from tipping into a complete Arctic catastrophe.

Chermside learned all of this when he returned to Sorgfjorden and saw *Polhem* anchored alongside *Diana*. Prior to a brief exploration northwards before his return to Sweden, Nordenskiöld had come to thank Leigh Smith. The other two Swedish vessels had made their way south on June 30, quickly departing the scene as soon as the ice had moved out of Mosselbukta. Potter wrote a hurried note home to his father, the M. P. Thomas Bayley Potter, saying that the *Polhem* had come in the night before and was leaving that day. "I must finish this letter," he concluded, "as we are just going on board the *Polhem* to say goodby." The contrast between the ordeal the Swedes had endured, and Leigh Smith's cruise, was apparent as Eaton finished his letter with a cheery postscript that he had "never enjoyed anything as much as this trip, and don't feel the cold a bit. It is warmer here than it has ever been, the temperature being 49 deg. We have plenty of sport with the birds, eider ducks, etc., and find them all capital eating..."[50]

The meeting between Leigh Smith and Nordenskiöld must have been brief indeed, or perhaps private, for Chermside makes no mention of it in his otherwise expansive journal. Both men were serious in their respective intent to explore the north, and the brief summer was already advanced. It was perhaps at this meeting that Nordenskiöld mentioned the possibility that the north cape of Svalbard was in fact an island and suggested naming it for the man who saved him, while the ever-humble Leigh Smith deferred the honor to his young aide-de-camp Chermside.

On July 7, *Diana* moved out of Sorgfjorden and down Hinlopenstretet to a group of islets called Fosterøyane that Nordenskiöld had explored in 1861.[51] There, Leigh Smith's expedition was itself beset by ice that stretched all the way across the strait from Kapp Fanshawe to Selanderneset. Retreating into Lomfjorden, Chermside found a "vertical pillar of flat stones covered with inscriptions in Norwegian."[52] Fishing smacks with ill or injured crew would call on *Diana* for a chance to be treated by Eaton, who passed for the only doctor north of Tromsø and served as a kind of replacement for Envall, who had performed much the same duty during the Swedish expedition's stay in Svalbard.

On July 14, they were hailed by Norwegian skipper Frederick (Fritz) Christian Mack (1837–1876), of whom *Place-Names* writes "several important geographical discoveries in the Arctic are due."[53] "Like all these Norwegian skippers that I have met," Chermside writes, "he is a well-informed superior class of man..."[54] Mack had also been sent by Nordenskiöld's agent in Tromsø with supplies for the beset expedition, but had arrived after the relief from Leigh Smith. (Three years later, Mack would write a plaintive letter to Leigh Smith complaining that he had been swindled by two putative foreign polar expeditions, had been without work for a year and a half, and was at the end of his tether.[55] He died a few months later.)

The next day, *Diana* escaped Hinlopenstretet and anchored back in Mosselbukta only a short distance from the Swedish camp. "How different on a calm sunny summer evening like this, it looks from the dreary icebound haven of 4 or 5 weeks ago."[56] With the way north blocked by ice pressing on the north coast, *Diana* maneuvered down Widjefjorden where Chermside and Potter went ashore for a bit of surveying while Leigh Smith initiated dredging and collecting operations. *Sampson* in the meantime had caught up with *Diana*, and for several days the two vessels—along with a collection of Norwegian fishing smacks—anchored in Lomfjorden to await favorable conditions.

Leaving its escort on the 14th, *Diana* passed beyond Kapp Petermann and anchored near a small island in Austfjorden that Leigh Smith named after *Diana*. On the newly christened Diana Island, they went on a reindeer hunt, returning with no less than sixty-eight reindeer, two or three seals, and seven white-fronted geese, which Chermside described as a "new species to Spitzbergen."[57] Chermside noted that the hunting was a very trying experience, as the "excitable sailors are not trained gillies, will not keep line, do not know when to stand, when to show, when to squat, or act in concert at a distance..."[58]

The expedition remained anchored near Diana Island for several days, surveying the island and a nearby shoal, while the crew performed chores of painting the vessel, flensing and salting skins, and preparing heads for mounting. On the 25th, they continued dredging operations and the following day began a running survey as *Diana* retreated toward the mouth of the fjord in another attempt to get north.

Dredging Operations and a Visit to Moffen

On July 25, the expedition sighted ice-covered Lågøya and did their best to avoid the seas around it, which Chermside noted were foul with "black, wicked rocks & reef."[59] The following day they took up dredging in 120 fathoms with a whale line and a heavy iron dredge weighing one cwt (112 lbs/50.8 kg).

Chermside described several other aspects to this dredging operation. First, a "swab" to gather up starfish and seaweed, echinoderms and stones from the bottom was fashioned from a "long broom of oakum used to clean the decks." To this was added a small tin to form a scoop to capture stones or gravel, along with a small hoop net and a sounding lead to gauge the character of the bottom. Deep-sea thermometers were attached at intervals of fifty fathoms. Deploying the dredge many times in a single day was impossible, as it was so heavy it required all hands to haul it out of the depths. The device likewise took a toll on the specimens collected. As Chermside wrote, any animals unlucky enough to be caught in this fashion were of little value, since they had been "pounded to fragments or reduced to become amateur jellyfish..."[60] Hemmed in by ice on the 29th, the crew nevertheless managed to sound and dredge with their "swab" when possible and retrieve what Chermside described as "some fine specimens of coral."[61]

Diana pushed through the ice and into Lomfjorden on the 30th, there to be met by four Norwegian fishing smacks. Increasingly unable to maneuver, Leigh Smith made for Heclahamna to gain an elevated view of the ice conditions. Still station-keeping after several days, Chermside wrote that "Macawber-like, we wait for something to turn up."[62] On August 5th, *Diana* made its way offshore to Moffen Island, where a year earlier Leigh Smith and his team had discovered an enormous whale skeleton buried in the shingle and covered with inscriptions and ship's names carved into the bone by earlier visits by hunting expeditions. Chermside described Moffen as a low-lying mass of shingle just a few feet above sea level, with a scattering of granite boulders apparently carried onto the island by the movement of ice grounding on the shoal ground. One group began walking around the lagoon in one direction while a second group walked the opposite way. An hour and a half later the two parties completed a circuit that Chermside estimated to be about seven miles and met "at opposite points of a large natural bay or harbour which occupies the greater part of the interior, & which had still a great deal of land ice in it."[63]

The entrance to the harbor lay on the northwest corner of the island, not along the north as shown on their charts. They had been looking for walrus but found only carcasses of walrus killed along the inner edge of the lagoon. "We also saw, which was far more interesting, the bone-covered sites of what at one time must have been glorious engagements with large numbers of sea-horse. There were three of the sites, white with immense quantities of bones of the animals evidently all killed at one time, & the largest of these was over an acre in extent, & I am rather sorry I did not count the number of skulls."[64] If Leigh Smith took this chance to seek out the skeleton he had found the previous year Chermside does not mention it. He does mention a painting hanging in the dining room of the Admiralty which depicted Moffen Island and an attack on a large number of walrus by men of the Phipps expedition, so it seems clear that the island's connections to British polar history were the subject of discussion amongst the exploring parties.

After two days around Moffen, *Diana* returned to Lågøya and then continued southeastwards down Hinlopenstretet, arriving at Augustabukta on August 10th. Here Chermside does mention a phenomenon associated with the whale skeleton from Moffen, which had apparently died naturally and been raised by isostatic uplift. The shingle beaches at Augustabukta showed the remains of whales "killed perhaps 100 years ago, the dates indeed have been in some cases determined, then the skeletons too heavy to drag above it must have lain in the water & now you find skeletons almost complete at six, eight & ten feet above the sea level."[65] Chermside does not describe the skeletons in any detail, or how he arrived at a relative death date, so it is not possible to know if the whales died naturally or showed evidence of butchering, but his assumption that the skeletons were left in shallow water being too heavy to be dragged makes it clear that he believes the whales were killed by humans. Such evidence at the remote Augustabukta could mean that, absent a shore tryworks, the whales were pursued into shallow water by the occasional visit of a pelagic hunting expedition or an expedition operating away from the established whaling stations on the west coast of Svalbard. There they could have been flensed alongside a ship, the blubber boiled down on board and the largely intact carcasses left to rot, as opposed to being set adrift as in deep-harbor or open-ocean whaling.[66]

Chermside's estimate of a death date of 100 years earlier seems to be confirmed by later geological research. Using an uplift rate of approximately 1 cm per year, posited for a Russian hunting hut in Nordaustlandet,[67] Chermside's whale skeletons would have died or been killed at some point between the years 1569 and 1691, an acceptable range that brackets the discovery and shore-based exploitation of right whales around Svalbard and its transition into pelagic whaling. The greater the annual uplift, such as that reported by Jahn[68] from southern Svalbard, the more recent the death of the whales. Jahn's data applied to Augustabukta would produce a death range very close to Chermside's supposition. Of course, the beach matrix itself is not an absolute dating of the whale since, as Blake[69] points out, such whales did not necessarily die right at the shore. Blake's radiocarbon dating of whalebone fragments produced dates of more than 6,000 BP.[70] A host of other factors, from storm surge to human agency, could also factor into the timing, location, and elevation of the remains. The only certainties are that the higher the whale's elevation and/or the slower the rate of uplift, the more likely such skeletons predate human activity on Svalbard. The low elevation of Chermside's whale skeletons at Augustabukta, like the palimpsest whale found at Moffen, both suggest human agency. Perhaps it was while contemplating these maritime mysteries that Leigh Smith repeated his comment about Svalbard to Chermside: "[T]he owner always says it is like Switzerland with the sea."[71]

Returning to familiar waters east of Tumlingodden, near where Leigh Smith had circumnavigated Wilhelmøya in 1871, Chermside recorded a humorous scene when the entire watch gathered around the heavy dredge for a deep-sea sounding in heavy ice, only to find that the water was but seventeen fathoms deep. *Diana* continued around the southern edge of Nordaustlandet to Vibebukta, which the crew also dredged. Chermside notes that Leigh Smith wanted to explore "Wiches Land" [Kong Karls Land], just visible across the pack, but the expedition was stopped by ice near Kapp Mohn. They returned northwestwards up Hinlopenstretet, emerging from the strait on August 23 with hopes of reaching the Sjuøyane.

To Chermsideøya and a Furthest North for the Summer

With ice still fast to the north, Leigh Smith led the expedition east towards Kapp Platen. There *Diana* anchored in shallow water in an ice-covered "Parry's 'Beverly Bay,'" which Leigh Smith—perhaps with an earlier inference from Nordenskiöld, correctly believed to be a channel, or sound, and the land north of the channel to be an island.[72] Nordenskiöld[73] (Intro-Fig. 5) in his chart of Svalbard named this 5.4 m²/14 km² island after Chermside, and it is now known as Chermsideøya.[74]

Intro-Fig. 5. Further detail from Nordenskiöld's 1875 chart of Svalbard, showing that the island north of Nordaustlandet had already by 1875 been named 'Chermsides ö' for the then-twenty-three-year-old Herbert Chermside (from Kjellman 1875)

This is apparently the only new place name added to the nomenclature of Svalbard during the 1873 expedition, and that it was named for the young expedition member speaks volumes for the impression he made upon Leigh Smith—and perhaps Nordenskiöld as well.

With the ice under heavy pressure, the crew spent its time moving from one anchorage to another. In his unpublished account of Leigh Smith's expeditions around Svalbard, Chermside provides an extensive discussion of the ice-breaking methods employed by the crew of *Diana*. Since leaving the southeastern end of Hinlopenstretet, the vessel had made only slow progress in the heavy ice. On the 28th, with the pressure of the ice lessening, the crew, as Chermside related, "set to work in earnest to extricate ourselves from the pack, as further progress E seemed impossible and the pack was more or less cemented together by early and very sharp frosts."[75]

On August 8, *Diana* had managed 15 miles progress in nine hours through heavy pack, but as Chermside wrote, this required that the ice have some 'play'. Now, with three miles of heavy ice separating the ship from sailing water, the steam engines were run up and the crew warped the ship from its temporary ice dock. Several of the crew were put onto the ice armed with boathooks. Once the vessel had some maneuvering space and the likeliest spot for the ice to crack was spied from the deck, the ship was rammed into that spot. Loose ice was then pushed aside by the men with the boathooks, or they hopped onto the broken ice and punted the ice out of the path of the ship —appropriate work for Cambridge men. While this was carried out, *Diana* was backed to the farthest corner of the cleared area, sometimes as much as 200 yards and "then she comes ahead full steam & jumping right on to the ice, succeeds in cracking it. At it she comes again & again & several large pieces are by this time smashed. On to these men jump with drills & boring holes ice hooks and warps are made fast from the ship & they are towed out of the way. The boathook men (or numbers as we say in drill, removing the smaller pieces) if the floes be nipping fast little is gained but, if not, the passage is slowly and gradually cleared…"[76]

The men bobbing about on the ice were then picked up by the dinghy and the ship proceeded to back once more to the farthest end of the cleared area. The bell was rung and the ship lurched forward again, this time with all hands running to one side of the deck to the other to create a rolling motion, "until by the time her head is straight at the obstacle and the Captain shouts "Steady" from the crow's nest, she is going at full speed and rolling almost to the rail—a crash—and we are through and looking forward to the next obstacle."[77]

When the ice would not crack, Chermside was often at work blasting it with explosives, with varying degrees of success. Yet another method was to jam the bow of the ship into the ice, then place an ice anchor ahead of the ship. A line was run from the ice anchor to the ship, the engines run up, and then the helm thrown hard over, the bow of the ship held in place while the stern forced the ice aside, thereby wedging the floe open further. "When there is no more to be gained over goes the helm the other way, and with the immense leverage gained by the length of the ship, the steam of course, besides giving her the power of pushing ahead gives her also the power of swinging, and thus we force our way through inch by inch, not however without breaking three new warps."[78]

Chermside admitted that often each of these techniques would fail in turn and the expedition had to give up and try to find another place to breakthrough. When all the hard work paid off, however, "a grand sight it is as one stands at work on the ice to watch the ship rush bravely at the obstacle at full speed, and how satisfactory as her bow leaps out of the water on to the mass, ere she falls back to feel it groan and quiver and crack under your feet, a dead dull muffled sound perhaps, and a new dark crack telling you that your blast has also done its work."[79]

By August 30th, *Diana* was hooked to ice between Parryøya and Phippsøya, which Chermside saw as a collection of "isolated hills connected by low sea beaches [with] immense quantities of driftwood and some whalebones,"[80] a similar description to that in Leslie's expeditionary biography of Nordenskiöld, wherein Leslie describes Nordenskiöld's landing on August 5, 1861, at Phippsøya, "several isolated mountains about 1,800 feet high, connected by a low land covered with driftwood and fragments of ships" together with remains of whale skeletons found lying high above the present level of the sea both on the east side of Parryøya and on the promontory of Martensøya.[81]

Nordenskiöld, in fact, had cached his small boat and a quantity of supplies on Phippsøya in the spring of 1873 after his aborted attempt to get north of Sjuøyane. Not finding the boat, Chermside wanted to cross to Martensøya and search for it but was stopped by fear of becoming trapped on Martensøya and having to be rescued. On August 31, Chermside climbed two peaks, presumably on Phippsøya, of 1,100 and 1,220 feet (335 and 372 m) above sea level, for a view to the north. It was not encouraging. On September 1, before a southeast gale, *Diana* made her farthest north for the summer at 80 ° 54'N, just beyond Sjuøyane, before being forced to retreat.

The expedition sailed westwards in a heavy gale until the 4th. Finding no possibility of getting further north, they ran for Magdalenefjorden where they rode out the gale for three more days. The seas were so violent that, even as she sheltered in the fjord, *Diana* dragged and broke both of her remaining anchors.

On the morning of September 8, *Diana* rendezvoused with *Sampson* at Grønfjorden. Captain Walker's crew on August 29 had found a coal mine there at the northeast point of the harbor and taken two boatloads on board. Before they parted for the year, *Sampson* passed a spare anchor to *Diana* while Leigh Smith decided on one last late-season attempt to get around the eastern edge of Svalbard.

As *Sampson* headed for home, *Diana* sailed between Edgeøya and Hopen and met pack ice just east of Hopen. The weather was fine and the pack loose enough for *Diana's* steam power to shoulder it aside. By September 14, however, severe frost had placed two inches of solid new ice in front of them and cemented the older pack ice together. After a brief exploration of Hopen, *Diana* turned and set sail for Dundee, which the expedition reached on September 26. The slower *Sampson* returned to Humber Dock early on the morning of Sunday, October 5. "All well," wrote Walker. "Thus ends the voyage."[82]

Aftermath

After the expedition, Chermside returned to England to study coastal defense. He would see service for several years in Turkey, Egypt, and Kurdistan, before returning to Constantinople for seven years as military attaché.[83] He then commanded British troops on Crete and fought in the Boer Wars before being appointed governor of Queensland in 1902. Despite Lady Tennyson's description of him as "a very short plain little general with a biggish moustache," his range of interests and travels and his genial personality earned him considerable popularity.[84]

Although Chermside never returned to the Arctic, his efforts there with Leigh Smith in 1873 must have made a considerable impression. When Nordenskiöld published his revised map of Spitzbergen in 1875, Chermside's name had been attached to the island discovered north of Beverlysundet. A valley that crosses Chermsideøya was also named for him in the 1920s,[85] but by this point his reputation would have been long secured.

For Leigh Smith, the conclusion of the expedition began a period of evaluation. By the fall of 1873, he had demonstrated a range of skills for geographic exploration in the waters around Svalbard, and his relief of Nordenskiöld had conferred upon him the knighthood of the Order of the Polar Star from the King of Sweden and Norway. As noted earlier, with his reticence to accept public accolades for his work, the medal and brooch were simply mailed to him.[86]

But his desire to reach the North Pole was strong as ever. Leigh Smith clearly wanted to push further north and put the theory of an open polar sea to the ultimate test of steam power. David Gray wrote to him in the fall of 1876 with the wish that Leigh Smith would soon "have another chance of getting a ship and that you will try once more to get north if you do not receive favorable news from the Arctic expedition this month [the disastrous government-sponsored Nares expedition, which would return to England in November having fallen far short of the Pole]. I think it would be well worth your trouble to try again."[87] Gray advised Leigh Smith that if the *Windward* came up for sale he would let him know.

In the summer of 1876, Leigh Smith offered 4,500 GBP for the vessel *Norvegen*, owned by the Deutsche Polar Schifffahrts Gesellschaft in Hamburg.[88] For unknown reasons the transaction did not come to pass. Other contacts would send letters to Leigh Smith whenever a potential vessel for Arctic exploration came onto the market. In 1877, Gray sent along a note that new engines installed on his *Eclipse* were giving more than nine knots and that he was ready "for a dash at the Pole should a favorable opportunity occur..."[89]

However, it would be seven years before Leigh Smith returned to the polar seas. When he did it would not be in someone else's vessel but in his own specially designed and Peterhead–constructed polar research vessel, *Eira*. His field of operations would shift from Svalbard to the newly-discovered Arctic lands of Franz Josef Land. There, Leigh Smith's adaptive method of polar exploration would find both its greatest success in the summer of 1880 and its greatest test, on August 21, 1881, when *Eira* would sink amid ice floes at Cape Flora. Following the wreck of the *Eira*, Leigh Smith and his crew were forced onto a stony shore to survive an Arctic winter. The following spring, in small boats salvaged from *Eira* before she went down, Leigh Smith would lead his men through hundreds of miles of ice as they successfully escaped to the south.

Conclusions

For Leigh Smith, the 1873 expedition marked the conclusion of his extensive explorations around Svalbard. During these three expeditions, the British explorer had gained vast experience with high latitude sailing, ice navigation, deep-ocean scientific research, and the infinitely complex geography and geographic nomenclature of the archipelago. He had added dozens of new names to the gazetteer of the islands, discovered several new islands and delineated Svalbard's eastern limits. He had come to the rescue of a major Swedish expedition and in the process gained a first-hand sense of the requirements for surviving an overwintering in the north. This invaluable knowledge would be put to the test during his own forced overwintering in Franz Josef Land in 1881–82.

Leigh Smith's nimbleness in adapting his plans to changing conditions was in marked contrast to the overweening government-sponsored expeditions of the era. In this sense, he can be seen to have initiated private oceanographic, biological, and geographic research in polar and sub-Arctic waters, work that would be continued in the twentieth century with, among others, Alexander Forbes' expeditions to Labrador in *Ramah* in the 1930s, and Jacques-Yves Cousteau's Antarctic expedition in *Calypso* in the 1970s. Without the national pressures of pole-seeking and without scores of seemingly expendable sailors, Leigh Smith had pioneered ways to achieve remarkable results by allowing ice and weather conditions to dictate his areas of operations.

P. J. Capelotti

Notes

1. See: https://www.jnytt.se/article/gavan-till-grenna-museum-en-polarhistorisk-skatt/ (accessed April 25, 2018). Throughout these introductory chapters, the editor has followed modern spellings of Arctic place names as standardized in *Place Names of Svalbard*, and not Chermside's original spelling of, for example, 'Spitzbergen' or other 19th century place names in and around Svalbard. A list of the two variants, both modern and Chermside's 1873 usage, is in Appendix 1
2. Kjellman FR (1875) Svenska Polarexpedition År 1872 + 1873 under ledning af A.E. Nordenskiöld. P.A. Norstedt + Söner, Stockholm

3. Capelotti PJ (2013) Shipwreck at Cape Flora: The Expeditions of Benjamin Leigh Smith, England's Forgotten Arctic Explorer. University of Calgary Press, Calgary
4. Krusenstjerna PM "Letter published in the *Official Gazette of Sweden*, July 28, 1873." From "Translated excerpts from newspapers re: 1872–1873 Swedish expedition." Cambridge: Scott Polar Research Institute Archive (hereafter cited as SPRI), MS 301/36/1–2
5. "Private letter written on board *Polhem* at Mossel Bay, June 19, 1873, and published in Stockholm newspapers, July 21, 1873." From "Translated excerpts from newspapers re: 1872–1873 Swedish expedition." SPRI MS 301/36/1-2
6. Leslie Alexander (1879) The Arctic Voyages of Adolf Erik Nordenskiöld, 1858–1879. Macmillan, London p. 406
7. Milum J (1881) "Report of the Evening Meetings, Third Meeting, December 12, 1881." Proc Roy Geogr Soc 4 (1882):60
8. Swedish Legation in London. Letter to Leigh Smith, May 6, 1874, SPRI MS 301/35
9. Chermside HC Journal on board of the SS Diana. SPRI, Vol. II, June 28, 1873. All quotes from this unnumbered journal are hereafter cited by their volume and entry date. The journals themselves are part of the general reference at SPRI: GB 15 Herbert Chermside, with the specific reference: MS 300/4/1-2-3
10. Anonymous (1875) Memoir to Illustrate the origin and foundation of the Pollock medal. Boddy and Co., Military Publishers, Woolwich
11. Bruce Anthony (1980) The Purchase System in the British Army, 1660-1871. Royal Historical Society, London
12. Wilson PD (1979) "Chermside, Sir Herbert Charles (1850–1929)." Australian Dictionary of Biography, vol. 7. Melbourne University Press, Melbourne, 631–632
13. Chermside HC no date [1874]. "Recent explorations in the Spitzbergen seas by Leigh Smith" (unpublished manuscript). Edinburgh: Edinburgh University Library special collections: MS Gen 77 (hereafter referred to as: Chermside. n.d. [1874])
14. Petermann A (1872a) Geographie und Erforschung der Polar-Regionen, Nr. 58: Die Englisch-Norwegischen Entdeckungen im Nordosten von Spitzbergen, Nordfahrten von Smyth, Ulve, Torkildsen, 19. Juni–27 Sept. 1871. Mittheilungen aus Justus Perthes' Geographischer Anstalt über Wichtige Neue Erforschungen auf Dem Gesammtgebiete der Geographie von Dr. A. Petermann 18 Band: 101–106. (Gotha: Justus Perthes); Petermann, A. 1872b. Geographie und Erforschung der Polar-Regionen, Nr. 59: Gillis-Land, König Karl-Land und das Seeboden-Relief um Spitzbergen, nach dem Standpunkte der Kenntniss im Jahre 1872. Mittheilungen aus Justus Perthes' Geographischer Anstalt über Wichtige Neue Erforschungen auf Dem Gesammtgebiete der Geographie von Dr. A. Petermann. 18 Band: 111–112. (Gotha: Justus Perthes); Petermann, A. 1872c. Originalkarte zur übersicht der Reisen von Smyth, Ulve, Torkildsen, 1871. Mittheilungen aus Justus Perthes' Geographischer Anstalt über Wichtige Neue Erforschungen auf Dem

Gesammtgebiete der Geographie von Dr. A. Petermann. 18 Band: Tafel 5 (Gotha: Justus Perthes); Petermann, A. 1872d. Smyth' & Ulve's Reise im Nordosten von Spitzbergen und ihre Aufnahmen im Nord-Ost-Lande, Aug. & Sept, 1871. Mittheilungen aus Justus Perthes' Geographischer Anstalt über Wichtige Neue Erforschungen auf Dem Gesammtgebiete der Geographie von Dr. A. Petermann. 18 Band: Tafel 6. (Gotha: Justus Perthes); Capelotti, P.J. 2006. "Benjamin Leigh Smith's first Arctic expedition, Spitzbergen, 1871." Polar Record 42(1):1–14

15. Wells JC (1873) *The gateway to polynia: a voyage to Spitzbergen from the journal of John C.Wells*. London: Henry S. King and Company; Capelotti, P.J. 2008. "Benjamin Leigh Smith's second Arctic expedition: Spitzbergen and Jan Mayen Land, 1872." *Polar Record* 44(3):255–264
16. Chermside HC n.d. [1874]; see also: Rawlinson HC (1874) "First meeting, November 10, 1874: opening address." Proc Roy Geogr Soc London 19(1):1–17
17. Chermside Journal, SPRI, MS 300/4/1-2-3
18. Norsk Polarinstitutt (1991) *The place names of Svalbard*. Oslo: Norsk Polarinstitutt (skrifter nr. 80, 112): 111 (hereafter cited as: Norsk Polarinstitutt 1991)
19. *Geographic Journal* (1913) "Obituary: Sir James Lamont." 42(3):301–302; Chermside. n.d. [1874]: 6
20. Chermside 1873/1: May 10
21. *Ibid*
22. *Ibid*
23. *Ibid.*, May 12
24. *Ibid.*, May 23
25. *Ibid.*, May 26
26. *Ibid.*, May 31
27. *Ibid*
28. *Ibid.*, June 7
29. Walker William (1873) "Logbook of the yacht *Sampson*, kept by Capt. William Walker, April 29-October 5, 1873." SPRI M 300/3; SL, hereafter referred to as: Walker 1873
30. Chermside 1873/2: June 13
31. Chermside n.d. [1874]: 8
32. Chermside 1873/2: June 13
33. *Ibid*
34. Leigh Smith, Benjamin. 1872. "To the Editor of the Times." *The Times* (27538): col. E, Nov. 19, 1872
35. *Ibid*
36. New York Times (1873) "The Spitzbergen Drama." August 3, 1873: 5
37. Kish G (1973) North-east passage: Adolf Erik Nordenskiöld, his life and times. Nico Israel, Amsterdam 111–112
38. Wijkander, A. 1873. Letter. June 22, 1873 (published in a Stockholm newspaper on July 18, 1873. In: Translated excerpts from newspapers re: 1872–1873 Swedish expedition. SPRI MS 301/36/1–2
39. Chermside 1873/2: June 13
40. *Ibid.*, June 14

41. Eaton to Bodichon, Shrove Tuesday 1873, SPRI MS 300/9 D
42. Chermside 1873/2: June 22
43. *Ibid.*, June 28
44. *Ibid*
45. *Ibid.*, June 29
46. Walker 1873: SPRI M 300/3; SL
47. Chermside 1873/2: July 2
48. *Ibid.*, July 6
49. Kish G (1968) "Adolf Erik Nordenskiöld (1832–1901): polar explorer and historian of cartography." Geogr J 134(4):487–500
50. "Mr. Leigh Smith's Arctic Expedition," The Times, August 26, 1873, 12
51. Norsk Polarinstitutt 1991: 140
52. Chermside 1873/2: July 11
53. Norsk Polarinstitutt 1991: 277
54. Chermside 1873/2: July 14
55. Mack letter to Leigh Smith, February 3, 1876, SPRI MS 301/17/1-3
56. Chermside 1873/2: July 15
57. *Ibid.*, July 21
58. *Ibid*
59. Chermside 1873/2: July 27
60. *Ibid.*, July 28
61. *Ibid.*, July 29
62. *Ibid.*, August 4
63. *Ibid.*, August 5
64. *Ibid*
65. *Ibid.*, August 10
66. See for example the description of harbor whaling by British whalers in the 1660s in: Conway, W.M. 1906. *No man's land: a history of Spitzbergen from its discovery in 1596 to the beginning of scientific exploration of the country*. Cambridge: Cambridge University Press. Chermside's estimate of a death date of 100 years earlier seems to be confirmed by later geological research. Using an uplift rate of approximately 1 cm per year, posited for a Russian hunting hut in Nordaustlandet (Olsson and Blake, Jr., 1962; see Note #67), Chermside's whale skeletons, resting between six and ten feet above sea level in 1873, would have died or been killed between 182-304 years earlier, or at some point between the years 1569-1691, an acceptable range that brackets the discovery and shore-based exploitation of Greenland whales around Spitzbergen and its transition into pelagic whaling. The greater the annual uplift, such as the 2.99 cm per year reported by Jahn (1959: 157; see Note #68) from Hornsund in southern Spitzbergen, the more recent the death of the whales. Jahn's Hornsund data applied to Augustabukta would produce a death range of 1764-1807, with a median year of 1786 —and very close to Chermside's supposition. Of course, the beach matrix itself is not an absolute dating of the whale since, as Blake (1961: 137; see Note #69) points out, such whales did not necessarily die right at the shore. Blake's radiocarbon dating of whalebone fragments found at an elevation of 7.8m asl (25.59') produced dates of more than 6,000 BP

67. Olsson IU, Blake W Jr (1962) "Problems of radiocarbon dating of raised beaches, based on experience in Spitzbergen." Norsk Geografisk Tidsskrift 18:47–64
68. Jahn Alfred (1959) "The Raised Shore Lines and Beaches in Hornsund and the Problem of Postglacial Vertical Movements of Spitzbergen." Polish Geographical Review, Vol. XXXI, Supplement: 157
69. Blake W Jr (1961) "Radiocarbon dating of raised beaches in Nordaustlandet, Spitzbergen." In: G.O. Raasch, ed. *Geology of the Arctic*. Toronto: University of Toronto Press: 133–145:137
70. *Ibid*, 141
71. Chermside 1873/2: August 10
72. *Ibid*., August 29
73. Kjellman FR (1875) Svenska Polarexpedition År 1872 + 1873 under ledning af A.E. Nordenskiöld. P.A. Norstedt + Söner, Stockholm
74. Norsk Polarinstitutt 1991: 96
75. Chermside n.d. [1874]: 19
76. *Ibid*: 21
77. *Ibid*: 22
78. *Ibid*: 23
79. *Ibid*: 23–24
80. Chermside 1873/3: August 30
81. Leslie Alexander (1879) The Arctic Voyages of Adolf Erik Nordenskiöld, 1858-1879. MacMillan and Co., London 79
82. Walker 1873: SPRI M 300/3; SL
83. Owen CV (2004) "Chermside, Sir Herbert Charles (1850-1929)," in *Oxford Dictionary of National Biography*. Oxford: Oxford University Press, 2004. Online edition: http://www.oxforddnb.com/view/article/32390, accessed July 6, 2009
84. Wilson Paul D (1979) "Chermside, Sir Herbert Charles (1850-1929)." Australian Dictionary of Biography, Volume 7. Melbourne University Press, Melbourne pp 631–632
85. Norsk Polarinstitutt 1991: 96
86. Ltr to Leigh Smith, May 6, 1874, SPRI Archives MS 301/35
87. Gray to Leigh Smith, September 23, 1876, SPRI Archives MS 301/13/2
88. Ltr to Leigh Smith, June 28, 1876, SPRI MS 301/9/1
89. Gray to Leigh Smith, January 19, 1877, SPRI Archives MS 301/13/2

List of Illustrations

(taken from a bit further North in Kokerihamna, compared to Chermside's location), the originally Dutch, now Russian, settlement of Barentsburg covers the shore under Grønfjordfjellet nowadays, while Chermside's two images show the area still in its pristine state, with fresh snow in the late summer season. His two images can be stitched together, though this leads to a doubling of the *Diana*, indicating that Chermside did not take the two photos straight in a row from the same position. Recent image with sailing vessel *Antigua*: 2017-06-03, Andreas Umbreit.

Figs. 9.9 and 9.9.1: Cape Fanshawe on the east side of the entrance of Lomfjorden (Lomme Bay). This area was visited twice by the expedition, for about 4 days following 11 July, and again on 30 July, and in both cases, they met Norwegian hunting vessels, as visible on the images. The different snow situation on the images in the slopes of Cape Fanshawe allows us to allocate this image to the first visit, with Chermside's position most likely on Hingstsletta a bit south of the ice front of Valhallfonna on the western shore of the entrance of Lomfjorden. The matching recent image with sailing vessel *Antigua* in about the same position as the *Diana* 125 years earlier was taken on 2018-07-04 by Andreas Umbreit.

Figs. 9.10 and 9.10.1: Recognizable scenery details on this image are sparse, but the situation with the vessel off a low ice cliff off what looks like a long and smooth sea front of an ice cap fits very well to the situation on the south side of the Valhallfonna ice cap on the western side of the entrance of Lomfjorden, where Chermside was on shore with his camera on the occasion of the first 1873 visit of the *Diana* for some days following 11 July. The accumulation of snow and fallen ice pieces all along the visible foot of the ice cliff indicates very shallow water, which matches the situation at Valhallfonna well, and contrary to most of the ice cliffs of the ice caps terminating in Augustabukta and Vibebukta, where the Diana visited as well in 1873. The low perspective of the image indicates that Chermside must have been either on a beach or on a safe ice resting on a shallow sea floor near the shore with his clumsy photography gear. Chermside mentions the hunting vessel *Sampson* anchoring very close to the glacier front in Lomfjord and this fits very well with the photography. As Valhallfonna has lost thickness and a bit of extension since 1873, the depicted glacier front is no help for trying to determine the position of the photographer more precisely. Recent image with front of Valhallfonna and sailing vessel *Antigua* taken by Andreas Umbreit, 2017-08-22.

Figs. 9.11 and 9.11.1: Hecla Cove (Heclahamna) in Sorgfjord (earlier: Treurenberg Bay) was visited by the *Diana* from 2–6 July and on 2–3 August 1873. For a British expedition to Svalbard, this was a famous site, as its name goes back to the expedition vessel *Hecla* of the 1828 expedition by Parry, which anchored here 45 years before the visit of Leigh Smith. Chermside's image shows the mountain Heclahuk (Hecla Hook, 434 m), ascended by a Swedish expedition (Torell, 1861) just 11 years before the arrival of the *Diana*. Heclahuk was also the name-giving mountain for the rocks of the older geology of Svalbard. Chermside used the opportunity of this visit also for taking the images 013, 014 and 015 from the area of this bay. The recent image with an almost identical view by Andreas Umbreit, 2011-08-31.

Figs. 9.12 and 9.12.1: View southwards from Hecla Cove/Heclahamna to the lower end of Sorgfjord with Dunérbreen at its end—taken by Chermside probably on 2 August 1873. The recent image taken by Andreas Umbreit (2011-08-31) from Crozierpynten on the north side of Heclahamna. Compared to 1873, where its clear ice-core moraines reached partly all the way to the shore, Dunérbreen in the background and its moraine have melted considerably.
Figs. 9.13 and 9.13.1: View from above the eastern shore of Heclahamna to Crozierpynten on its northern side, photography by Chermside probably on 2 August 1873. Twenty-five years later, this location became the main Swedish base of the Swedish-Russian Arc-of-Meridian Expedition (1898–1902), which erected a number of buildings here, along with scientific measuring instruments, to determine the exact shape of Earth so far north—with the first precise triangulation maps of much of the archipelago as a side result. The recent image (Andreas Umbreit, 2011-08-31) shows almost the same perspective, with some remains of the Swedish base visible as ruins, while Chermside's image shows the same location still in its pristine condition.
Figs. 9.14 and 9.14.1: Heclahamna, with the Norwegian hunting vessel *Sofia*, as depicted by Chermside, probably during their second anchorage here from 2–3 August 1873. Observe the little dot on the hilltop to the right, which is probably the base of the flagstaff set up here by the Parry expedition in 1828 (see also images 012-014) as the only clear human sign in an otherwise pristine scenery. The corresponding new image (Andreas Umbreit, 2011-08-31) shows almost the same perspective, 138 years later—now with parts of the ruins of the 1898–1902 Arc-of-Meridian Expedition Swedish main base, which in the meantime has been both set up and then given up again.
Figs. 9.15: Cape Fanshawe, seen from Hingstsletta on the west side of the entrance of Lomfjord (Lomme Bay) during the first visit of the *Diana* (ship to the right) from 11–14 July 1873, with two Norwegian hunting vessels in the picture.
Figs. 9.16 and 9.16.1: View from Hingstsletta on west side of northern Lomfjord (Lomme Bay) southwards over the inner part of the fjord. Chermside took this photo during the first stay of *Diana* in Lomfjord from 11–14 July. Most of the bay is still covered by dense drift ice or winter ice, with the ice border visible in the water a few hundred metres into the picture on the left side. Much of the lower parts of the gravel-covered plain behind the beach wall is still covered with ice and snow from the winter, while the corresponding image from 22 August 2017 by Andreas Umbreit, about 6 weeks later in the season, is free of ice and snow both in the fjord and on the lowland.
Figs. 9.17 and 9.17.1: View across sea ice pushed into shallow DeGeerbukta (DeGeer Bay) on the northern west side of Lomfjord onto Faksefjellet mountains and further mountains in innermost Lomfjord. This image was probably taken by Chermside during the second visit of the *Diana* in this area at the end of July, where she anchored in almost the same position off Hingstsletta as during her first stay here from 11–14 July 1873. Some of the mountains in the background are the same as on Fig. 9.16, but with less

snow. The recent image shows a similar perspective, though without packed sea ice in the bay, taken on 22 August 2017 by Andreas Umbreit.

Figs. 9.18, 9.18.1, and 9.18.2: After an ill-fated attempt of sailing deeper southwards into Hinlopen Strait around mid-July, the *Diana* managed to reach the southern end of Hinlopen Strait with Augustabukta (10 August) and even Vibebukta (22 August, both in Nordaustlandet), navigating through fairly dense drift ice. These photos were taken by Chermside from on board *Diana* probably around 17 August, when they were in the ice up to somewhere east of Thumb Point (Tumlingodden, on eastern Wilhelm Island (Wilhelmøya)). The drift ice helps calm the waves, thus minimizing the movements of the ship, which was essential for sharp photographies from on board with the long exposure times required. Stitched together, both images show the complete northeastern scenery of Wilhelmøya from northeast, from Kapp Breusing (right) to Tumlingodden (left). However, different weather conditions on the two photos indicate that they were taken not immediately one after the other. The recent image by Andreas Umbreit, taken on 30 June 2012 from Torellneset, is somewhat more northwest than Chermside's corresponding positions in 1873, but shows the same coastline of Wilhelmøya from a slightly different angle—and with no drift ice at all in this part of Hinlopen Strait, contrary to 1873. However, ice conditions in Hinlopen Strait vary also nowadays a lot in summer. Not in every summer, but this area is often free of ice in late June nowadays as here in summer 2012.

Figs. 9.19 and 9.19.1: Unidentified areas of ice fields.

Fig. 9.20: *Diana*, with signal flags flying.

Figs. 9.21 and 9.21.1: The *Diana* visited Mosselbukta (Mosselbay, northern Wijdefjord) and the Swedish scientific wintering station there at that time of the Nordenskiöld expedition repeatedly during its 1873 voyage. The first time was for several days in mid-June when the three Swedish vessels there were still locked in solid winter sea ice and the Swedish expedition team was running short of provisions and accordingly welcomed a substantial food donation by Benjamin Leigh-Smith from the *Diana*. This image was taken by Chermside on a later visit, probably on 16 July, when the winter ice had left the bay. Unfortunately, the quality of the image is somewhat poor, probably taken from the ship with its movements, but in the picture, the wintering house Polhem and some of the scientific installations north of it are just about discernable in dark to the left of the middle of the photo. As Nordenskiöld largely relied on drawings for the documentation of his expedition, this photography by Chermside is a rare photographic account of the Polhem base at the end of its operation. The recent images by Andreas Umbreit show the Mosselbay area from a similar angle, though from shorter distance from the shore, on 5 July 2018, with the Swedish constructions collapsed—and seen in the detail picture (2017-08-24) by Andreas Umbreit depicting the remains of the main Polhem building as it looked 144 years later, after parts of it had also been reused by trappers and others.

Fig. 9.22: Another fine view of *Diana*.

Figs. 9.23 and 9.23.1: From 15–27 July, the *Diana* explored 100 km long Wijdefjord over almost its complete length. Though this is one of the biggest fjords of the archipelago, it was (and is) rarely visited both nowadays and in nineteenth century. After the call at Mosselbukta, the next anchorage was in Dirksbukta (just south of Dirksodden) on the east side of the fjord on 17 July, viewing across Wijdefjord onto northern Andrée Land between Våtnedalen (valley almost to the left) and Gråhuken (Grey Hook, right, flat northern end). The corresponding modern picture by Andreas Umbreit (2017-08-24) shows about the same scenery, but from above Vassfarbukta, around 8 km further north, which means a different angle of view.

Figs. 9.24, 9.24.1, and 9.24.2: By 21 August, *Diana* had reached East Fjord (Austfjorden), the innermost eastern branch of Wijdefjord, and anchored on its west side on the height of what they named Diana Island after their expedition vessel and nowadays called Bjørnnesholmen after a Norwegian trapper who had his main station on nearby Austfjordneset for seven years from 1924 onwards. Here, the Leigh Smith expedition spent five days on hunting and exploration, and Chermside took these photographs from Diana Island onto the glacier-polished rocky hills of Austfjordnes and behind it the mountains from Rutherfjordfjellet (left on Fig. 9.24, up to 1329 m) via the glacier Schrödingerbreen to Grossfjellet (right on Fig. 9.24.1, up to 1172 m), the tops hidden in a cloud layer starting at about 1100 m above sea level. Some 12 km inland, behind these mountains, mount Perriertoppen (1712 m) is hidden as the second-highest peak of the archipelago. The recent image by Andreas Umbreit was taken from the hilltop of Austfjordnes on 6 June 2006, looking into easterly direction onto the same mountain scenery as the two images by Chermside, 133 years earlier. Both his and the recent pictures show remarkably little snow and ice coverage of this area despite the partly high elevations, which is due to the extremely low annual precipitation (meaning first of all very little snowfall) in this sheltered area of innermost Wijdefjord deep within the main island.

Fig. 9.25: This image by Chermside probably taken around 17 July from on board the *Diana* as it approached its southernmost reach in Hinlopen Strait, show the northeastern side of Wilhelmøya between Moltkeneset and Tumlingodden (Thumb Point) from Northeast.

Figs. 9.26 and 9.26.1: In its first attempt of 1873 to go deeper south into Hinlopen Strait, *Diana* met heavy drift ice around the Foster islands (visited also by Parry in 1827, and named after a member of his expedition), combined with strong winds, which kept her in this area for some days around 7 July, close to Sværingen, the main island of these small rocky islands in the middle of Hinlopen Strait. An intended landing was not possible under these conditions, but Chermside took this image from on board the Diana, showing the south side of Sværingen. The corresponding recent picture by Andreas Umbreit shows the south side of Sværingen, too, from a slightly more eastern position, on 5 July 2011, without any drift ice around it. However, also nowadays drift ice is very much possible in these waters in July, varying strongly from year to year.

Fig. 9.27: Cape Fanshawe, depicted by Chermside from ashore at Hingstsletta on the west side of the entrance of Lomfjord when the *Diana* visited this bay for the second time at the end of July, anchoring at almost the same position off Hingstsletta. Thick drift ice had entered the bay again, but snow cover on the slopes of Cape Fanshawe is less than during the first visit about 20 days earlier.

Fig. 9.28 and 9.28.1: Between 10 and 22 August, *Diana* reached its southernmost positions at the southern end of Hinlopenstretet (further south, the sea was completely blocked by very dense drift ice), between Bastianøyane and Vibebukta. There she did some scientific dredging of the sea floor as far as possible regarding ice conditions almost into Vibebukta, but with only very few landing possibilities. This image was taken by Chermside from on board the *Diana* in this period along the coast of Nordaustlandet (Northeast Land), and shows low Karl Alexander Island (Karl Alexanderøya) to the right and, partly hidden behind it, the bigger but equally low and rocky Franzøya with its western end to the left, seen from North, both islands situated south of western Ulvebukta, where *Diana* was struggling with the ice between these islands and the shore at Kapp Oetker around 17–18 August. The corresponding recent image by Andreas Umbreit shows these two islands from north as well, but from a slightly more western position, taken on 7 July 2011, without any drift ice around them, which is not a given in early July even nowadays.

Figs. 9.29, 9.29.1, and 9.29.2: These images were taken by Chermside on or about 17–18 August, when the *Diana* struggled with the ice conditions between Karl Alexander Island/Franz Island to the north and Kapp Oetker at the shore of Nordaustlandet east of Kapp Torell. Both images show neighbouring stretches of the coastline a bit east of Kapp Oetker. As a comparison with the contemporary corresponding see the image by Andreas Umbreit (taken 4 July 2011) that shows that the shore was far more covered by permanent snow accumulations back in 1873 than nowadays, even more so as the Chermside photographs were taken six weeks later in the season, close to the average minimal annual snow coverage. In the background behind the coast and the partly snow and ice-covered ascent, the gentle rise of the huge ice cap Vegafonna starts (hidden in low clouds on the Chermside images).

Fig. 9.30: Washing day on board *Diana*. This image was taken near Low Island (Lågøya).

Fig. 9.31: Crew of the *Diana* towing the skins of a recently killed bear (left) and a walrus.

Fig. 9.32: *Diana*, draped with dozens of slaughtered birds and at least one bear.

Fig. 9.33: This image, with the hunting party on drift ice in the foreground, shows the Seven Islands (Sjuøyane) from south, with Parryøya to the left (mountains Øykollen and—with the big snow field—Gullberget, and then left, from the middle, western Phippsøya with Phippsøyhorga in the distance, right of the middle eastern Phippsøya with rampart-like Høgberget and Trollenykjen in front of it more to the right, and finally to the right of the image, western Martensøya). The photo appears to have been taken during the second visit to Sjuøyane, when the expedition explored also the area between Chermsideøya, Kapp Platen, and Sjuøyane. It also seems to reflect Chermside's difficulty in getting the crew to stay still in one place while he composed his images.

Figs. 9.34 and 9.34.1: At the end of August 1873, *Diana* approached the Sjuøyane again, this time with open water west of the islands. This image by Chermside shows the west side of Phippsøya (left, to the north) and Parryøya (right, to the south), with the southernmost tip of Martensøya looming behind to the far left, seen from southwest. The corresponding contemporary image by Andreas Umbreit was taken (2016-07-02) from southwest, as well, but from a closer position, therefore showing only the southwestern coastline of Parryøya (with small Nelsonøya (Nelson Island) in front of it, left of the middle of the picture), while Phippsøya and Martensøya (except of the snow-covered top of Solvberget in the background) are hidden behind.

Figs. 9.35 and 9.35.1: Between 15 and 29 June, Leigh-Smith tried to reach the Sjuøyane for the first time during the 1873 voyage, but the *Diana* was stopped by dense sea ice off the west side of the islands, with only small Tabular Island (Tavleøya) accessible without ice. This image taken by Chermside shows a view from the *Diana* to high Høgneset as the westernmost steep cliff of Parryøya to the right and behind it the western coastline northwards of northern Phippsøya), then Tavleøya, and furthest to the left in the distance Little Tabular Island (Vesle Tavleøya), with Ross Island as the northernmost point of the Spitsbergen archipelago hidden just behind it. The contemporary matching image by Andreas Umbreit shows almost the same perspective, taken on 2 July 2016, with no drift ice—and this is not common so early in the season even today.

Figs. 9.36 and 9.36.1: Southern Phippsøya in Sjuøyane, with Granittberget and Hytteberget, seen from west during the first 1873 attempt to reach the islands in second part of June, with the hindering dense drift ice visible in the foreground. A similar perspective (from a slightly more southerly position) was depicted as a comparison by Andreas Umbreit on 2 July 2016.

Figs. 9.37 and 9.37.1: Tryggve Granfjellet as the northern end of Phippsøya (right) and Tavleøya (left) depicted by Chermside in the second part of June 1873 from a southerly direction with the dense drift ice blocking the west side of Sjuøyane in the foreground. A matching recent perspective by Andreas Umbreit shows the same scenery on 2 July 2016.

Figs. 9.38 and 9.38.1: Høgneset as the westernmost steep cliff of Parryøya, seen by Chermside from the *Diana* across the hindering thick drift ice from North during the first attempt to reach Sjuøyane during the second part of June 1873. A similar recent perspective by Andreas Umbreit from 7 August 2013 shows the same area, then free of ice. Nelsonøya is situated just southwest of Parryøya.

Figs. 9.39 and 9.39.1: Tiny but high Nelsonøya (middle) and the south of Parryøya, depicted by Chermside from the *Diana* across the heavy drift ice from Westsouthwest during the first attempt to reach Sjuøyane in second part of June 1873. The contemporary image (from 2 July 2016) by Andreas Umbreit shows much of the same scenery from a more southwesterly direction.

Figs. 9.40 and 9.40.1: Small but high Nelsonøya (middle) and the south of Parryøya, depicted by Chermside from *Diana* across the heavy drift ice from the West during the first attempt to reach Sjuøyane in second part of June 1873. The contemporary image (from 2 July 2016) by Andreas Umbreit

shows much of the same scenery from a more southwesterly direction, but closer.

Figs. 9.41 and 9.41.1: Here, Chermside took a photo of southernmost Phippsøya with Hytteberget (left) and Utkiksnosa (middle) through Straumporten and onto southernmost Martin's Island (misspelled by Chermside for what is actually Martens Island (Martensøya)) in the background, far right and looming through between the mountains in the middle. The picture was taken from the Diana across the hindering dense drift ice from the west during the first attempt to reach Sjuøyane in second part of June 1873. A contemporary image (from 2 July 2016) by Andreas Umbreit shows parts of this scenery from a somewhat different position.

3. Other illustrations

Frontispiece: Svalbard/Spitsbergen with prominent place names from the 1873 Leigh Smith Expedition. Map created by As the Crow Flies Cartography, Canada. © P.J. Capelotti.

Intro-Fig. 1: Herbert C. Chermside. Long after his summer in the Arctic and after a long military career, Chermside was Governor of Queensland from 24 March 1902 to 10 October 1904. He appears here in military dress for this portrait (courtesy State Library of Queensland).

Intro-Fig. 2: Benjamin Leigh Smith, from a newspaper illustration around the time of his final expedition to Svalbard (courtesy of Hancox Archive).

Intro-Fig. 3. Nordenskiöld's 1875 chart of Svalbard, showing his route from Sjuøyane southwestwards to Nordauslandet and back to Mosselbukta (from Kjellman 1875).

Intro-Fig. 4. Detail from Nordenskiöld's 1875 chart of Svalbard (from Kjellman 1875).

Intro-Fig. 5. Further detail from Nordenskiöld's 1875 chart of Svalbard, showing Chermsideøya (from Kjellman 1875).

Chapter 2-Fig. 2.1. Adolf Erik Nordenskiöld, in a famous painting by Georg von Rosen (1886; Source: https://www.ub.uit.no/northernlights/eng/northeast01.htm).

Chapter 2-Fig. 2.2. The site of Edward Sabine's "pendulum observatory" on the southern tip of Indre Norskøya in Fair Haven, NW Svalbard (Courtesy of Urban Wråkberg).

Chapter 2-Fig. 2.3. Nordenskiöld's ships frozen in at Mosselbukta (from Kjellman, 1875).

Chapter 2-Fig. 2.4. Nordenskiöld's winter base at Mosselbukta (from Kjellman, 1875).

Chapter 2-Fig. 2.5. The house of Oscar Dickson and A.E. Nordenskiöld's polar station for combined mining and scientific purposes is still standing today at Icefjord on Svalbard (photo in 2011 and courtesy of Urban Wråkberg).

Chapter 2-Fig. 2.6. The house of the Swedish polar station in Mossel Bay has not been cared for and is today a heap of planks (image from 4 September 2000). For the informed and careful visitor, it is still rich in signs of the drama it once witnessed (photo by and courtesy of Urban Wråkberg).

Chapter 3-Fig. 3.1. Axel Enwall (1844–1912), doctor and polar photo pioneer. From Svenskt porträttgalleri XIII, Läkarekåren, Stockholm 1899.

Chapter 3-Fig. 3.2. A photograph by Axel Enwall of Benjamin Leigh Smith's ship *Diana* with members of the crew from *Gladan* transferring provisions for the Swedish expedition, 13 June 1873. (courtesy Grenna Museum: Andréexpeditionen Polarcenter).

Chapter 3-Fig. 3.3. Letter from Axel Enwall to August Wijkander, 29 August 1874. (courtesy Regional state archives, Gothenburg).

Chapter 3-Fig. 3.4. Newspaper article from *Kristinehamns-Tidningen* from 11 June 1896 telling that a person in Stockholm had become mad, thinking he was the discoverer of the North Pole and that the balloonist S. A. Andrée was going to strip him from the honor. The person was sent to Dr. Enwall, but he gave no hope because 5–6 other cases of insanity in the persons family indicated hereditary predispositions.

Chapter 4-Fig. 1. Alfred Eaton (courtesy of the Royal Entomological Society).

Contents

Contributors

Susan Barr Oslo, Norway

P. J. Capelotti Penn State University, Abington, PA, USA

C. L. Devlin Penn State University, Abington, PA, USA

Håkan Jorikson Grenna Museum, Grenna, Sweden

Anders Larsson Gothenburg University Library, Gothenburg, Sweden

Tyrone Martinsson Valand Academy, University of Gothenburg, Gothenburg, Sweden

Charlotte Moore Hancox, Whatlington, UK

Andreas Umbreit Munich, Germany

Urban Wråkberg Dep. Tourism and Northern Studies, UiT The Arctic University of Norway, Kirkenes, Norway

List of Figures

The Nineteenth Century Exploration of Svalbard

1

Susan Barr

Introduction

The seventeenth and eighteenth centuries had seen the development of the exploitation of the natural resources of the Svalbard archipelago and thus the availability of information about this High Arctic island group. Given that the ice conditions of the period were mostly more severe than today, and that the only access to the islands was by ship—with sails as the main method of propulsion—the economic prospects were obviously great enough to sustain a relatively wide interest in this uninhabited and climatically harsh area.

Exploitation leads to exploration

The strain on the natural resources of the area began with excessive walrus and whale hunting in the early seventeenth century, which led to the near extinction of both species. British and Dutch were the main perpetrators and the slaughter only stopped when it became non-economic as the stocks diminished. Hunting then turned to the land and on a much smaller scale, where the first Russian Pomor (= from the northwest corner of Russia) hunters and later Norwegian and other trappers overwintered in small huts on the islands to reap the foxes, polar bears, reindeer and any remaining walrus for skins, blubber and tusks.

Mapping and descriptions of natural conditions had followed the hunters to a certain extent and provided a baseline for the scientists to come, but it was during the nineteenth century that scientific exploration blossomed. At the very beginning of the century, it was a British whaling expedition with William Scoresby, Sr., and Jr., on board that can be said to have provided a transition between the commercial and the scientific at this time. In 1806, in *Resolution*, they reached the highest latitude then obtained by ship (81°30′ N19°E), north of Spitsbergen, which was not surpassed before the Finnish–Swedish polar scientist A.E. Nordenskiöld reached 81°42′N 62 years later.[1] Although not working on Svalbard itself, Scoresby, Jr. (1789–1857), made invaluable contributions to the knowledge of the geographical and natural conditions of the European Arctic, particularly through his 1820 publication *An Account of the Arctic Regions*, hailed as "undoubtedly the foremost book of its kind."[2] Scoresby was already elected Fellow of the Royal Society of Edinburgh in 1819 and was elected Fellow of the Royal Society in 1824. Although he maintained his interest in science, his last Arctic voyage was in 1822, after which he turned to studies in divinity, later becoming a clergyman.

British naval expeditions

Perhaps the first entirely scientific observatory in Svalbard was established for a 3-week period on Indre Norskøya, north of Spitsbergen. In 1823, a British naval expedition was sent from London at

S. Barr (✉)
Oslo, Norway
e-mail: snbarr@gmail.com

P. J. Capelotti (ed.), *The Coldest Coast*, Historical Geography and Geosciences,
https://doi.org/10.1007/978-3-030-67880-7_1

the request of the Board of Longitude to enable the astronomer and pioneer of earth magnetism observations Edward Sabine (1788–1883) to conduct pendulum measurements in the north of Svalbard and East Greenland in the quest to determine the exact shape of the earth. Sabine was an advocate of conducting a magnetic survey of the entire globe in order to understand the magnetic forces and variations and he was highly honoured internationally for his lifelong work.[3] During the period of June 30–July 22, Sabine carried out his observations on Indre Norskøya before the expedition continued to Greenland for further work.[4]

Four years later, the British Navy again sent an expedition to Svalbard, but this time, the goal was the North Pole. William Edward Parry sailed to the northernmost Svalbard islands—Waldenøya and Vesle Tavleøya—to lay depots before establishing a base in Sorgfjorden, Northwest Spitsbergen. The North Pole attempt, pulling sledge boats over the ice, reached the furthest north record of c. 82°45′ N, which was to stand for almost 50 years (being beaten in 1876 by another British Naval expedition northwest of Greenland). It was the southward drift of the ice that defeated Parry's attempts to push northward. However, while the attempt was being made another part of the expedition conducted surveys in Hinlopenstretet and general natural history observations were made.[5]

Keilhau and the first true scientific expedition

This same year, 1827, saw what has been called the first truly scientific expedition to Svalbard. Norwegian geologist Balthasar Mathias Keilhau (1797–1858) had set as his goal the compilation of a complete overview of the geology of Norway. In 1826, he was appointed as an associate professor in geology at the University of Christiania (Oslo), becoming a professor in mineralogy and geology in 1834. In these positions, he was expected to carry out field studies and he started with a study tour to Finnmark, Bjørnøya and other parts of the Svalbard archipelago in 1827.[6] His resulting publication in 1831, *Reise i Øst-og Vest-Finmarken samt til Beeren-Eiland og Spitsbergen, i Aarene 1827 og 1828* (Travels in East and West Finnmark and to Bjørnøya and Spitsbergen in the Years 1827 and 1828) took place in August–September, 1827, and contained, in addition to pioneering descriptions of geology, botany and palaeontology, valuable observations concerning the Russian and Norwegian hunters in Svalbard as well as sketches of what today are important cultural heritage sites.

British gentleman explorers

The mix between tourism, hunting and scientific expeditions was a fine art indulged by nineteenth century gentry in England and Scotland who could afford the ways and means of exploring remote areas. Despite the pith hat and khaki shorts image, this might raise in our minds, there were also those who chose to go north, to the fog and ice and the possibility of game at least as exotic (and dangerous) as lions and tigers.

Lord Dufferin—*aka* F.T. Hamilton-Temple-Blackwood, Marquis of Dufferin and Ava—was probably the first. In 1856, he took his own yacht *Foam* to Jan Mayen and the northwest coast of Spitsbergen. His travel account, *Letters from High Latitudes* (1857), received great and international success at the time and most certainly inspired others to follow, including James Edward W.T. Butler, Marquess of Ormonde, who sailed in *Mirage* to Svalbard in 1874 after reading Dufferin's book. Scottish James Lamont (1858, 1859, 1869, 1871) and English Benjamin Leigh-Smith[7] (1871, 1872, 1873) were gentleman sportsmen and explorers, with the latter certainly providing the greatest scientific output even though any descriptions from the region at this time added new knowledge.

The wealthy who could afford to travel in their own or a hired yacht inevitably came to be followed by the slightly less well-off who joined organised cruise trips to the exotic north. The first such tourist expedition to Svalbard was apparently organised in 1881 on *Pallas* from Tromsø to the west coast of Spitsbergen.[8] Hunting and tourism were the main purpose of the voyage, but British naturalists A. Chapman and A.H. Cocks recorded and published observations of the wildlife. Organised cruise tourism has since become big business in Svalbard.

Cocks returned to Svalbard in 1882 on a British–French zoological expedition with Charles Rabot (1856–1944). Rabot was a French geographer, explorer and mountaineer and he and Cocks particularly studied the avifauna, collecting specimens and leading to a publication by Cocks of the birds and mammals collected during the voyage. They also visited the Swedish IPY station at Kapp Thordsen where Rabot took interesting photographs (see later).

La Recherche in Svalbard

An unusually international expedition worked in Svalbard in 1838 and 1839, leaving as its most glorious legacy a series of 27 hand-coloured lithographs of Svalbard scenery interspersed with scenes from the expedition itself. This was a French naval expedition led by Paul Gaimard and with the corvette *La Recherche*. The task was to make extensive studies in the European Arctic region, which encompassed northern Scandinavia as well as Svalbard. Also participating were Norwegian, Danish and Swedish scientists. Recherchefjorden, Madalenefjorden and Bjørnøya were visited and partly surveyed, and the usual geological and zoological observations and collections were made. French medical officer, botanist and meteorologist Ch.F. Martins made the first scientific study of glaciers on Spitsbergen. The expeditions were well documented both by sketches and paintings, and by author and historian Xavier Marmier.[9] The expedition was even mentioned by French science-fiction author Jules Verne in his book *Voyage au centre de la Terre* (Journey to the Centre of the Earth) in 1864. On another curious note, one of the artists, François Auguste Biard, smuggled his fiancé Léonie d'Aunet (1820–1879) on board in 1839, dressed in men's clothing. According to her own narrative published in 1854, *Voyage d'une femme au Spitzberg*, this was arranged with Gaimard beforehand as a way to get Biard to join the expedition.

Swedish explorations in Svalbard

The last half of the nineteenth century was definitely a pinnacle of Swedish scientific expeditions to Svalbard, with names that have become immortal not only through scientific achievements but also by being placed on geographical features around the archipelago. In 1993, the Swedish Polar Research Secretariat celebrated a long history of Swedish science in the polar regions with the publication of *High Latitudes, A History of Swedish Polar Travels and Research* by professor of meteorology and polar veteran Gösta H. Liljequist (1914–1995). Much of the information below is taken from this valuable book.

Liljequist mentioned that Sven Lovén (1809–1895) is often referred to as "the father of Swedish polar exploration" after his visit to the west coast of Spitsbergen in 1837, which resulted in zoological and geological collections, although Liljequist would rather give this honour to Otto Torell.[10] Supporting Lovén's candidacy, Clive Holland states in his chronicle of Arctic expeditions[11] that Lovén's expedition was the "direct forerunner of the long series of important Swedish expeditions starting 21 years later" as he went on to encourage Torell and others to continue the important work to be found there.[12]

Otto Torell (1828–1900) is described by Liljequist as: "A student of the Ice Age and an initiator of scientific polar exploration."[13] The idea of an Ice Age that had covered southern Scandinavia with glaciers at one time was revolutionary and not entirely acceptable in the mid-nineteenth century, but Torell was encouraged by Sven Lovén's studies of marine invertebrates that indicated a once-Arctic climate in southern Sweden. In 1858, he carried out a reconnaissance expedition along the west coast of Spitsbergen accompanied by Finnish–Swedish mineralogist Adolf Erik Nordenskiöld (1832–1901) and zoologist August Quennerstedt (1837–1926). Torell financed the expedition himself. The expedition on *Frithjof* visited Hornsund, Bellsund, Grønfjorden, Amsterdamøya, Norskøyane and Isfjorden, collecting geological and fossil specimens as well as zoological and botanical examples. Together with other expeditions to Switzerland, Iceland and Greenland, Torell was able to work out the latest Pleistocene glaciation in Europe, which earned him world fame despite his unwillingness to write about the results of his expeditions and investigations.[14]

Torell led one more expedition to Svalbard; this time, in 1861, it was on a completely different scale. Supported by the Royal Swedish Academy of Sciences and Crown Prince Oscar, the plans were threefold: to carry out interdisciplinary studies of Svalbard on land and the marine near-area, to reconnoitre for the establishment of a triangulation network (an extension of Sabine's work in 1823 and a precursor of the Swedish-Russian Arc of Meridian expedition at the end of the century—see later), and to attempt to sledge north from Sjuøyane to achieve a furthest north and maybe discover new land. An impressive array of scientists was assembled, including Nordenskiöld, Christian Vilhelm Blomstrand, Nils Christofer Dunér, Jakob K.E. Chydenius and others, requiring two ships to transport them all: *Aeolus* and *Magdalena*. The crews of the ships were mostly Norwegian, 15 on *Aeolus* and 11 on *Magdalena*. The sledge journey was abandoned, but extensive investigations were carried out both off the west coast of Spitsbergen and northeast to Hinlopenstretet, the north coast of Nordaustlandet, and Sjuøyane.

Liljequist explains in *High Latitudes* how the use of small boats for investigations along the coasts became the pattern for Swedish expeditions to come. The boats were used both for transport and living quarters with a ridge-roofed tent fitting over the boat for lodging, when they slept during the day and worked and travelled during the night (of course with the continual daylight at this latitude). Meals were augmented with fresh meat from reindeer, seals and birds.[15] Interesting for us today are descriptions and sketches of cultural heritage sites and objects such as Pomor crosses and huts that have long since disappeared.

The effort and the resources put into this expedition reaped extremely successful results, both regarding scientific material and improved mapping of a large part of Svalbard. Deep-sea soundings were conducted down to 2,500 m/8,200 ft and a knowledge of the paleoclimate of Svalbard and an outline of the geological structure of the northern part were obtained.

A small follow-up expedition in 1864 was sent out by the Swedish state and included Nordenskiöld and originally Chydenius (who died that year and was replaced by Dunér) and the zoologist A.J. Malmgren, but not Torell—which greatly offended him. However, he went on to further develop his scientific career and to see later scientists, including Nordenskiöld, follow the course he had staked out.

Adolf Erik Nordenskiöld

The 1864 expedition was to continue surveys for the measurement of an arc of meridian. Surveys were carried out at Bjørnøya, Isfjorden and Van Keulenfjorden, before visits were made to Hornsund, southwest Edgeøya, Agardhbukta and western Barentsøya, before returning to the west coast of Spitsbergen. At Prins Karls Forland, they rescued the shipwrecked crews of three Norwegian hunting vessels, which had been caught in the ice off Nordaustlandet. This could almost be seen as a preview of events that were to occur during Nordenskiöld's 1872–1873 expedition described in the next chapter.

In 1868, Nordenskiöld returned to Svalbard with the steamer *Sofia,* which he hoped would be able to penetrate the ice north of the archipelago. On the way north, scientific work was carried out on Bjørnøya and in Grønfjorden and Adventfjorden. The expedition moved north to Kongsfjorden and then left a group on Danskøya while the ship was sailed on the north. The ice conditions sent them eastwards to Liefdefjorden before they were again stopped by the ice near Sjuøyane. A return was made to Amsterdamøya where they met a small supply vessel, which could transport most of the scientific staff home. *Sofia* was again turned towards Nordaustlandet and on 19 September, she reached 81°42′N, which beat Scoresby's record for a ship (81°30′N in 1806). One more attempt was made, but *Sofia* was damaged by ice and despite Nordenskiöld in mid-October still wishing to press northeast to search for a supposed land area east of Svalbard called 'Gillis Land' the ice forced them to return south.

In 1870, a small expedition with the two young scientists, geologist and botanist Alfred Gabriel Nathorst (1850–1921) and engineer and geologist Hjalmar Mathias Wilander (1844–1891) was privately sponsored to investigate phosphorite deposits that Nordenskiöld had discovered at Kapp Thordsen in 1864. The aim was to see whether the deposits were economically worthwhile extracting for fertilisers. This was Nathorst's introduction to polar research and he was to go on to inherit Nordenskiöld's position as the "grand old man" of Swedish polar science.[16] The report the two compiled led to the plan to establish a colony at Kapp Thordsen for mining the phosphorite, but apart from a large house being erected in 1872 together with a railed track between the shore and the house, the project was abandoned. The house, now known as "Svenskhuset" (the Swedish House), is still standing and is maintained as important cultural heritage by the Norwegian authorities. Nordenskiöld's expedition of 1872–1873 is described in the next chapter. This was his last expedition to Svalbard.

The First International Polar Year 1882–1883

The International Polar Year (IPY) was a plan to turn national efforts from the aims of conquering geographical points such as the Northwest Passage and the North Pole to an international network of Arctic geophysical observatories that could coordinate efforts to measure and register the same phenomena in the same way and at the same time, which would give a far more useful scientific foundation than the more haphazard observations that previous expeditions had obtained. Austrian naval officer Karl Weyprecht initiated the idea, which led to 11 nations running 12 stations in the Arctic/sub-Arctic and two in the sub-Antarctic for c. 13 months from August 1882.[17] On the basis of their experience in Svalbard, the Swedes chose to make their contribution here. The government only provided the naval ship to transport the expedition, while the cost otherwise was covered through private donations. Nils Gustaf Ekholm (1848–1923) led the expedition, which consisted of five other scientists and six assistants. Among the scientists was civil engineer Salomon August Andrée who was to become most famous for his attempt to fly to the North Pole by hydrogen balloon in 1897—see later.

The original plan was to spend the year at Mosselbukta where Nordenskiöld had stayed 10 years before, but ice conditions stopped progress northwards and it was decided to use the house at Kapp Thordsen from 1872. The Polar Year passed by extremely successfully with the fixed observations and measurements being carried out according to the programme and the stay in Svalbard finished with all the participants in good health. This Polar Year was later followed by a second (1932–1933), third (1957–1958) and fourth (2007–2008).[18]

Also in 1882, Torell, who was now head of the Geological Survey of Sweden, organised a small Swedish geological expedition to be carried out by A.G. Nathorst and Gerard De Geer (1858–1943) to study geology and the glaciers of southwest Spitsbergen from Hornsund to Isfjorden. De Geer was to become a worthy successor to Torell in glacial geology and the study of Ice Age complexities. Nathorst was to be recognised for his studies of plant palaeontology. The small yacht *Bjona* of Tromsø was hired with a crew of eight. Once they had left the shelter of land, the rolling of the ship swilled the bilge water around and revealed that the earlier cargo of fish had not been cleaned out sufficiently and unbearable fumes of decomposed fish made it almost impossible to stay under deck.[19] In mid-July, they rendezvoused in Grønfjorden with the *Urd* and *Verdande,* which were transporting the Swedish IPY expedition northwards. Just before parting the news arrived that the ice was too thick further north for the ships to reach the planned base in Mosselbukta.

This small expedition managed to achieve an impressive amount of field studies leading to important advances in the knowledge of Svalbard's palaeontology and glacial geology. It was also to boost the careers of De Geer and Nathorst towards becoming recognised as outstanding scientists in their fields.

Sweden and the 1890s

In the last decade of the nineteenth century A.E. Nordenskiöld's active Arctic period was over, but new young Swedish scientists were developing with his support and advice, and still with the financial patronage he had enjoyed from Baron Oscar Dickson and the royal support from King Oscar II of Sweden and Norway. At this time, too, the Antarctic was beginning to interest scientists, following in the wake of the movement of whaling from the northern to the southern regions. Nordenskiöld was interested in a proposed expedition to Antarctica in cooperation with Australian interests and, being himself now 58 years old, planned to send his son Gustaf (1868–95) as the leader. To gain experience, Gustaf was to lead an expedition to Svalbard in 1890. This he did, but he contracted tuberculosis that same year and the Antarctic expedition never materialised.[20]

Gustaf Nordenskiöld's Svalbard expedition was financed mainly by Oscar Dickson with the aim of making geological, zoological and botanical observations and collections—an aim of just about all expeditions of the time. They were unable to land on Bjørnøya owing to fog but landed in Hornsund from where Gustaf and three others skied to Recherchefjorden to study the glaciers en route and reconnoitre for a possible arc measurement in the area. Their equipment was improvised, and conditions were worse than they had imagined with ice, fog, difficult snow and steep mountain sides, and they almost lost botany student Johan Alfred Björling on the way. This was probably the first time a longer inland journey was made in Svalbard. The expedition then sailed north and east to Hinlopenstretet doing field studies on the way and again a Swedish scientific expedition returned from Svalbard with rich results.

In 1892, a French naval expedition made a short scientific visit to Svalbard during which Charles Rabot and G. Lancelin attempted to cross Spitsbergen from Sassenfjorden to Agardhbukta. They walked up Sassendalen and climbed a mountain at its head, probably Wallenbergfjellet, before turning back.[21] This was perhaps the second longer inland journey in the archipelago. In 1896 and 1897, English explorer and mountaineer Sir Martin Conway made two expeditions to Svalbard with the express purpose of exploring the interior—probably the first to have this as their primary aim. With two companions, Conway crossed from Sassendalen to Agardhbukta in 1896 and that year and the next made further explorations inland on Spitsbergen. Conway's books *The First Crossing of Spitsbergen* from 1897 and *No Man's Land, a History of Spitsbergen from its discovery in 1596 to the beginning of the Scientific Exploration of the Country* from 1906 are classics within Svalbard literature.

Nathorst returned to Svalbard in 1898 on a scientific expedition again sponsored in part by King Oscar II. They successfully surveyed Bjørnøya before moving north to Bellsund to chart the Sound and accompanying fjords. Kong Karls Land, Kvitøya and the northern islands were visited with the expedition unfortunately missing the fact that Andrée's last camp was on Kvitøya. The mystery of the fate of the balloon expedition could then have been solved only a year after its disappearance instead of the 33 years it actually took. Considerable scientific results were obtained within geophysical observations and natural history collections. The following year Johan Gunnar Andersson led a small expedition to Bjørnøya where they spent nearly 2 months documenting the geology, flora and fauna.

The final Swedish expeditions this century were part of the ambitious Swedish–Russian undertaking to measure an arc of meridian in Svalbard in connection with determining the shape of the earth (i.e., whether it narrowed near to the poles). A preliminary expedition in 1898 reconnoitred triangulation stations and erected signals in the northeast of the archipelago for the main party to follow. In 1899–1900, the Swedish Arc of Meridian expedition worked out from Sorgfjorden in northeast Svalbard while the Russian counterpart was based on Gåshamna,

Hornsund. The time spent allowed both expeditions to carry out further surveys and scientific observations.

Other expeditions

Swedish scientists were not alone in Svalbard during the last half of the nineteenth century. In addition to Norwegian sealers and walrus hunters and Pomor and Norwegian wintering hunters, there were other scientific expeditions to the archipelago but not with the dedicated concentration to Arctic science seen from Sweden.

German expeditions went to Svalbard in the summers of 1868, 1869, 1870, 1886, 1889, 1891 and 1898,[22] often with the combined aims of hunting, exploration and various specimen collections, although the expedition in 1889 led by Willy Kükenthal was sponsored by the Geographische Gesellschaft in Bremen for marine zoological studies around Svalbard. Wilhelm Bade's expedition in 1891 was primarily a commercial tourist cruise with some minor scientific work also being carried out. Finally, Theodor Lerner's expedition med *Helgoland* in 1898 was a success with extensive scientific results, not least in zoology and the first accurate mapping of Kong Karls Land. The expedition also made the first recorded anticlockwise circumnavigation of Svalbard.[23]Also in 1891 and 1892, Prinz Heinrich von Bourbon's sporting expedition to Svalbard in the yachts *Fleur de Lys I* and *II* added hydrographical observations and sketch maps of all the harbours visited to the diverse knowledge of the archipelago.

North Pole attempts

During the 1890s and over the turn of the century, there were spectacular—and failed—attempts to travel to the north pole from the north of Svalbard. These certainly added to the general interest for Svalbard and generated an upswing in tourist traffic but added little on the scientific side apart from the remains they left behind. Today, these are important cultural heritage sites and objects. American journalist Walter Wellman tried first by ship in 1894 in the same area as Parry almost 70 years earlier. His later attempts by airship from Virgohamna on Danskøya took place a decade later and were equally a failure. Swedish engineer S.A. Andrée, who had been a member of the Swedish IPY expedition at Kapp Thordsen, tried twice—in 1896 and 1897—to fly a hydrogen balloon from Virgohamna and he and his two companions' remains were found 33 years later on Kvitøya in the far northeast of Svalbard. Their contribution to Svalbard science was mainly on the archaeological side, although Andrée's 1896 expedition ship provided transport for a small Swedish geological expedition led by Gerard De Geer, which examined the Isfjord region.

Norway moves into the field

From 1814 to 1905, Norway was in a personal union with Sweden under the Swedish king and although the Svalbard area was "in Norway's backyard," it is not surprising that national scientific expeditions were not sent to the archipelago before the rise of nationalism and the craving for independence became manifest towards the end of the nineteenth and beginning of the twentieth century. Hunting and trapping in Svalbard were, however, regarded as regular activities and Norwegian sailors were known for their skills in icy waters, where they also made important geographical discoveries and contributed to the scanty knowledge of the natural conditions in the European High Arctic. Fridtjof Nansen's Arctic expeditions across Greenland and with the *Fram* in the period 1888–1896 greatly helped to raise national awareness, but it was Prince Albert I of Monaco, a respected oceanographer, who helped to bring Norwegian scientists into Svalbard on what became a regular basis. Albert organised and led two ship-based expeditions to Svalbard in 1898 and 1899 on which he invited international scientists. On a new expedition in 1906, he invited among others, Gunnar Isachsen, a Norwegian cavalry captain and topographer who had participated on Otto Sverdrup's Fram Expedition to northeast Canada in 1898–1902. Isachsen's group started the topographic and scientific expeditions, which became an annual occurrence, and which led to the establishment in 1928 of the precursor of what is today the Norwegian Polar Institute.[24]

Conclusion

The nineteenth century saw the development of scientific expeditions to Svalbard with Swedish scientists among the most consistent in the last half of the century. In addition to their publications and collections, many place names today are living memorials to the scientists who made their names and careers in the archipelago.[25] Among other actors, British gentleman explorers also contributed to the sum of knowledge of the Svalbard archipelago both through their own mapping work and scientific collections and through the popular books they published after their voyages.

Notes

1. Holland, Clive. 1994. *Arctic Exploration and Development c. 500 b.c. to 1915*. New York & London: Garland Publishing Inc.: 171 (hereafter referred to as 'Holland 1994').
2. Stamp, Tom and Cordelia. 1993. *William Scoresby, Arctic Scientist*. Whitby: Caedmon of Whitby Press: 73.
3. Encyclopædia Britannica: https://www.britannica.com/biography/Edward-Sabine (Accessed May 2019).
4. Holland 1994: 94.
5. *ibid*: 200.
6. *Store norske leksikon* (encyclopaedia): https://nbl.snl.no/Balthazar_Keilhau (Accessed May 2019).
7. See elsewhere in this book.
8. Holland 1994: 325.
9. Mamier, Xavier. 1840. *Lettres sur le Nord*. Paris: H. L. Delloye.
10. Liljequist, Gösta H. 1993. *High Latitudes, A History of Swedish Polar Travels and Research*. Stockholm: The Swedish Polar Research Secretariat: 16 (hereafter referred to as 'Liljequist 1993').
11. Holland 1994.
12. *ibid*: 213.
13. Liljequist 1993: 22.
14. *ibid*: 24.
15. *ibid*: 31.
16. *ibid*: 73.
17. Barr, Susan & Cornelia Lüdecke (Eds.). 2010. *The History of the International Polar Years (IPYs)*. Springer.
18. *ibid*.
19. Liljequist 1993: 188.
20. *ibid*: 200.
21. Holland 1994: 374.
22. *ibid*.
23. *ibid*: 409.
24. Barr, Susan. 2003. *Norway—A Consistent Polar Nation?* 2nd edition. Oslo: Fram Museum: 61–62.
25. Norsk Polarinstitutt. 2003. *The place names of Svalbard*. Tromsø: Norsk Polarinstitutt (skrifter nr. 80, 112).

2 Adolf Erik Nordenskiöld and His Svalbard Expedition of 1872–73

Urban Wråkberg

Introduction: Northern confluence of interests

For most nineteenth-century Western explorers and scientists, participating in one Arctic voyage was enough. However, this was not so for Benjamin Leigh Smith (1828–1913) and Adolf Erik Nordenskiöld (1832–1901). The latter led the Swedish overwintering expedition on the northern coast of Svalbard in 1872–73. Nordenskiöld later offered both Axel Enwall (1844–1912), the medical doctor of that undertaking, and his geophysicist August Wijkander (1849–1913) an opportunity to participate in new expeditions he planned to Arctic Siberia and Svalbard. Although they had been keen volunteers for the 1872–73 overwintering, both Enwall and Wijkander declined.[1] As this chapter will show, for most, the tribulations of the 1872–73 overwintering could be endured but once.

The Swedish Arctic expedition of 1872–73 was one in a series of scientifically and partly industrially motivated tours that were dispatched from Sweden to Svalbard, Greenland, and the Siberian Arctic in the second half of the nineteenth century. They were based on ships borrowed from the Swedish navy or sealing or whaling ships chartered with crews from Northern Norway. Nordenskiöld had participated in these voyages since 1858; by 1872, he had led three major undertakings to Svalbard, the summer expeditions of 1861, 1864, and 1868. During the latter, a northern record at sea was set north of Svalbard, turning, out of the grip of the drift ice, at 81° 42′ N.[2]

As discussed elsewhere in this volume, Svalbard remained an uninhabited and lawless *terra communis* throughout the nineteenth century. The archipelago was beneficially situated under the influence of the comparatively warm southwesterly current of the North Atlantic Drift. It was favored logistically since the fishing communities of the towns of Hammerfest and Tromsø in northernmost Norway dispatched a number of sealing and hunting expeditions to the islands each summer. The Ebeltoft shipbroker family in Tromsø arranged re-supply in that port to expeditions on their way north. They could offer North Norwegian sailing ships for charter, including a crew with experience from the unchartered and ice-infested seas around Svalbard. This enabled any bold geographer with economical means to go north into what was little-known scientific territory at the time.[3]

In the first half of the nineteenth century, the British Admiralty and the Royal Geographical Society set the scene for a kind of open competition among industrializing states in providing the international scientific community and the colonially minded public with news on Arctic discoveries. Popular literature chronicling the efforts of polar explorers crossing ice fields,

U. Wråkberg (✉)
Dep. Tourism and Northern Studies, UiT The Arctic University of Norway, Kirkenes, Norway
e-mail: urban.wrakberg@uit.no

P. J. Capelotti (ed.), *The Coldest Coast*, Historical Geography and Geosciences,
https://doi.org/10.1007/978-3-030-67880-7_2

sighting new land, and attempting to attain the North Pole arose in affordable editions. The Royal Navy expedition under John Franklin (1786–1847) ordered to explore a North-West passage from the Atlantic to the Pacific Ocean disappeared in the Canadian Arctic archipelago after its departure from Portsmouth in 1845. The Franklin mystery ignited the public's interest and demonstrated how Western virtues of masculinity, heroism, and personal sacrifice could be found in the dramas of polar exploration. The reports and travelogues produced from the series of search expeditions on both sea and land for Franklin and his men created a literary genre still thriving today among armchair travelers and polar tourists.[4]

Small and colonially insignificant nations like the Nordic ones could earn international acclaim by organizing expeditions to the High North. This motivated the lobbyists who were instrumental in raising funds for Nordenskiöld's 1872–73 Arctic undertaking, which was to include an attempt at the highest north by marching on the sea ice north of Svalbard. The record of 82° 45′ N, along what was then called "the Spitsbergen route to the North Pole," had been set by man-hauling sledge parties of Edward Parry's (1790–1855) Arctic summer expedition in 1827. The logistical innovation of Nordenskiöld's new expedition was to overwinter in the far north and then start sledging in the spring, when daylight had returned and the sea ice was less broken, using Scandinavian domesticated reindeer as draft animals. This required building a winter camp in northern Svalbard, and to this purpose, the materials of a prefabricated house had to be brought north onboard what was finally no less than three ships: the brig *Gladan* of the Swedish Navy, an ice-strengthened rigged steamer named *Polhem,* also on loan from the Swedish state, and the merchant steamer *Onkel Adam* chartered at the last moment in Gothenburg to transport the draft animals and their fodder of reindeer moss. All three ships transported various amounts of provisions and supplies, while scientific instruments and storage for geological samples and biological specimens were transported on the *Polhem,* which was the sole ship intended to winter in the north.[5]

The Royal Swedish Academy of Sciences was successful on several occasions during the nineteenth century in lobbying for Arctic expeditions to be based on extra state funding voted by the Swedish Parliament. Not all parliamentarians were in favor of this manner of spending state funds, and similarly, the engagement of the Swedish Navy varied. Of crucial importance were the private funds raised for the 1872–73 expedition, mostly from western Sweden and the port of Gothenburg, which saw significant growth at that time in shipping, trade, and industry. Foremost among Nordenskiöld's private sponsors was the timber magnate Oscar Dickson (1823–97), who, given his tainted reputation for rough business methods in the lumber industry of northern Sweden, liked to be associated with the altruistic cause of exploring the unknown Arctic for the benefit of all humankind. Nordenskiöld and the Academy of Sciences enjoyed patronage from the Swedish Royal Court, which comprised the monarchy of both Norway and Sweden since 1814. The union of these two nations weakened over time until it was peacefully dissolved in 1905. In 1872, upon the death of Carl XV, King Oscar II (1829–1907) ascended the double throne. He was, among much else, a university-trained mathematician and keen theater lover. He engaged in culture, academic work, and foreign politics throughout his reign. He publicly endorsed and personally contributed economically to several Arctic expeditions that were launched from Sweden and Norway until his death in 1907, including the famous first crossing of the central Arctic by the Norwegian *Fram* expedition led by Fridtjof Nansen (1861–1930) in 1893–96.[6]

Nordenskiöld was a baron of the Swedish speaking nobility of Finland. Once in union with, or rather a part of Sweden, after unsuccessful wars with Russia at the beginning of the nineteenth century, Finland became a Grand Duchy of Imperial Russia. After peace, members of the Finnish nobility often made successful careers in the Russian administration. Nordenskiöld's

Fig. 2.1 Adolf Erik Nordenskiöld, in a famous painting by Georg von Rosen (1886; *Source* https://www.ub.uit.no/northernlights/eng/northeast01.htm)

father Nils Gustaf (1792–1866) was, like Adolf Erik, who in this regard followed his example, a mineralogist trained at Helsinki University. N. G. Nordenskiöld traveled widely in Russia in the service of the Czars, researched the mineralogy of the Ural Mountains, and later became superintendent of the Mining Board of Finland. Adolf Erik, in contrast, had to leave his country because of his involvement in liberal political agitation as a post-graduate student. He migrated to neighboring Sweden, where many regarded him a heroic political refugee, and embarked on a successful career in science. He was appointed as the professor of mineralogy at the Museum of Natural History in Stockholm in 1858 (Fig. 2.1).[7]

A. E. Nordenskiöld, among many skills, possessed the ability to communicate successfully and form alliances with people of all nations, and every walk of life, apparently. His networks extended far beyond the academic world, certainly including royalty and nobility and also landed gentry such as Leigh Smith. They comprised naval officers, politicians, and businesspersons, as well as ships-brokers in remote Tromsø, Arctic skippers, ice-pilots, and Sami indigenous hunters from the tundra of sub-Arctic Norway and Sweden. In the latter half of his career, he got along well with diplomats, leading scientists, investors, and merchants in Russia, where he also found a new major Arctic sponsor, the gold mine owner and Irkutsk philanthropist Aleksandr Sibiryakov (1849–1933).[8]

The socio-economic range of interests in the network of Nordenskiöld and his closest Arctic allies, Oscar Dickson, Georg Lindhagen (1819–1906), the permanent secretary of the Swedish Academy of Sciences in 1866–1901, and King Oscar II, is apparent in the several Swedish engagements that happened at Spitsbergen around 1872, in addition to the overwintering expedition of 1872–73. Dickson and Nordenskiöld had established a small Arctic mining company for extracting coprolite at Cape Thordsen [Kapp Thordsen] in Icefjord [Isfjorden], on western Spitsbergen, and introduced it to adventurous investors like themselves. Coprolite, i.e. fossilized guano, was mined, especially in England at the time, for use as fertilizer because of its high phosphate content. Nordenskiöld had discovered the deposit during previous expeditions. The first attempt to mine this mineral was during the summer of 1872, and near the site, a spacious house was constructed and stockpiled with supplies.[9]

Nordenskiöld and Dickson were the main agents not only of the 1872–73 expedition and this mining enterprise, they also lobbied successfully for an official Swedish request to be sent to all governments of nations with a possible interest in the terra nullius of Spitsbergen, to get their approval of an annexation making the islands part of Norway. The declared intents were to extend national legislation to protect the land claim and real estate of the Swedish mine at Cape Thordsen and facilitate scientific research on the islands in the future by using its mining camp as a scientific station. This idea was unpopular in Norway, where politicians generally disliked having the country's foreign relations guided or run from Stockholm and the national parliament was suspicious of the costs this seizure might entail in the long and short run, without denying that sealing and later trapping by North Norwegian hunters were crucial businesses on the islands. Later on, this diplomatic push stranded on Russian opposition.[10]

Nevertheless, the idea of mining at Cape Thordsen did not leave Nordenskiöld's mind. In 1888, he communicated with Leigh Smith and two other British Arctic explorers, Allen Young (1827–1915) and Henry Gore Booth (1843–1900), to engage them in forming a new company to restart mining at Spitsbergen. Nordenskiöld sent samples of the coprolite from Icefjord to London, Leigh Smith brought gold mining sites into the discussions, and Gore Booth considered undertaking an expedition to Spitsbergen that summer to prospect.[11] One may identify this and Nordenskiöld and Dickson's previous mining attempt as colonial moves toward so-called effective occupation resulting in land claims typical of Western geopolitics at the time, were it not for the lack of any permanent population on Svalbard, indigenous or otherwise. This often proved a crucial local component in any successful undertaking within the global colonial system of exploiting resources, people, and markets in non-European areas of the world.

The crisis of nineteenth-century polar exploration

The Swedish expedition and wintering on Svalbard in 1872–73 were the result of a confluence of scientific, mining, and geopolitical interests, orchestrated by Nordenskiöld and Dickson. Because of the endorsement by the Swedish Parliament and public support from Oscar II, leaders of the Swedish Navy decided to participate with two state-owned ships, two commanders, and crews selected from navy volunteers.

The goals and programs of the expedition were closely related to an international discussion of the time on the "polar question." It involved three main issues. First, what conditions should be expected in the completely unknown central Arctic based on new or old theories and conjectures from single remarkable observations and/or series of recorded data on the environment of adjacent, more accessible regions? Second, what should be regarded of most interest to science and humanity to explore in the little-known rim and center of the Arctic? The third issue was the perennial debate on which geographical route and mode of transportation were the most promising among the innovative alternatives proposed by different individuals and institutions.

After Parry's northern record by man-hauling sledging on the ice north of Svalbard in 1827, polar commentators in or associated with the British Admiralty promoted the alternative route north along the west coast of Greenland, arguing that traveling on or close to land for as long as possible was the most promising alternative for anyone intent on reaching the North Pole.[12]

Nordenskiöld had the sailing steamer *Sofia* at his disposal when setting the northern record at sea in 1868. The *Sofia*, like the *Polhem* of the 1872–73 expedition, was designed with an ice-strengthened hull, making it suitable for its regular service of transporting mail across the Baltic Sea between Finland and Sweden during winters. The ice formed there was light in comparison with the ice rim of the Arctic Ocean during summers. After 1868, Nordenskiöld and Dickson held no enthusiasm for further attempts to navigate far north at sea anywhere in the Arctic, not least because the *Sofia* sprung a leak during its ice encounters in the fall of 1868 and only made it back south after makeshift repairs on Svalbard.[13] As documented in other chapters of this book, Leigh Smith was more persistent in his numerous attempts at steam-assisted navigation of dense drift ice fields north and east of Svalbard during the following years.

In circles around the leading geographical journals on the European continent, enthusiasm was stirred in the 1860s by the prospects of applying steam propulsion in Arctic exploration. The old idea of the (seasonal) existence of an open polar sea in the central Arctic had many adherents. This "school" regarded it likely that large expanses of open sea would form in the central Arctic because of seasonal melting at the end of summer. The problem, they argued, was finding navigable leads into this more open center.[14]

This hypothesis was partly inferred from observations made off the coast of Siberia of so-called *polynyas*, wide expanses of open water in the sea ice that occurred even in winters. These occurrences were investigated in detail, interpreted scientifically much later, and found to result from local bathymetric and oceanographic conditions. The idea of a "penetrable ice wall" expanding/drifting outwards/south from the central Arctic because of the sea ice melting in the summer was "built" on a lack of knowledge at the time of the sea currents that we now know transport sea ice around and across the Arctic Ocean all year.

Nevertheless, following the German school of Arctic logistics, Julius Payer (1841–1915), who had participated in an attempt in 1868 to go north against the south-going stream of sea ice along the desolate eastern coast of Greenland, led two Austro-Hungarian expeditions searching for leads in the summer ice east of Spitsbergen in 1871 and in 1872–74 together with Karl Weyprecht (1838–81).[15] During the latter expedition, the steam barque *Admiral Tegetthoff* was beset in the ice and drifted through two winters within sight of an unknown icy coast on which some cartographic works were conducted. The new discovery was named Franz Josef Land, after the emperor of Austro-Hungary. It was revealed to be an Arctic archipelago rather than the tip of a previously unknown Arctic continent in subsequent expeditions led by Leigh Smith. Following this, little enthusiasm remained for further attempts to find open sea around the North Pole in summers. At the close of the nineteenth century, the first direct observations of ice conditions very far north, including soundings of the considerable depth of the central Arctic Ocean, were made on Nansen's Arctic crossing in the *Fram*.[16]

Despite the widely acclaimed endurance, successful return, and fine geographic discoveries of the 1872–74 Austro-Hungarian expedition,[17] its leader at sea, Karl Weyprecht, afterward joined the critics of polar exploration and promoted the concept of stationary polar research based in a safe and comfortable station as a future model for the systematic gathering of scientific data on the Arctic. He proposed that the station would preferably be part of a network of stations, which could then collaborate in making synoptic meteorological and geomagnetic observations. Such data would be more valuable to geoscience than single observations made on odd occasions, published in hard-to-find reports in the generic scientific literature. Middle-range excursions could be made from a well-supplied base camp. Polar stations would be built in the far north, but accessible from the south during the summer/fall season. Weyprecht's system of polar exploration in terms of network conception was realized during the so-called International Polar Year in 1882–83, where 12 nations participated by installing a scientific station along the rim of the high Arctic. August Wijkander, the young geophysicist and astronomer on the 1872–73 expedition, acted as the Swedish delegate in the planning of this joint undertaking.[18]

When pondering the significant difficulties, the Swedish expedition of 1872–73 encountered during its overwintering, it is clarifying to view them in the contemporary context of less successful polar exploration. Charles Francis Hall's (1821–71) *Polaris* expedition that aimed for the North Pole in 1871–73 was undoubtedly extraordinary in that Hall was killed during the expedition by arsenic, probably administered by one or more of the many expedition members critical of his failing leadership. The British Arctic Expedition of 1875–76 under George Nares (1831–1915), which managed to increase the northern record incrementally to 83°20′26″N, was plagued by scurvy and heavily criticized after its return home, causing the national interest for polar research to turn toward the Antarctic.[19]

John W. Danenhower (1849–87) was one of 12 survivors of the crew of 33 of George De Long's (1844–81) expedition in 1879–81 toward the Pole along the Bering Strait route, in the steam yacht *Jeanette*. Danenhower later stated, "I unhesitatingly record myself as opposed to further exploration of the central polar basin with our present resources. The gradual extension of observatory stations in the interest of meteorology, magnetism, and other scientific branches should be made, but national support should not be given to another polar expedition."[20] While not uncommon, at the time, this view was not the general opinion of the Western public of armchair travelers and potential sponsors of polar exploration. In 1909, Robert Peary (1856–1920), with considerable logistic support from his indigenous helpers, claimed to have reached the North Pole, and in 1911, Roald Amundsen (1872–1928) undoubtedly reached the South Pole using a modified system of dog team transport he had learned and adapted from the indigenous people of northern Canada.

Polar exploration at the time entailed long sea voyages in uncharted and ice-infested waters. It is difficult to imagine the challenges ice navigation presented even to experienced mariners during the age of sail. In remote Arctic or Antarctic seas, steam propulsion provided no immediate solutions to ice navigation. Away from the main shipping routes of the world, coal had to be used sparingly, and early steam engines consumed coal heavily. Moreover, there was no network of ports to enter and bunker coal in the polar regions. Having extra transport capacity to carry coal in a tender ship, like Nordenskiöld had in 1872 and Leigh Smith had the following year, enabled laying depots of it in strategic positions. Some of the coal stored at the Swedish station during the 1872–73 expedition still remains on the site today. In contrast to later diesel engines, steam propulsion offers no emergency start option. It took many hours to fire the boiler and get steam up before the engine and propeller could be put in motion. If you suddenly lost your mooring and went adrift, which happened more than once to the ships of Nordenskiöld's expedition, steam was no immediate helper. Travel under steam had to be planned according to the amount of coal remaining in the bunker and always took several hours to prepare.

Two of the ships used in the 1872–73 expedition had steam engines as well as sailing rigs. The brig *Gladan* was a sailing ship. The *Gladan* and the *Onkel Adam* were intended to return in the fall when the wintering station had been built, while the *Polhem* was to remain frozen in sea ice in a sheltered location close to the base camp, to be used for the return at the end of the following summer.

Nordenskiöld had visited Greenland in 1870 for scientific fieldwork, to enquire about dog team sledging, and to acquire dogs. The latter failed because of an infectious disease among the dogs, as a consequence of which almost none were for sale. Nordenskiöld then distributed a questionnaire in northern Norway and Sweden, probing the informants' experiences of reindeer as draft animals, and asked for advice on their usability under arctic conditions. The feedback was deemed positive.

Based on this, Nordenskiöld and Dickson's answer to the "polar question" was the following: no attempt at sailing or navigating by steam far north. By raising funding for a 2-year campaign, the first summer season would be used to find a good location for an on-land base camp and the erection of a prefabricated house, preferably somewhere on the Seven Islands [Sjuøyane], the northernmost islets of the Svalbard archipelago. After overwintering, much earlier in the season than when the north would be accessible by ship, sledge teams would assist and lay depots along a return route for one advance party that would sledge to the highest north in the early spring. This was the time of the year when the sea ice was expected to be less broken by pressure and open water than it would be later. Domesticated Scandinavian reindeer were to be used as draft animals. Their preferred fodder, reindeer moss, does not grow on Spitsbergen, so it had to be brought on the ships going north in 1872 for the winter and the ensuing sledge tours. Compared with meat for dogs, the ratio of food value to weight was regarded to be in favor of reindeer moss.[21] (This was before Amundsen developed his Antarctic transport system of starting out with very large dog teams and then using some of the dogs successively as food for the remaining ones while traveling). A scientific program in meteorology and geomagnetism was to be conducted at the base camp, from which a series of short biological excursions could be made into the surroundings.[22]

A winter on the dark side of Earth

One may be prone to say that most things went wrong on Nordenskiöld's 1872–73 expedition. The main cause for this was the unusually early arrival of large amounts of drift ice in northern Svalbard in mid-September 1872. During the 1868 Swedish expedition, the sea to the north of the archipelago was still navigable in October. The summer of 1872 was likewise characterized by larger than normal amounts of drift ice in northern Svalbard, hindering navigation earlier that year as well. However, ice conditions around Svalbard were known to change quickly with the wind direction; therefore, the *Polhem* and the *Gladan,* which carried the main parts of the expedition's house, after reaching Svalbard from Sweden, waited at anchor in a sheltered coastal area called Fair Haven on Northwest Spitsbergen for the ice to move in the northeast. The *Onkel Adam* arrived later with further supplies, coal, stores of reindeer moss, and the reindeer, which had been bought in Kautokeino in northern Norway (Fig. 2.2).[23]

Time was used productively in Fair Haven for research onshore and along the coast. August Wijkander repeated pendulum measurements of the strength of the Earth's gravitational field at a place on Inner Norwegian Island [Indre Norskøya] where the British army officer Edward Sabine (1788–1883) had undertaken measurements of high quality as a member of Douglas Clavering's (1794–1827) naval expedition in 1823. The Swedish expedition's botanist/biologist Frans Kjellman (1846–1907) dredged many places to investigate the fauna and algal flora on the coastal seafloor.

Leigh Smith's ship, the *Samson,* was sighted at Fair Haven on the evening of 29 August. He was returning south from a hunting and exploration voyage to Svalbard. He had been locked in by drift ice for 5 weeks in Wijdefjorden, which opens on the northern coast of Svalbard not very far from Fair Haven. Nevertheless, the trip from there had taken 12 days because of prevailing ice conditions, instead of the 1 or 2 days on an open sea. The Swedish scientists and officers visited Leigh Smith's ship and vice versa; they dined and exchanged information on ice conditions. Sharing, among others, a great interest inthe scientific exploration of the archipelago and its surrounding seas, they had much in common. Leigh Smith informed them of his plans for

Fig. 2.2 The site of Edward Sabine's "pendulum observatory" on the southern tip of Indre Norskøya in Fair Haven, NW Svalbard (Courtesy of Urban Wråkberg)

another expedition the following year and volunteered to call upon his arrival at the Swedish winter quarter to ensure that everyone was fine. Frans Kjellman ended on this meeting in his travelogue of the 1872–73 expedition: "After receiving the most heartfelt well wishes, we parted from this man, who was to care in the future for the expedition and whom is forever tied to its participants with the bonds of gratitude and affection."[24]

The Swedish flotilla of three vessels did not make it to the Seven Islands that fall. The sea ice was clinging to the coast. Days were spent scouting for leads or waiting for winds that would disperse the ice fields. Four attempts were made to reach Parry Island [Parryøya], one of the islets of the Seven Islands. Nordenskiöld and the two naval officers, the commander of the *Polhem* Louis Palander (1842–1920) and Gerhard von Krusenstjerna (1840–1907), who was in charge of the *Gladan,* did not always see eye-to-eye about the tactics, but Nordenskiöld had the final word when needed. The ice remained too dense for navigation toward the east as well, where alternative sites to those on the Seven Islands could be found. Murchison Bay [Murchisonfjorden] and Hinlopen Strait [Hinlopenstretet] could not be reached. Even Treurenberg Bay [Sorgfjorden], between Fair Haven and Hinlopen, which the Swedes knew well from previous expeditions, remained blocked.

On 5 September, after further reconnaissance, Mossel Bay [Mosselbukta] was chosen as the winter campsite. This bay is situated on the coast of the Neu-Friesland [Ny-Friesland] region of northern Svalbard. It is 7 kilometers wide and approximately 6 kilometers long and opening to the northwest. It is fairly exposed to northern winds and drift ice, but on its eastern shore, there is a narrow spit of stony land 5–10 meters above sea level, stretching half a kilometer in the north–south direction. The winter station was built on its elevated end facing the bay, and sheltered inside the natural breakwater it formed it was deemed safe for the *Polhem* to freeze in, situated a short walk from the main quarters. The position of the station was 79°53′N, 16°04′E.

The expedition experienced another major setback on the morning of 16 September. The station was now equipped with supplies, and the reindeer were on the shore close to it, where, assisted by their herders, they attempted to graze on the stony tundra. The *Onkel Adam* and the *Gladan* were ready to lift anchor since 15 September but had orders from Nordenskiöld to stay longer to give the wintering party time to complete letters and reports that had to be brought back home. However, since a north-westerly wind suddenly rose to a storm, none of the ships could head out to sea. This wind persisted unexpectedly for many days. This resulted in sea ice filling the entire bay. The temperature fell below freezing, and the sea began to freeze and cement ice blocks together. It was initially believed that the wind and temperature would soon change for the better and break up the ice, enabling the tide to flush the bay open. However, there was no such luck. The shocking prospect of having to winter at Mossel Bay dawned on those present who were not intended to take part in the overwintering according to the initial plan of the expedition.

An emergency plan was conceived. The winter station was spacious and suitable for those who intended to winter there, but it was not regarded desirable or possible to try to refit it to house more people. New lodgings had to be partly built in the holds of the *Onkel Adam* and the *Gladan* to accommodate their crews. The party initially intended for overwintering numbered 22 persons, including the scientific team and the officers and crew of the *Polhem*. They had ample stores of provisions, including antiscorbutics, for 16 months. The *Gladan* carried provisions that would be sufficient only until December for its officers and crew of 25. The merchant captain of the freighter *Onkel Adam* Leonard Clase (1839–83) had been wise enough to buy extra provisions on his own initiative en route to northern Svalbard, albeit only salted food. His crew of 13 included the female cook Amanda Wennberg, who became the first woman known to history by name to winter on Spitsbergen, but not much more.[25] Taking stock of everything, the stores were recalculated, and it was found that the provisions would last 66 persons until April the following year on full rations, and maybe until August or September on two-third rations. The anti-scorbutic in the store was harder to calculate. It was the kind common in Scandinavia at the time: mainly lingonberry or cloudberry jam and some sauerkraut. Cloudberries were later found to be the most effective, but because of its shortage, the jam would have to be prescribed by the MD Axel Enwall to those in need of it, rather than being distributed along with food rations.

The next mishap of the 1872–73 expedition occurred as collateral damage from the gales that filled the bay with ice. The reindeer on the shore were tended by experienced Sami herders; they were not inside an enclosure. Fences are not used in regular reindeer husbandry in northernmost Scandinavia. The entire herd of reindeer fled from the wind exposed coast, seeking shelter in the mountainous interior. None of the animals, except one, was ever seen again, despite extensive searches. A possibly overlooked factor that might have played a role in this is that Scandinavian reindeer do not inhabit the barren tundra during winter. Their natural seasonal behavior is to wander from their summer grazing grounds on the mountains down to sheltered forested regions in the fall. In any case, the second part of the expedition's innovative logistic concept was lost in this mishap. However, adaptability was then, as is now, part of productive scientific field research in the polar regions; we will return to this in the following section.

At the end of September, more bad news reached the people in Mossel Bay. A small group of Norwegian seal hunters suddenly arrived on foot and informed that no less than four of their ships were beset at Grey Hook [Gråhuken], approximately 30 km west across Wijdefjorden, and two more ships at Welcome Point

[Velkomstpynten], approximately 10 km farther west. Like Nordenskiöld, they had all been caught off guard by the sudden northern winds and the early arrival of ice. In those days, North Norwegian sloops and yachts engaged in Arctic seal and walrus hunting carried no emergency provisions. Opportunities to hunt for food would soon be gone with the daylight and, in contrast to the rich winter cod fisheries off Northern Norway, the shrimp and the tiny polar cod in the waters around Svalbard, were insufficient to sustain humans marooned anywhere on the islands.

Overall, 58 Norwegian sailors were trapped. Their representatives enquired about receiving support from the camp at Mossel Bay; Nordenskiöld and Palander suggested they join them at Mossel Bay and offered to share their rations after 1 December. They also informed the sailors that if any of them could drag their deck boats over the shore-fast ice to open sea and sail southwest to Icefjord, they could survive comfortably until the next summer using the fine house and the ample stores at Cape Thordsen, which were left by the Swedish mining company. The sailors' representatives returned west on foot.

Given what was known at the beginning of October, Nordenskiöld had to recalculate the provisions again. The addition of 58 men would mean reducing rations to an untenable degree, making a loss of life likely. Nordenskiöld disliked having to this, but reason and duty now compelled him to ask for rescue of the expedition. He entrusted the Norwegian hunters returning from Mossel Bay with a copy of a telegram that briefly described the situation to be sent from Tromsø to Oscar Dickson, should any of them succeed in reaching there subsequently. If most of the seal hunters were to join the camp at Mossel Bay from December, more than 100 men would have to be fed. Considering the available supplies, Nordenskiöld concluded, "With the necessary restrictions [in rations], this is only enough until the beginning of July; thus, we hope the government will dispatch a rescue vessel early next spring. We are healthy, in good spirit, face the future with calm, and hope for success in our undertaking."[26] As things turned out, none of the sealers returned to share stores at Mossel Bay. A revised, more optimistic calculation of the stores was then made, and the emergency interpretation of the situation was dropped from the expedition's agenda, which was now malleable and dependent on Nordenskiöld and Louis Palander's decision-making.

In December 1872, Palander chose to answer in writing to a suggestion made by Nordenskiöld to raise the daily rations of canned meat, given the latter's opinion that many had lost weight over the last three months. Palander, after consulting with Enwall, replied that the current rationing should be maintained because according to him, signs of scurvy among some winterers did not result from small food rations, and that it was unwise, as a consequence of increasing rations, to reduce the time the total supplies would suffice (from the beginning of September to the end of July). He added that if anyone at the scientific station and the *Polhem* provided a good reason and requested for more food, he would allow for it on an individual basis.[27] It indicates, among other things, that both Nordenskiöld and Palander wanted to ensure that the people wintering in the *Polhem* and the land station were in shape for the scientific field program in the coming spring (Figs. 2.3 and 2.4).

Later that winter, Nordenskiöld briefed the situation in another draft report/letter that he hoped somebody passing outside the ice of Mossel Bay would convey to Oscar Dickson: "Everyone at Polhem [the name then used for both the ship and the scientific station on shore] are well and fit. Hopes for the future are high. Provisions and the supplies of coal are sufficient. Rescue is not necessary."[28] This may strike an observer as an optimism verging on recklessness. Value statements aside, it demonstrates the complete commitment of Nordenskiöld (which was likely one shared by all participants in the expedition who were originally intended to winter) toward producing some prestigious scientific results from the expedition, despite all setbacks. Nordenskiöld and the *Polhem* team were prepared to push everyone at Mossel Bay for outcomes that would be considered successes

Fig. 2.3 Nordenskiöld's ships frozen in at Mosselbukta (from Kjellman 1875)

Fig. 2.4 Nordenskiöld's winter base at Mosselbukta (from Kjellman 1875)

at home, in line with the expedition's original goals.

Daily programs were put in place for all crew members and overseen by the ship's masters. Nordenskiöld stated, "Crews have been kept in steady work during weekdays, with useful things if such where to be had, else with unnecessary or purposely invented tasks."[29] The scientists and naval officers took turns to fulfill the repetitive diurnal and hourly observations required for the meteorology and geomagnetism scientific programs. This appears to have kept most minds occupied. The Sami team members took charge of the marine biological dredgings, which they executed regularly from small boats or openings they made in the ice on the bay close to the wintering station.

The first death to occur in the Swedish 1872–73 expedition was in December, when a 23-year-old *Gladan* boatswain surnamed Svan succumbed, probably from a lung ailment he had contracted before joining the expedition that was exacerbated by early stage scurvy.[30] Signs of scurvy increased and Enwall was completely occupied with checking on everyone's health, diagnosing scurvy in varying degrees, and administering antiscorbutics.

The situation was strained, among others, because of the guilt over the decision in mid-September, when Nordenskiöld had instructed the commander of the *Gladan* and the captain of the *Onkel Adam* to wait for reports and letters to be written, compelling them to stay those fateful 24 hours that resulted in them and their crews having to winter. The sun remained under the horizon between 21 October and 20 February. Because of the mountains beyond the coast of Mossel Bay, not even twilight at noon was visible in December in the south. Under an overcast sky, in the absence of moonlight or auroras, there was complete darkness. Seabirds and land-living animals were all gone, only Ptarmigans were noticed further inland but couldn't be hunted due to the dark.

During the first half of the winter, the temperature was occasionally above zero, which we now know to be normal on Svalbard. Episodes of southerly winds occurred, and on a few occasions, to everyone's surprise, the ice broke all the way to the beset ships and the bay largely opened. Nordenskiöld and Palander strongly urged the captains of the *Onkel Adam* and the *Gladan* to stand out to sea and leave. On 29 January, the *Onkel Adam* had steam up. In the dark, facing deteriorating weather, the ship lost its moorings and became dangerously adrift inside the bay. Both ships' crews and commanders battled the wind that changed direction toward land and rose to a storm in the evening. By assisting the *Onkel Adam* with the *Gladan's* longboat, despite all, the steamer's main anchor and valuable length of chain were salvaged. Sea ice reappeared and drifted into the bay, which soon froze over.[31] After these dramatic events, the temperature continued dropping.

Today, we need to imagine something like a moon landing without radio communication and media cover if we are to consider the isolation, stress levels, and the limited information on which plans had to be revised and decisions had to be made at Mossel Bay in the winter of 1872–73. August Wijkander noted in his diary that Nordenskiöld was able to rebuild a positive atmosphere despite all ills and setbacks.[32] Against all odds, people of various kinds in their various quarters returned to speaking terms.

Among the tidings that were unknown at Mossel Bay before the summer of 1873 were the fate of the Norwegian sealers at Welcome Point and Grey Hook. They were sturdy mariners in good physical shape. On 7 October, before scurvy had weakened any of them, 17 of the men at Welcome Point managed to pull two deck boats across the ice to open water and sailed and rowed successfully south, into Icefjord and to Cape Thordsen, to stay in the Swedish house for the winter. A month later, open water reached the beset ships at Welcome Point. Apart from two men who chose to winter at Grey Hook to watch over their uninsured ship and its catch of blubber and seal skins, the sealers from Grey Hook joined those at Welcome Point and, after hard sailing, returned to Tromsø; thus, 39 men were saved. When passing the mouth of Icefjord, an easterly gale prevented them from entering and rescuing the men at Cape Thordsen (Fig. 2.5).

The two men at Grey Hook succumbed to scurvy. Between January and April 1873, so did the 17 staying at Cape Thordsen, one after the other, according to their journal. The house of the mining company held ample stores of provisions, antiscorbutics, clothes, fuel, guns, etc. The cause of this sad outcome was much discussed at the time and is still debated today. Before the invention of the term "vitamin" and a stable medical consensus on the causes and remedies of scurvy, many different countermeasures were advocated, including not only various types of food but also physical exercise, outdoor activity, fighting lethargy or sloth with handiwork, and incessant indoor activities. Although the party at Cape Thordsen was young and fit, the most senior among them died early on, and he might have known the importance of organizing hunting and cooking routines during a high Arctic wintering. As seems to be the case with all notorious Arctic overwinterings with fatal outcomes, various theories on what really killed

Fig. 2.5 The house of Oscar Dickson and A.E. Nordenskiöld's polar station for combined mining and scientific purposes is still standing today at Icefjord on Svalbard (photo in 2011 and courtesy of Urban Wråkberg)

these men have been proposed over the years. The most recent one is lead poisoning.[33]

Rescue expeditions were dispatched north from Tromsø after the 39 sealers returned from Welcome Point, but darkness, rough weather, and dangerous icing on the riggings of the ships compelled all to turn. Experience told the experts in northern Norway that the sealers at Cape Thordsen would survive until the following summer, as would the scientific expedition in Mossel Bay. Of course, already in September 1872, the fact that the sealers had not returned as expected to Tromsø and the news that neither the *Gladan* nor the *Onkel Adam* had reached their home ports as planned made clear to Leigh Smith, Oscar Dickson, and the promoters of the 1872–73 expedition, that things were not going altogether as planned. Nevertheless, only Leigh Smith, true to his word, prepared a rescue operation to bring supplies to the Swedish winter camp as early as possible in the spring of 1873. By the end of 1872, at least everyone knew it to be located in Mossel Bay, not anywhere on the Seven Islands.

Running the revised scientific program of 1873

The winter and spring seasons of 1873 followed what is now known to be the usual weather pattern on Svalbard. The cold intensified, and there were no episodes of mild weather. Daylight reappeared and increased swiftly. Scurvy was afflicting many, especially among the crews wintering on board the *Onkel Adam* and the *Gladan*. Conditions in the house on shore were better.

Nordenskiöld and Palander developed a revised sledging program. Given that they were far south of the Seven Islands and would have to rely on man-hauling, they were still able to use the early spring season. Three teams were set up, two supporting teams carrying supplies for part of the distance and laying depots, and one advance party making a push across the sea ice and hopefully reaching beyond the Seven Islands. This was based on food provisions that had been reserved for the purpose, mainly pemmican. Sledging began on 24 April and continued mainly as planned initially, given that bad weather frequently halted progress. One man was lost despite security routines by walking alone on the sea ice between two groups engaged in pitching camp. Searches were undertaken to no avail. For anyone unfortunate enough to slip through a weak spot in the ice and into seawater, even the ability to shout for help is taken by hypothermia within minutes. The man-hauling continued, pulling sledges during the night to counter snow blindness and camping in one common tent during the day to keep warm. The physical strain and speed varied with the unevenness of the sea ice. The occurrence of pressure ridges increased as they proceeded north.

They reached the Seven Islands and climbed a mountain on Phipps Island [Phippsøya] on 17 May. Looking north, as far as the eye could see, there was nothing but a chaotic expanse of piled-up ice. It was too late to try to reconnoiter a route in this maze. Plans were now changed. They would sledge eastward, close to the northern shore of North East Land [Nordaustlandet], the easternmost part of the Svalbard archipelago. Moving in this direction would bring them to poorly mapped regions, beyond the furthest east reached by the Swedish expedition of 1861. This would be valuable in the construction of the new map of the archipelago that Nordenskiöld and his colleagues were preparing. However, open water now occurred between the sea ice and the shore. As the coast became steeper toward the east, they chose to abort plan B as well. They could not reach the cape, later named after Leigh Smith, which was a point to the east of which many Arctic geographers expected unknown land to exist. This included Nordenskiöld, who had made disputed observations from mainland Spitsbergen in 1864 of a hazy coastline of what he identified as the old Dutch whaler's enigmatic Giles Land, far in the pack ice east of Barents Land [Barentsøya] on southern Svalbard.[34]

The advance team innovatively turned southward on to land, climbed the crevasses on the side of the ice dome covering most of North East Land, and, much hampered by fog, but based on Nordenskiöld's experiences of marching on the Greenland ice sheet in 1870, moved far south before turning west. After several days of traversing the dangerous crevassed inland ice, they reached Wahlenberg Bay [Wahlenbergfjorden], which opens on the east coast of the Hinlopen Strait. They continued north along the shore of the Hinlopen Strait, and on 23 June, they reached Shoal Point [Langgrunnodden] in Murchison Bay, where they had laid a small boat on their outward tour. A part of the team used it, while the rest were offered a lift by a passing Norwegian sealing vessel. Thus, they returned to Mossel Bay. Here, the situation was very different from when they had left.

Trying months at Mossel Bay

Although daylight, milder temperatures, and many species of land and sea birds returned to Mossel Bay, the ice did not loosen its grip on the coast during the months of April and May. Subsisting on sub-sized portions of salted and preserved meat had taken its toll on everyone. Many had scurvy with anemia, bleeding gums, loose teeth, and lack of lower limb mobility caused by swelling and pain. It was clear that the amount of game the best hunters among the winterers could hunt in one day was insufficient to help them, even if reserved for those worst hit by scurvy. Smaller groups made two attempts to sledge to Grey Hook carrying a small boat along, but man-hauling conditions on the melting sea ice across the mouth of Wijdefjord proved extremely demanding and dangerous, so both excursions soon had to turn back.

The sea ice outside the bay moved in earnest only at the end of May, and open water was sighted through binoculars from the base camp on 29 May, still far from the beset ships. On 6 June, two sealing yachts from Hammerfest and Tromsø called at the ice rim and the winterers had their first update with news from the outside world. One of these ships carried some provisions, bread, butter, and ale sent by the shipbroker Ebeltoft in Tromsø. This was very welcome but made little difference to the overall situation. As stated previously, Norwegian sealers carried no extra provisions that they could spare, and plenty would have been needed for 60 starving people. Despite the arrival of summer, many at Mossel Bay worried that they might perish despite enduring the hard winter, as no appropriately stocked rescue missions seemed to be forthcoming from Sweden. Spirits were low at the camp; everyone was afflicted by scurvy, and the stores of preserved meat, which would be depleted within a month or two, were way past their "best before" date and were almost inedible.

It was only on the evening of 12 June, with the appearance of Leigh Smith's ship, the whaler *Diana,* on the horizon, and after its mooring along the ice outside Mossel Bay, that an end was put to the crisis of the 1872–73 Swedish overwintering expedition. All the officers and the two scientists Kjellman and Wijkander were present at Mossel Bay to greet Leigh Smith, to exchange information with him "and three young Englishmen Mr. Chermeside, Reverend Eaton, and one whose name we are unable to recall."[35] Palander, Nordenskiöld, and nine men were still away on their sledge tour.

Leigh Smith carried provisions of excellent quality and, most importantly, in ample quantities and freely bestowed a lot of it to the starved winterers. They presented a sorry sight to him and his fellow travelers. Kjellman later stated, apparently misinterpreting the mix of feelings, "We were received most kindly on-board the *Diana* and could read in the sincere earnestness of the ship's commander's face an appreciation over the way we had fought through the dangers of the Arctic winter, with which he was by no means unfamiliar."[36] The supplies included fresh potatoes, preserved vegetables and soups, and canned meat of various brands, not to mention wine, cognac, and tobacco.[37] As we know today, the many bottles of lemon juice were the most important. After many thanks and an exchange of well wishes, Leigh Smith unmoored from the ice edge at Mossel Bay and continued his journey north towards the Seven Islands.

Before its terminal stage, the ailments produced by scurvy can be swiftly cured by supplying the body with vitamin C. The sick at Mossel Bay cast off their crutches and spirits improved. On full food rations, many were soon able to undertake heavy physical work. Therefore, ice saws were used, and inch by inch, a canal was cut from the *Onkel Adam* and the *Gladan* to the melting ice edge.

Thus, on 29 June, when Nordenskiöld, Palander, and the men of the northern sledge excursion assembled at Mossel Bay, the *Onkel Adam* and the *Gladan* had already been pulled along the channel sawed in the ice, and were anchored in the open bay, reloaded and ready to leave. The *Onkel Adam* left that evening, and the *Gladan,* the next day.

Nordenskiöld and the scientists, who were to present the results of the expedition to its sponsors and the public back home, stayed on with the crew of the *Polhem* to produce what was to be an almost complete first annual series of meteorological data, and tidal and geomagnetic records from a scientific overwintering on Svalbard. The *Polhem* left Mossel Bay, first to find and bury the dead at Grey Hook, then to make an extensive summer excursion at sea on northern and western Svalbard, including biological dredgings, among others, at the Seven Islands. Nordenskiöld, Palander, and everyone else then onboard the *Polhem* got occasion to meet and thank Leigh Smith on 3 July when they found him in Treurenberg Bay onboard the *Diana,* and the *Samson,* which served as the coal tender.

Nordenskiöld undertook geological and paleontological research in the Icefjord and Bellsund regions until the *Polhem* returned north to collect the last people at the base in Mossel Bay, who were shutting it down on 18 July. They departed Svalbard on 1 August and reached Tromsø on the sixth. The *Polhem* experienced engine trouble that delayed their return to Sweden, but they reached Gothenburg on the morning of 29 August. They were greeted, among others, by Captain Clase, who informed them that both the *Gladan* and the *Onkel Adam* had returned to Sweden without further trouble. At this point, the 1872–73 Swedish overwintering expedition to Spitsbergen was dissolved (Fig. 2.6).

Homecoming

The 1872–73 expedition to Spitsbergen failed to set any spectacular geographical record. The march on the ice dome of North East Land was a novel feat but was not too productive in scientific

Fig. 2.6 The house of the Swedish polar station in Mossel Bay has not been cared for and is today a heap of planks (image from 4 September 2000). For the informed and careful visitor, it is still rich in signs of the drama it once witnessed (photo by and courtesy of Urban Wråkberg)

terms. The 1873 summer excursions in Icefjord, where paleontological collections completing his earlier fieldwork were made, were actually more fruitful, as they enabled Nordenskiöld to publish on the former climate of Svalbard.[38] This was long before Alfred Wegener (1880–1930) proposed his initially disputed theory on continental drift. Oscar Dickson covered the extra costs incurred because of the overwintering of the *Onkel Adam* and the *Gladan*. Nordenskiöld had to rely on the moral support of the Swedish Academy of Sciences, which acknowledged the value of the basic research produced by the expedition, foremost of which were an almost complete annual record of weather data and observations of life conditions through the polar night on northern Svalbard.

It is clear that several winterers would have succumbed at Mossel Bay, had they not been helped by Leigh Smith. This was acknowledged more openly in reports made public by the naval officer and captain leading the crews of the *Gladan* and the *Onkel Adam* than by Nordenskiöld.[39] In addition, the research conducted in the summer of 1873 would not have been possible without the provisions from Leigh Smith.

Nevertheless, the 1872–73 wintering did not hurt Nordenskiöld's Arctic career. Like Oscar Dickson, the new King Oscar II remained supportive of all his future Arctic endeavors. His focused, albeit callous, leadership and ability to cheer people up under adverse conditions remain impressive. Impressive as well was his ability to share leadership with Louis Palander in 1872–73 to make many decisions, including some where none of the alternatives was morally attractive.

As mentioned initially, most of the winterers at Mossel Bay were happy never to return to Svalbard. For Nordenskiöld, it became his final expedition to the islands but not his last Arctic voyage. He and his main sponsor Oscar Dickson directed their combined business and scientific interests toward the unknown interior of Greenland as a possible region of forestry (this was before Nansen's first crossing of the Greenland icecap in 1888–89), and toward Siberia and the opportunity to open trade along the northern sea route. Two members of the 1872–73 wintering did travel far north again with Nordenskiöld. They were the naval commander of the *Polhem*, Louis Palander, who later reached the rank of admiral in the Swedish Navy, and the field botanist and expert in Arctic algae, Frans Kjellman.

Kjellman, who was not even intended to join the wintering team in 1872–73, proved scientifically productive during it. He enlisted in Nordenskiöld's 1875 expedition to the Kara Sea and the mouth of the Yenisei River, where he assumed command of the sea party, while Nordenskiöld returned on land through Russia, meeting people and thus extending his scientific and economic network. Kjellman, with an experienced Norwegian crew in the *Prøven,* attempted to round the north cape of Novaya Zemlya, but had to turn back because of the sea ice, and returned safely to Tromsø. Both Kjellman, as the botanist, and Palander, as commander at sea, subsequently participated in Nordenskiöld's much-celebrated navigation of the North-East Passage between the Euroarctic and the northwest Pacific. In the winter of 1878–79, they survived another Arctic overwintering on-board the expedition ship *Vega,* beset on the north coast of Eastern Siberia not far from the Bering Strait. Based on more successful planning and bartering for fresh food with the Chukchi indigenous people of the region, that overwintering was characterized by much superior conditions than those of the one endured at Mossel Bay in 1872–73.[40]

Notes

1. Axel Enwall (1878) to Adolf Erik Nordenskiöld, 16 March. Letters to A.E. Nordenskiöld, main series. A.E. Nordenskiöld Collection. Center for History of Science, Royal Swedish Academy of Sciences, Stockholm, Sweden (hereafter, AEN Coll. CHS, RSAS); Josef Edström (1961) *August Wijkander: Vetenskapsmannen, kommunalmannen, riksdagsmannen* (Gothenburg: Göteborgs Kungl. Vetenskaps- och vitterhets-samhälle), 27–28.
2. The main English biographies on Nordenskiöld are: Alexander Leslie (1879) *The Arctic Voyages of Adolf Erik Nordenskiöld* (London: Macmillan); George Kish (1973) *North-East Passage: Adolf Erik Nordenskiöld, His Life and Times* (Amsterdam: Nico Israel). Swedish biographies: Sven Hedin (1926) *Adolf Erik Nordenskiöld: En levnadsbeskrivning* (Stockholm: Bonnier); Henrik Ramsay (1950) *Nordenskiöld: Sjöfararen* (Helsinki: Schildt).
3. Einar-Arne Drivenes et al. (2006) *Into the Ice: The History of Norway and the Polar Regions* (Oslo: Gyldendal Akademisk), 13–42.
4. Beau Riffenburgh (1993) *The Myth of the Explorer* (Cambridge: Scott Polar Research Institute); Robert David (2000) *The Arctic in the British Imagination 1818–1914* (Manchester: Manchester University Press); Urban Wråkberg (2019) "A.E. Nordenskiöld in Swedish Memory: The Origin and Uses of Arctic Heroism," *Acta Borealia* 36, 2, 166–182.
5. Frans R. Kjellman (1875) *Svenska polarexpeditionen år 1872–1873 under ledning av A.E. Nordenskiöld* (Stockholm: P.A. Norstedt & Söner), 1–14.
6. Urban Wråkberg (1999) *Vetenskapens Vikingatåg: Perspektiv på svensk polarforskning 1860–1930* (Stockholm: Royal Swedish Academy of Sciences), 18–27.
7. Leslie, *Arctic Voyages,* 4–20.
8. Wråkberg, "A.E. Nordenskiöld in Swedish Memory," 173–178. On Nordenskiöld's Russian network, see Andrey Skrydlov (2019) "N.A. Nordenskiöld's Polar Expeditions and the Russian Society," *Earth and Environmental Science 302: 4th International Scientific Conference "Arctic: History and Modernity"* (Bristol: IOP Conference Series) https://doi.org/10.1088/1755-1315/302/1/012003.
9. Urban Wråkberg (1999) "Politik och vetenskap i A.E. Nordenskiölds ockupationsförsök av Spetsbergen år 1871–1873," in Urban Wråkberg (ed.) *Arktisk gruvdrift: Teknik, vetenskap och historia i norr* (Stockholm: Jernkontoret), 39–52.
10. Swedish Ministry of Foreign Affairs, [memorandum] "Spetsbergen Maj 1892," Archives of the Swedish Ministry of Foreign Affairs, 1902 dossiersystem, vol. 308, file 70:I, Swedish National Archives, Stockholm.
11. Benjamin Leigh Smith (1888) to Adolf Erik Nordenskiöld, 2 May and 3 June. Letters to A.E. Nordenskiöld, main series. AEN Coll. CHS, RSAS.

12. Clements R. Markham (1865) "On the Best Route for North Polar Exploration," *Proceedings of the Royal Geographical Society* 9, 138–163; Clements R. Markham (1872–73) "On Discoveries East of Spitzbergen, and Attempts to reach the Pole on the Spitzbergen Meridians," *Proceedings of the Royal Geographical Society* 17, 97–107.
13. Gösta H. Liljequist (1993) *High Latitudes: A History of Swedish Polar Travels and Research* (Stockholm: Streiffert), ch. 9; Leslie, *Arctic Voyages,* 128–152.
14. August Petermann (1865) "Die Eisverhältnisse in den Polar-Meeren und die Möglichkeit des Vordringens in Schiffen bis zu den höchsten Breiten," *Petermanns Geographische Mitteilungen* 11, 136–146.
15. August Petermann (1872) "Geographie und Erforschung der Polar-Regionen, Nr. 60: Die neue Österreichische Nordpolar-Expedition unter Commando von Weyprecht und Payer und sechs andere neue Expeditionen, Rückblick auf die Geschichte arktischer Entdeckungen, die Walfische und das offene Polarmeer," *Petermanns Geographische Mitteilungen* 18, 145–150.
16. Drivenes, *Into the Ice*, 57–117.
17. Johan Schimanski & Ulrike Spring (2015) *Passagiere des Eises: Polarhelten und arktische Diskurse 1874* (Wien: Böhlau Verlag).
18. Susan Barr & Cornelia Lüdecke (eds.) (2010) *The History of the International Polar Years (IPYs)* (Heidelberg: Springer); Urban Wråkberg (2009) "IPY Field Stations: Functions and Meanings," in Jessica M. Shadian & Monica Tennberg (eds.) *Legacies and Change in Polar Sciences: Historical, Legal and Political Reflections on the International Polar Year* (Farnham: Ashgate Publishing), 47–71.
19. Pierre Berton (2000) *The Arctic Grail: The Quest for the Northwest Passage and the North Pole 1818–1909* (New York: Lyons Press).
20. John W. Danenhower (1885) "The Polar Question," *Proceedings of the United States Naval Institute* 11, 633–699, on 685.
21. "Svar på förfrågningar angående möjligheterna att till arktiska trakter medföra renar," [Answers to questioner regarding the possibilities of bringing reindeer to the Arctic] dossier 10, box F02b:1; "Samlade inbundna handlingar rörande de svenska polarexpeditionerna 1872–1873" [Collected documents regarding the Swedish polar expeditions 1872–1873] (hereafter just F02b:1), AEN Coll. CHS, RSAS.
22. Adolf Erik Nordenskiöld (1872) "Account of an Expedition to Greenland in the Year 1870," *The Geological Magazine* 9, 97, 289–306; Kjellman, *Svenska polar-expeditionen år 1872–1873,* 1–14.
23. Where no reference is provided in the following, and for several details of the execution of the 1872–1873 expedition, see Kjellman, *Svenska polar-expeditionen år 1872–1873* and Liljequist, *High Latitudes,* 75–93.
24. Kjellman, *Svenska polar-expeditionen år 1872–1873,* 82.
25. Susan Nicol (2007) "Women Over-winterers in Svalbard, 1898–1941," *Polar Record* 43, 224, 49–53. During the first half of the nineteenth century, Pomor fur trappers from the White Sea region of Russia wintered on Spitsbergen with families, including women and even children, e.g., in huts on two islands in Murchison Bay, not very far from Mossel Bay. See Urban Wråkberg (2006) "In the Days of Pioneers and Amateurs: On the Early History of Arctic Archaeology," in Vadim Starkov (ed.) *Spitsbergen in the History Research Works* (Moscow: Russian Academy of Sciences), 40–87 on 73.
26. Adolf Erik Nordenskiöld (1872) to Oscar Dickson, 1 October. Item 92, F02b:1, AEN Coll. CHS, RSAS. Original: "Med nödiga inskränkningar räcker detta endast till början af Juli, hvadan vi hoppas att

regeringen utsänder undsättningsfartyg tidigt nästa vår. Vi äro friska, vid godt mod, gå framtiden med lugn till mötes och hoppas på framgång i vårt företag."

27. Louis Palander 1872) to Adolf Erik Nordenskiöld, 29 December. Item 99, F02b:1, AEN Coll. CHS, RSAS.
28. Adolf Erik Nordenskiöld (1873) to Oscar Dickson, 15 March. Item 105, F02b:1, AEN Coll. CHS, RSAS. Original: "Alla Polhems innevånare raska. Framtids förhoppningarna goda. Proviant och koltillgången tillräcklig. Undsättning ej nödig."
29. Adolf Erik Nordenskiöld (1873) to Oscar Dickson, 25 January. Item 112, F02b:1, AEN Coll. CHS, RSAS. Original: "manskapet har under hvardagarne hållits till träget arbete, nödigt om sådant funnits, i annat fall onödigt eller för ändamålet påhittat."
30. Axel Enwall (1872) 22 December. "Intyg om båtsman Svan" [Death certificate boatswain Svan] Item 97, F02b:1, AEN Coll. CHS, RSAS.
31. Leonard Clase (1873) 1 February. "Utdrag ur Journalen ombord å Onkel Adam" [Excerpt from the logbook of *Onkel Adam* (for 29 January)] Item 114, F02b:1, AEN Coll. CHS, RSAS.
32. Edström, *August Wijkander*, 17–18.
33. Torill Ustad Stav (2012) "Derfor døde 17 i dette huset," NRK Troms & Finnmark, see: https://www.nrk.no/tromsogfinnmark/avslorer-hvorfor-17-fangstmenn-dode-1.8407675 (accessed May 9, 2020).
34. Urban Wråkberg (2002) "The Politics of Naming: Contested Observations and the Shaping of Geographical Knowledge," in Michael Bravo & Sverker Sörlin (eds.) *Narrating the Arctic: Collective Memory, Science, and the Nordic Nations, 1800–1940* (Canton, Mass.: Science History Publications), 155–197 on 163–166.
35. Kjellman, *Svenska polar-expeditionen år 1872–1873*, 268. Original: "jämte några unga engelsmän, Reverend Eaton, Mr Chermeside och en, hvars namn vi ej erinra oss".
36. Ibid. 268. Original: "Vi mottogos mycket vänligt om bord och läste i den allvarlige fartygschefens drag den sannaste tillfredsställelse öfver det sätt, på hvilket vi bekämpat den artiska vinterns faror, om hvilka han ingalunda var okunnig."
37. "Uppgift på den proviant som af Hr. Smitt skänkts till Svenska polar expeditionen" [List of the provisions bestowed by Mr. Smith to the Swedish polar expedition] (1873) 16 June. Item 120, F02b:1, AEN Coll. CHS, RSAS.
38. Kjellman, *Svenska polar-expeditionen år 1872–1873,* 334–345; Adolf Erik Nordenskiöld (1875) "On the Former Climate of the Polar Regions," *The Geological Magazine*, 525–532.
39. P.J. Capelotti (2010) "Benjamin Leigh Smith's Third Arctic Expedition: Svalbard, 1873," *Polar Record* 46, 239, 359–371, on 363–364.
40. Leslie, *Arctic Voyages,* 278–387; Liljequist, *High Latitudes,* ch. 15–16; Urban Wråkberg & Gunilla Lindberg-Wada (eds.) (2002) *An Arctic Passage to the Far East: The Visit of the Swedish Vega Expedition to Meiji Japan in 1879* (Stockholm: Royal Swedish Academy of Sciences).

References

Barr S, Lüdecke C (eds) (2010) The history of the International Polar Years (IPYs). Springer, Heidelberg

Berton P (2000) The Arctic grail: The quest for the Northwest Passage and the North Pole 1818–1909. Lyons Press, New York

Capelotti PJ (2010) Benjamin Leigh Smith's third Arctic expedition: Svalbard, 1873. Polar Rec 46(239):359–371

Clase L (1873) Utdrag ur Journalen ombord å Onkel Adam, 1 February, item 114, Samlade inbundna handlingar rörande de svenska polarexpeditionerna 1872–1873 [Collected documents regarding the Swedish polar expeditions 1872–1873] box F02b:1 [hereafter, F02b:1], A.E. Nordenskiöld Collection.

Center for History of Science, Royal Swedish Academy of Sciences, Stockholm, Sweden [hereafter, AEN Coll. CHS, RSAS]
Danenhower JW (1885) The polar question. In: Proceedings of the United States Naval Institute, vol 11, pp 633–699
David R (2000) The Arctic in the British imagination 1818–1914. Manchester University Press, Manchester
Drivenes E-A et al (2006) Into the ice: The history of Norway and the polar regions. Gyldendal Akademisk, Oslo
Edström J (1961) August Wijkander: Vetenskapsmannen, kommunalmannen, riksdagsmannen. Göteborgs Kungl. Vetenskaps- och vitterhets-samhälle, Gothenburg
Enwall A (1872) Intyg om båtsman Svan, 22 December, item 97, F02b:1, AEN Coll. CHS, RSAS
Enwall A (1878) to AE Nordenskiöld, 16 March, Letters to AE Nordenskiöld main series, AEN Coll. CHS, RSAS
Hedin S (1926) Adolf Erik Nordenskiöld: En levnadsbeskrivning. Bonnier, Stockholm
Kish G (1973) North-East Passage: Adolf Erik Nordenskiöld, his life and times. Nico Israel, Amsterdam
Kjellman FR (1875) Svenska polar-expeditionen år 1872–1873 under ledning av A.E. Nordenskiöld. P.A. Norstedt & Söner, Stockholm
Leigh Smith B (1888) to AE Nordenskiöld, 2 May, Letters to AE Nordenskiöld main series, AEN Coll. CHS, RSAS
Leigh Smith B (1888) to AE Nordenskiöld, 3 June, Letters to AE Nordenskiöld main series, AEN Coll. CHS, RSAS
Leslie A (1879) The Arctic voyages of Adolf Erik Nordenskiöld. Macmillan, London
Liljequist GH (1993) High latitudes: A history of Swedish polar travels and research. Streiffert, Stockholm
Markham CR (1865) On the best route for north polar exploration. Proc Roy Geogr Soc 9:138–163
Markham CR (1872–73) On discoveries east of Spitzbergen, and attempts to reach the pole on the Spitzbergen meridians. Proc Roy Geogr Soc 17:97–107
Nicol S (2007) Women over-winterers in Svalbard, 1898–1941. Polar Rec 43(224):49–53
Nordenskiöld AE (1872) Account of an expedition to Greenland in the year 1870. Geol Mag 9(97):289–306
Nordenskiöld AE (1872) to Oscar Dickson, 1 October, item 92, F02b:1, AEN Coll. CHS, RSAS
Nordenskiöld AE (1873) to Oscar Dickson, 15 March, item 105, F02b:1, AEN Coll. CHS, RSAS
Nordenskiöld AE (1873) to Oscar Dickson, 25 January, item 112, F02b:1, AEN Coll. CHS, RSAS
Nordenskiöld AE (1875) On the former climate of the polar regions. Geol Mag, 525–532
Palander L (1872) to AE Nordenskiöld, 29 December, item 99, F02b:1, AEN Coll. CHS, RSAS
Petermann A (1865) Die Eisverhältnisse in den Polar-Meeren und die Möglichkeit des Vordringens in Schiffen bis zu den höchsten Breiten. Petermanns Geographische Mitteilungen 11:136–146
Petermann A (1872) Geographie und Erforschung der Polar-Regionen, Nr. 60: Die neue Österreichische Nordpolar-Expedition unter Commando von Weyprecht und Payer und sechs andere neue Expeditionen, Rückblick auf die Geschichte arktischer Entdeckungen, die Walfische und das offene Polarmeer. Petermanns Geographische Mitteilungen 18:145–150
Ramsay H (1950) Nordenskiöld: Sjöfararen. Schildt, Helsinki
Riffenburgh B (1993) The myth of the explorer. Scott Polar Research Institute, Cambridge
Schimanski J, Spring U (2015) Passagiere des Eises: Polarhelten und arktische Diskurse 1874. Böhlau Verlag, Wien
Skrydlov A (2019) N.A. Nordenskiöld's polar expeditions and the Russian Society. In: Earth and environmental science 302: 4th international scientific conference "Arctic: History and Modernity". IOP conference series, Bristol. https://doi.org/10.1088/1755-1315/302/1/012003
Svar på förfrågningar angående möjligheterna att till arktiska trakter medföra renar [Answers to questioner regarding the possibilities of bringing reindeer to the Arctic]. Dossier 10, box F02b:1, AEN Coll. CHS, RSAS
Swedish Ministry of Foreign Affairs (1892) Spetsbergen Maj 1892 [memorandum]. Archives of the Swedish Ministry of Foreign Affairs, 1902 dossiersystem, vol 308, file 70:I, Swedish National Archives, Stockholm
Uppgift på den proviant som af Hr. Smitt skänkts till Svenska polar expeditionen [List of the provisions bestowed by Mr. Smith to the Swedish polar expedition] (1873) 16 June, item 120, F02b:1, AEN Coll. CHS, RSAS
Ustad Stav T (2012) Derfor døde 17 i dette huset. NRK Troms & Finnmark. https://www.nrk.no/tromsogfinnmark/avslorer-hvorfor-17-fangstmenn-dode-1.8407675
Wråkberg U (1999) Politik och vetenskap i A.E. Nordenskiölds ockupationsförsök av Spetsbergen år 1871–1873. In: Wråkberg U (ed) Arktisk gruvdrift: Teknik, vetenskap och historia i norr. Jernkontoret, Stockholm, pp 39–52
Wråkberg U (1999) Vetenskapens vikingatåg: Perspektiv på svensk polarforskning 1860–1930. Royal Swedish Academy of Sciences, Stockholm
Wråkberg U (2002) The politics of naming: Contested observations and the shaping of geographical knowledge. In: Bravo M, Sörlin S (eds) Narrating the Arctic: Collective memory, science, and the Nordic nations, 1800–1940. Science History Publications, Canton, Mass., pp 155–197

Wråkberg U (2006) In the days of pioneers and amateurs: On the early history of Arctic archaeology. In: V Starkov (ed) Spitsbergen in the history research works. Russian Academy of Sciences, Moscow, pp 40–87

Wråkberg U (2009) IPY field stations: Functions and meanings. In: Shadian JM, Tennberg M (eds) Legacies and change in polar sciences: Historical, legal and political reflections on the International Polar Year. Ashgate Publishing, Farnham, pp 47–71

Wråkberg U (2019) A.E. Nordenskiöld in Swedish memory: The origin and uses of Arctic heroism. Acta Borealia 36(2):166–182

Wråkberg U, Lindberg-Wada G (eds) (2002) An Arctic passage to the Far East: The visit of the Swedish Vega expedition to Meiji Japan in 1879. Royal Swedish Academy of Sciences, Stockholm

3 Axel W. Enwall (1844–1912), Medical Doctor and Polar Photo Pioneer

Anders Larsson

Introduction

In the newspaper *Nya Kristinehamnsposten* on 22 July 1912, one could read the following: "*Yesterday evening an unusual solemn burial ceremony took place at the local church, when the remains of the late Dr. Axel Enwall the former head physician at Kristinehamn psychiatric hospital were put to rest.*" The burial ceremony marked the end of a remarkable life. The work of Enwall within many areas of society and his travels to different parts of the world were unusual and ahead of his time.

Early life and preparation for a polar expedition

Axel Wilhelm Enwall was born in 1844 in the town of Enköping, a son of the shipmaster Pehr Ulrik Enwall. He completed his schooling at Katedralskolan in Uppsala and thereafter went on to study medicine at Uppsala University, and he eventually achieved his licentiate's degree at the Karolinska Institute in Stockholm in 1874. Already during his ongoing medical studies, Enwall in 1866 was recruited to the general military hospital in Uppsala (*Allmänna garnisonssjukhuset*) as an assistant physician. He also served as naval surgeon on the monitor battleship *John Ericsson* during the ship′s expedition to the Baltic Sea and the Gulf of Bothnia in 1868. From 1869, he served as a battalion doctor at the Royal Naval station in the town of Karlskrona. Around this time, he also served as a doctor on board the corvette *af Chapman* during an extensive journey to England, Spain and the Azores. Possibly, these travels and his experiences from the trips to the Gulf of Bothnia inspired Enwall to accompany on journeys to even more remote destinations later on (Fig. 3.1).

Hence, for some years, Axel Enwall had been working as a military doctor both at sea and land. For someone of his age, he had acquired broad and thorough working experience as a medical doctor under different conditions, which made him well equipped to face new challenges. These presented themselves already in 1872 when he was offered to participate as a medical doctor in the wintering expedition of Adolf Erik Nordenskiöld to Svalbard. This polar expedition, mainly sponsored by various patrons from Gothenburg, was the first Swedish expedition, which had the intent of wintering at the highest northern latitude possible in the polar region. It is not apparent as to how A. E Nordenskiöld became aware of the young doctor, but it is thought possible that contacts were channeled through the Royal Navy, as two of the expedition's ships were crewed by the marines, thus providing contacts that may have facilitated Enwall's path into the polar career.

A. Larsson (✉)
Gothenburg University Library, Gothenburg, Sweden
e-mail: anders.larsson@ub.gu.se

P. J. Capelotti (ed.), *The Coldest Coast*, Historical Geography and Geosciences,
https://doi.org/10.1007/978-3-030-67880-7_3

Fig. 3.1 Axel Enwall (1844–1912), doctor and polar photo pioneer. From Svenskt porträttgalleri XIII, Läkarekåren, Stockholm 1899

A doctor overwinters in Svalbard

A.E. Nordenskiöld had participated or headed four expeditions to Svalbard, the latest as late as 1868. During the latter expedition, doctor Carl Nyström was the responsible medical officer. After his homecoming, he wrote a report on the conditions of health and provisions on this expedition, which presumably was useful to Axel Enwall.

The history of the expedition is documented in other parts of this publication and shall therefore not be gone into here. As for Axel Enwall, he went ardently into his work as the responsible medical officer on the main ship of the expedition, namely the former mail steamer *Polhem*. As events unfolded, this proved to be even more of a challenge than Enwall could ever have anticipated.

The expedition was originally planned to head for the Seven Islands, which are located to the north of Svalbard at approximately 80 degrees north latitude. When this proved impossible, the expedition had to establish its wintering station at the so-called Mossel Bay on the Northwest Spitsbergen mainland. Here, a building was raised to house the expedition over the winter, with the intention that the two assisting ships, *Gladan* and *Onkel Adam*, would soon after unloading provisions and other materials return to Sweden.

This plan was not implemented until the 16th of September, when a severe storm brought massive drift ice into Mossel Bay and rendered the planned journey back to Sweden impossible. Enwall was now stranded, with the medical responsibility for 67 men and a female cook (Amanda Wennberg), rather than just over 20 men for whom the wintering was originally planned.

Events took turn for the worse when on the 30th of September, seven Norwegian seal hunters, sent out from some ice-trapped ships, arrived at Mossel Bay to ask Nordenskiöld's expedition for provisions assistance over the winter. The seal hunting crews amounted to 60 men, which together with the people at Mossel Bay would amount to 120 persons. This meant the rations had to be reduced at a minimum. As Nordenskiöld writes: "*As long as some provisions remained to be shared, it was not possible to let a fellow human in close proximity to face death through starvation, and we despatch therefore with the trappers a written confirmation, that we would to our utmost ability receive, feed and shelter them.*"[1]

As events developed, the promised assistance for provisions for the Norwegian seal hunters was not required, as some were sent on to Isfjorden (where the so-called Svenskhuset (Swedish House) was located), whilst others managed to free themselves from the ice and made their way to Tromsö in northern Norway. A relieved *Nordenskiöld* writes:*"Had that happened, none of us nor the seal hunters would likely have survived the winter."*[2].

The only ones who remained behind were a seal hunter by the name of Johan Mattilas and his aide Gabriel Andersen, who chose to pass the winter at Gråhuken, a couple of miles from Mossel Bay where they had their catch to guard. The two of them died during the winter and were found dead during the spring of 1873. Furthermore, all of the seal hunters who were sent to Svenskhuset passed away there during the winter, probably caused by lead poisoning. The narrative concerning this incident and their death is richly documented but will be left aside here as it was beyond Axel Enwall's medical responsibility.[3]

Even though the absence of the expected Norwegian seal hunters alleviated the anticipated burden on the expedition at Mossel Bay, Enwall still had enough on his mind concerning the health conditions and provisions prior to the impending wintering. Enwall writes: "*As afore mentioned, Gladan with its 25 and Onkel Adam with its 13 crew (among them a female cook) were confined here on the 16th September, and the expedition with its stock of clothing and other necessities, enough for 21–22 men for 18 months, found itself necessitated to feed no less than 67. Had the addition of 58 (57) trappers arrived and shared our bounties, the figure would have amounted to a respectable 125 instead of 22, a calamity that we now have avoided* (Fig. 3.3)."[4]

The accommodation of the two crews in addition to the regular expedition developed into a problem as all of them could not be housed on land but were forced to remain on board the ships. This entailed crowded living and problems of dampness. The observant doctor Enwall realized that these conditions could cause several problems, in particular with scurvy. Moreover, there was a psychological dimension, which darkened the atmosphere of those members of the expedition who had not anticipated remaining at the station over the winter. Enwall comments: "*A further aspect to be considered is the importance of the etiological argument relating to the development of scurvy on board these ships. It is indubitably that with this disease, anything that heightens or lowers the senses is of the utmost importance. We who have our accommodation on land were prepared to spend the winter here, but not so for the crew of the other two ships a state which more or less must evoke feelings of misery with the unforeseen separation from home and relatives, and incarcerated in darkness for 4 months with 2/3 rations.*"[5]

As early as the end of November and the beginning of December, the first cases of scurvy appeared. The causes of the disease were at the time not known, i.e., the lack of vitamin C (ascorbic acid), and Enwall tested different medical concoctions such as quinine and iron. It was nonetheless known that juice and jam made from fresh berries and fruits had a very positive effect in curbing the disease. "*Next to lime juice I believe cloudberries among all of the antiscorbutica that are known to me, to be the best. The effect achieved from these berries at a dose of 120–130 g daily was quite substantial.*"[6]

Throughout the winter season, the health condition of the expedition and of the involuntary hibernators was overall satisfactory with exception of the cases of scurvy. One man died, a boatswain named Svan on board the *Gladan*—albeit not of scurvy. According to Enwall, the diagnosis was "pleuritis dextra," i.e., the inflammation of the right lung lining (pleural). There were yet numerous cases of muscular rheumatism and digestion disorders and several cases of snow blindness and frostbite.

Enwall also carried out inspections of the other ships to keep an eye on health conditions. He thus met occasionally with the captain of *Onkel Adam,* Börje Leonard Case, originally from Onsala, to share a toddy. The rest of the winter season passed without further death, but time and the reduced rations took their toll. "*We are now surviving on 2/3 rations for the 9th month, my antiscorbutica will soon be depleted, the cloudberries are a long while ago consumed, and only half a jar of lime juice remains and still 15 patients are under my care. The iron, which was provided for in the well stock apothecary, is all consumed and I have now asked the sick to file iron filings that I have prepared with conserved milk and chocolate into trochisi (pills) which has a good effect.*"[7] Most men were anemic and suffered from dejection, "since [any] rescue ship from Sweden, long awaited all winter or at least during spring, will not arrive."[8]

Rescue

Rescue was, nevertheless, closer at hand than one had otherwise thought. On the evening of 12 June 1873, a steamship equipped for whaling purposes was observed at the edge of the ice. It proved to be *Diana*, the ship of the Englishman Benjamin Leigh Smith that had arrived at Svalbard and for the second time paid a visit to the Swedish expedition. Hurriedly the ship was met at the edge of the ice and the men were very well received on board. The hardships of the winter

Fig. 3.2 A photograph by Axel Enwall of Benjamin Leigh Smith's ship *Diana* with members of the crew from *Gladan* transferring provisions for the Swedish expedition, 13 June 1873. (Courtesy Grenna Museum: Andréexpeditionen Polarcenter)

and the critical situation of provisions were related to the visitors and thereafter Leigh Smith gave his assurance that he was well stocked with provisions and was willing to provide as much as he could to the Swedish expedition. One of the participants on the Swedish expedition, the botanist Frans Reinhold Kjellman writes "*It wasn't until late morning on the 13th that we returned home, well pleased by our visit on the Diana. Later during the same day Mr. Smith came ashore and was thoroughly furnished with information of our circumstances, looked at our little cabin and observatories, and had us give an account of our fate and work* (Fig. 3.2)."[9]

The provisions provided by Leigh Smith were very much appreciated and Enwall writes: "*In the month of June, however, the condition of the sick was such that I am almost certain that if the Englishman, Mr. Leigh Smith, had not kindly presented the expedition with a large quantity of preserved provisions, fresh potatoes, and other refreshments, we should not have got off with less than one or more deaths in addition.*"[10] Now, with the reinforcement of provisions from Leigh Smith, the future horizons looked far brighter for the Swedish expedition, and Enwall observes that it even had a positive influence on the general health situation, which earlier was negatively influenced because of the uncertainty to be rescued from the ice.

Leigh Smith's visit found A.E. Nordenskiöld away from the base camp, together with, among others, Louis Palander, the expedition's Sami participants, and the crew from *Polhem* on a few weeks research expedition at the nearby North East Land (incidentally the first comprehensive survey and study of this part of Svalbard). Upon his return to Mossel Bay, Nordenskiöld together with the other returnees were gladly surprised by the visit of *Diana* and the enforcement of provisions.

Nevertheless, not everything had gone according to Nordenskiöld's plans. As mentioned, a man on board the ship *Gladan* had already died and now Nordenskiöld reported another death. This time the death of a boatswain

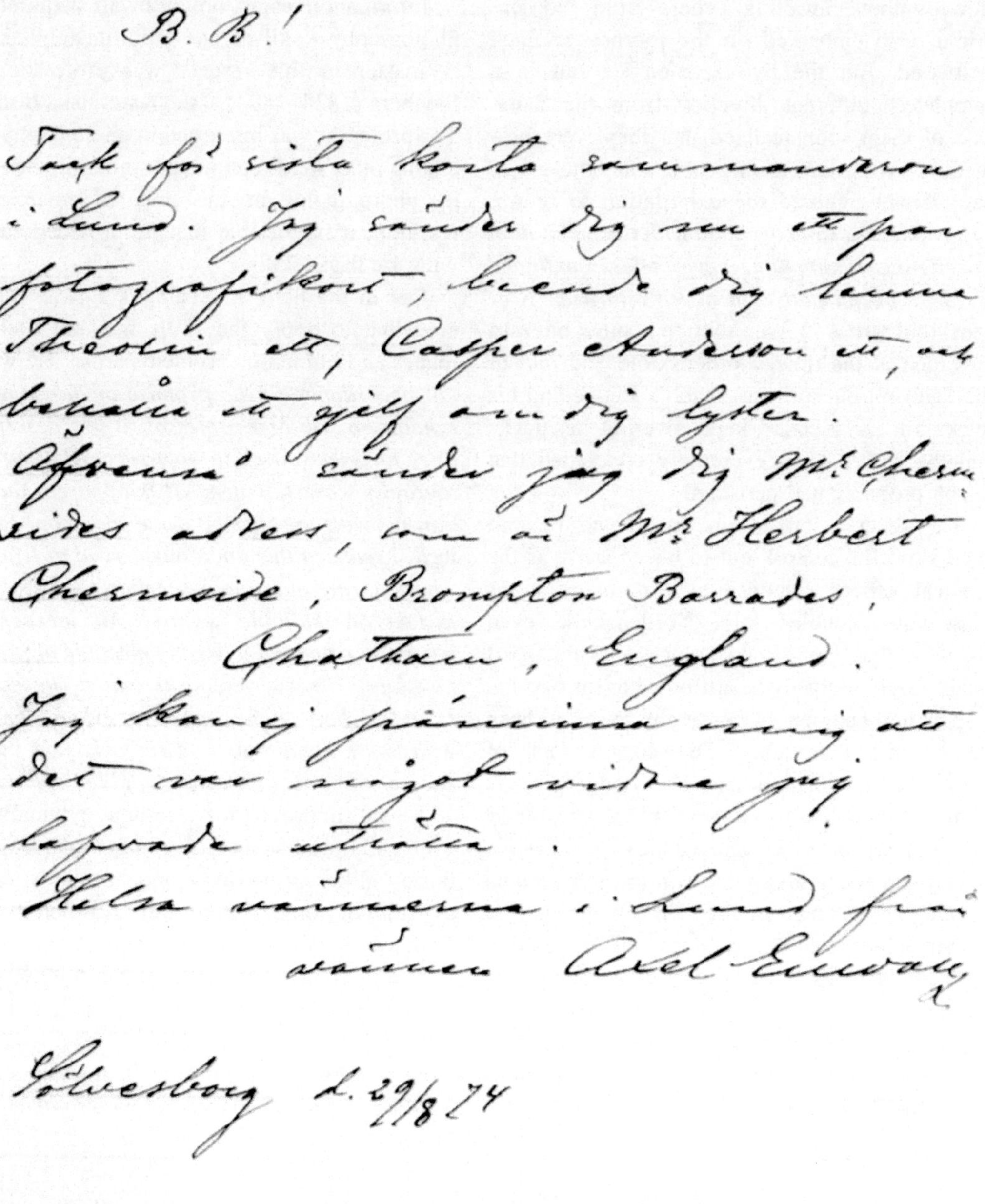
B. B!

Tack för sista korta sammanvaron i Lund. Jag sänder dig nu ett par fotografikort, beder dig lämna Theodor ett, Casper Anderson ett och behålla ett sjelf om dig lyster.
Äfvenså sänder jag dig Mr Chermsides adress, som är Mr Herbert Chermside, Brompton Barracks, Chatham, England.
Jag kan ej påminna mig att det var något vidare jag lofvade uträtta.
Helsa vännerna i Lund från vännen Axel Enwall

Sölvesborg d. 29/8 74

Fig. 3.3 Letter from Axel Enwall to August Wijkander, 29 August 1874. (Courtesy Regional state archives, Gothenburg)

named Snabb. According to Nordenskiöld's account of the events, he had sent two people from among the Sami group to look for driftwood. Upon their return, they reported that driftwood was found in the area. Nordenskiöld issued an order to fetch the driftwood on one of the sledges, which were among their equipment. By misunderstanding, two seamen among the crew, without awaiting orders from Nordenskiöld, also embarked on the journey to fetch driftwood. But the two seamen set out in a completely different direction from the Sami. One of them soon realized that they were proceeding in the wrong direction and returned to base. Snabb ignored the exhortation to return. *"This able, but stubborn man with the inclination to religious ruminations nevertheless continued his once begun path and never returned*" Nordenskiöld writes.[11] Even though a snow blizzard prevailed at the time, Nordenskiöld and one of the Sami set out to trace Snabb's sledge and his tracks but the blizzard kept covering his tracks and the search team eventually concluded that Snabb probably had perished.

The death of boatswain Snabb was beyond Axel Enwall's control, but in his capacity as the medical officer responsible, certain questions must have troubled him. Nordenskiöld even suggests that the disappearance and death of Snabb could probably be attributed to his psychic disposition and the event ought to have been looked into thoroughly. Due to the lack of additional information and knowledge Enwall writes; "*A man was lost during the ice-journey by going astray on the ice among the drifting snow. He was probably frozen to death when overcome by fatigue.*"[12] Later in life, Enwall acted as a doctor with psychiatric care responsibilities and the case of boatswain Snabb's disappearance—and its possible explanation being his mental predisposition—would possibly have come back to Enwall even though he could not have influenced the incident itself.

Arctic photography

Axel Enwall's interest in photography led him even to shoulder the responsibility of photographic documentation during the expedition of 1872–1873. Enwall of his own accord and voluntarily informed Nordenskiöld of his interest in undertaking this task. The usage of cameras within the discipline of polar research was somewhat of a new phenomenon, and Enwall obviously realized the possibilities that photographic techniques could provide to geographic research such as the expedition was now undertaking.

Information as to how Enwall acquired his photography skills is unfortunately scant. A master in this respect was professor Sten Stenberg (1824–1884) at the Karolinska Institute. The professor was interested in photography, and among other fields conducted studies in the usage of photography in the area of microscopic research; it is possible that he provided Enwall with the basic skills.

Out in the field, in Svalbard, Enwall discovered that to apply the skills was not a simple matter and numerous problems arose. He writes: "*In accordance to the promise made I am now sending to you Mr. professor, a few descriptive lines to brief you as to how successful my photography work has been. It is apparent that my parting with the master came too soon. Rather well, I realised this upon our arrival at Isfjorden where during our stopover I didn't even succeed to take one suitable negative. My inexperience left me ill equipped to swiftly meet the obstacles I now faced However on my arrival at The Norway Islands I had gained more experience to cope with harsh conditions and setbacks and thereby succeeded to acquire some suitable negatives.*"[13]

Enwall improved his technique gradually and eventually produced some 60 photographs, taken during different periods and locations of the expedition. Some of these were featured in Frans Reinhold Kjellman's travel book on the expedition, for instance some panoramic foldouts from Foul Bay and Skansbukten. Overall, Kjellman offers a pleasant picture of how a photographer's work could be depicted on such an excursion. "*In his camera, one view after the other paints itself, different from another and, as one could perceive, surpassing each other in magnificent wilderness and impressive grandeur. Some of these are chosen and let to be engraved on the silver bath dipped glass plate, set in the dark tent and preserved.*"[14]

Although it is difficult to obtain detailed information about the equipment used, Enwall does give us a list of photographic equipment in a letter to Nordenskiöld such as lenses, glass plates, a tent for developing negatives and also what Enwall calls *"Düben's camera."*[15] What this means is somewhat unclear but could possibly refer to a camera which was used by Lotten von Düben (1828–1915) while photographing her husband Gustaf von Düben's pictorial works *Lappland och lapparne, företrädesvis de svenske. Ethnografiska studier,* published in 1873.

In conjunction with Leigh Smith's visit at Mossel Bay, Enwall made an acquaintance with another of the English members of the expedition, Herbert C. Chermside (1850–1929). As we see evidenced in this volume, Chermside was an enthusiastic neophyte photographer and on his return home to England presented a comprehensive account of his photographic work published in *The Photographic Journal*. Chermside describes himself as an amateur who only had a few days of training in photography prior to the expedition's departure, and he emphasizes that the main purpose of the expedition was geographical research and therefore photography had no precedence over other work. Chermside writes: "*As, however, so little has hitherto been done in the way of Arctic photography, I trust that photographers may be able to, at any rate, deduce some conclusions of interest from my experiences. The only existing Arctic photographs that I know of, besides Mr. Leigh Smith's have been taken by Americans, Austrians and Swedes.*"[16]

After briefly mentioning William Bradford's photographs taken in 1869 and the pictorial material of the Austrian Payer-Weyprecht's expedition, Chermside expounds further about the photographs taken by the Swedish expedition. "*Those forming the Swedish collection were taken in 1872–3, and are plates about 6" by 8". I have seen the negatives and shortly expect some prints.*"[17] One can from the narrative deduce that Chermside probably received a collection of photographs from Enwall as so presumably did Enwall receive a set of Chermside's.

With the exception of the above-cited sources, and as mentioned earlier, Enwall did neither recount his methods nor did he describe the equipment he employed in his work with photography. Chermside's article in *The Photographic Journal* does, however, provide us with a relatively sound understanding of the then prevailing circumstances. "*On the 8th of June we reached King's Bay, Spitzbergen, lat. 79° N., long. 12° E.; and the jagged peaks of the mountains, with here and there dark patches of hard slate or quartzite showing through their snowy covering, the immense glaciers, like frozen rivers, winding down the gullies and ravines between them, all lying bathed in the sunlight of an Arctic summer night, promised a fair field for the camera. The next day was spent in futile attempts at photography; and perhaps an account of them may be the best way of illustrating the difficulties we encountered.*

The camera is first taken ashore about a quarter of a mile, the passage being at some times of tide rather obstructed by ice, and we at once meet the difficulty that the Arctic photographer continually encounters, namely that of locomotion. The ground is ragged and rocky, covered with snow from 1 to 3 or 4 feet deep, all very slushy from the fierce heat of the June sun, and serving to carefully conceal the holes between the stones, so that the greatest difficulty is continually occurring in selecting a suitable position."[18]

Chermside's article expounds the methodology and procedures of photography related to development, light conditions and so forth. *"Photography with either wet or dry plates was successfully carried on from June to September, between lat. 78°N. and 80°40'N., and at temperatures varying from 45° F. to 25° F. In cases where after exposure the films of wet plates appeared dry or frozen, it was only necessary to redip the plate in the bath for a short time before development."*[19].

Fog was constantly present which is taken as normal for this time of the year. But this was hardly the only difficulty facing Chermside. Another was the problem of mirage. "*But of adverse conditions, the most frequent and annoying is one almost invariably present in greater or less degree, and especially so in the brightest conditions of light—namely, the vibration of the atmosphere caused by the heated vapours that rises from the thawing soil.—From the unequal densities of the different layers of the air in calm and fine weather, when the temperature of the air is above that of the sea, the various effects of refraction are particularly noticeable, and inverted images in the air of various objects, especially of bergs and masses of ice, are excessively common; and as it by no means always occurs that the layers of air in which the phenomena of various degrees of refraction and the total internal reflection take place are horizontal, the images are often inclined at wonderful angles.*"[20] Although Axel Enwall provided somewhat meagre information on his pictorial documentation of Nordenskiöld's expedition, one can probably assume from the citation above that he faced similar challenges as Chermside.

End of the expedition and return

Subsequent to the timely visit of Leigh Smith, the situation for the Swedish expedition radically changed, nowhere more so than for Enwall in his role as the responsible medical officer. Health conditions of the Swedes swiftly improved after the delivery of the provisions from the Englishman. On Nordenskiöld's return from the expedition to North East Land, the expedition remained yet for a while at Mossel Bay while the two ships *Gladan* and *Onkel Adam* made their journey back to Sweden. The expedition's scientific team on board the *Polhem* journeyed first to Gråhuken to care for the two deceased trappers. After Enwall had confirmed that the two had died of scurvy and they were buried, *Polhem* continued its journey south.

During the summer months that followed, the team was engaged with scientific work along the west coast of Svalbard and eventually arrived at Tromsö on 6 August 1873; there the expedition was disbanded and the members were returned to Sweden.

On his return home in 1873, Axel Enwall resumed his work as a doctor at among other institutes The Military School at Carlberg and at the General Pediatric Hospital in Stockholm. Simultaneously, Enwall was engaged with the task of compiling the photographic material from the expedition. With his colleague from the expedition, August Wijkander, he corresponds on the subject and sent him a register of 60 or more photographs with the format of 12 × 18 cm, and also mounted on cardboard with the format of 24 × 32 cm.[21]

In the register, Enwall describes the motif and also the quality of the respective photographs by grading them on a scale from "*excellent*" to "*poor,*" the idea being that members of the expedition would be given the possibility of purchasing the pictures for one krona each. It is moreover apparent through the correspondence with Wijkander that within the expedition, there were plans to publish the photographs in the form of an album and to distribute it through bookshops. Enwall however had doubts over this scheme. He writes: "*I willingly agree, that your remarks to a certain extent could be correct, but your suggestion can, however, for a number of reasons not be realised. First and foremost, concerning the question of price, it is calculated by Jaeger as low as he can set it with the hope of attaining a rich amount of orders through me or Nordenskiöld. Should the 'Spitsbergen album' according to your suggestion be channelled through a (book)shop, it would necessitate first and foremost a publisher which would increase the costs for the enterprise. (Jaeger has no intention of becoming a publisher nor do I have the capital to be one.).*"[22]

The named Jaeger in Enwall's letter refers to the then well renowned court photographer Johannes Jaeger (1832–1908). Originally from Berlin, he established himself in Sweden in the 1860s, where he pursued his career as, in among other areas, museum photographer, and through his pictorial documentation of the Stockholm Exhibition of 1866. Johannes Jaeger even published what is to be regarded as the first Swedish photographic book, named *Molins fontän* (The fountain of Molin), in 1866. In this book, a technique of combining text and pasted original photographs was employed and it was possible with this background that Jaeger's name appears in the discussion between the participants of the expedition regarding the eventual publication of Enwall's photographs.[23] The matter was at length discussed in correspondence with Wijkander, but as this would entail the need to process and retouch the photographs, write a connecting text and add maps, the outcome was then that a commercial bookshop product never materialized. However, a number of albums or suites containing Enwall's photographs from the expedition of 1872–1873 are today preserved at the Regional State Archives in Gothenburg, the Polar Centre at Grenna Museum, The Royal Swedish Academy of Sciences and Uppsala University Library.

Subsequent life

The year after his homecoming, i.e.1874, Axel Enwall assumed the post of city physician in Sölvesborg, a position which according to a contemporary newspaper announcement carried with it a modest remuneration of 200 Swedish riksdaler per year, or as the newspaper put it "*Wonderful prospects: meagre salary but a greater amount of work for the same price!*"[24] His appointment in Sölvesborg was short lived. Already in the following year, he was appointed as a battalion doctor at the Kronoberg Regiment in Växjö, a post which he held up to 1866.[25]

Meanwhile, August Wijkander continued to maintain his interest in the pictorial material from the expedition and in 1876 he once again turned to Enwall with the question of the material. Enwall answers: "*I don't know if Jaeger still has the plates, but since Nordenskiöld is not at home, Jaeger would be the best to know where the plates are, in case he himself doesn't have them, and hence the man to turn to. A few months ago I donated all the prints I had left, among them 1 or 2 of the ones you wish to have.*"[26] He continues: "*I am now in private practice in Wexiö instead of Sölvesborg and I very much enjoy my appointment here. I rather have a great deal of work, a decent income and I am hoping for even better conditions in due course.*"[27]

During the following years, Axel Enwall continued working as a battalion doctor and as a private practitioner in Växjö. At the same time, Enwall was even engaged in political activities. He was a member of the city council and the city treasury. For a few years, Enwall also held the positions of assistant and head physician at the

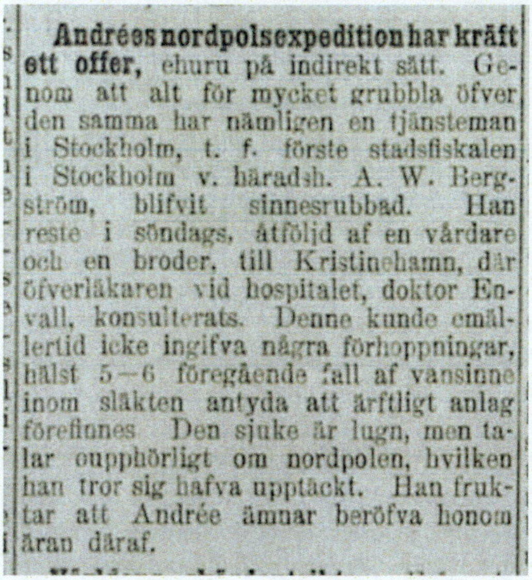

Andrées nordpolsexpedition har kräft ett offer, ehuru på indirekt sätt. Genom att alt för mycket grubbla öfver den samma har nämligen en tjänsteman i Stockholm, t. f. förste stadsfiskalen i Stockholm v. häradsh. A. W. Bergström, blifvit sinnesrubbad. Han reste i söndags, åtföljd af en vårdare och en broder, till Kristinehamn, där öfverläkaren vid hospitalet, doktor Envall, konsulterats. Denne kunde emellertid icke ingifva några förhoppningar, hälst 5—6 föregående fall af vansinne inom släkten antyda att ärftligt anlag förefinnes. Den sjuke är lugn, men talar oupphörligt om nordpolen, hvilken han tror sig hafva upptäckt. Han fruktar att Andrée ämnar beröfva honom äran däraf.

Fig. 3.4 Newspaper article from *Kristinehamns-Tidningen* from 11 June 1896 telling that a person in Stockholm had become mad, thinking he was the discoverer of the North Pole and that the balloonist S. A. Andrée was going to strip him from the honor. The person was sent to Dr. Enwall, but he gave no hope because five to six other cases of insanity in the persons' family indicated hereditary predispositions

Växjö Hospital. On 9 July 1880, Enwall married Agnes Finne from Tammerfors. The wedding was held in Finland. The marriage lasted up to the death of his wife in 1911.

Enwall also kept more sporadic contacts than earlier with the circle of the Polar researchers. Previously, during his stay at Sölvesborg, he was contacted by Nordenskiöld with the request of preparing a list of medical preparations, for Nordenskiöld's imminent expedition to the Kara Sea. Enwall, extremely occupied with other responsibilities, replies: "*Such a list, requested by you for others who are not in the medical profession is extremely difficult to achieve., nonetheless I have made an effort to provide you with the list which herein is attached..*" - - - "*Good luck with the journey wishing you to attain outstanding results in more fields than one.*"[28] Enclosed with the letter is a list of 40 different medical preparations of which Enwall considers to be important, but points out: "*Obviously a number of additions could have been made or altered qualitatively as well as quantitatively, but I think that ought to be about adequate* (Fig. 3.4)."[29]

Nordenskiöld continued to have great confidence in Axel Enwall. A few years later, he received a request from Nordenskiöld to act as a ship medical officer on the planned Vega Expedition around the Northeast Passage. Enwall was by now a well-established and an extremely busy doctor, and thereby found it difficult to accede to Nordenskiöld's request: *Although your offer to the post of a medical officer on your imminent expedition is very much appealing and however tempting the arctic region is to the one who has closely encountered it, my answer is, dictated by prevailing circumstances and dictated as a no. The reasons are apparent and clear. I am now shouldering the responsibilities of partly being the permanent army battalion doctor and an assistant doctor at Wexiö hospital (the latter appointment admittedly ceased in March but I am awaiting and hope for a renewal), and partly I have now under the given circumstances succeeded in establishing an income generating practice. Assuming, if I could obtain leave of absence from my appointment as a battalion doctor, of which I think improbable, it would inevitably lead me to abandon all to its fate, and upon my homecoming, should all go well, I would have to start from scratch and find myself looking for means of sustenance in another place as the present would be occupied by others. The creditors furthermore who are now set upon my pledging my debts would surely not allow me to let all this to go adrift.*

"As I said, as much as I would like to revisit the arctic regions, the obstacles that I am faced with this time are too many and too high, almost insurmountable, for me to answer your request with anything but a no. And you must also realise this, as the situation carries no optional preference other than to incur irreparable losses should I now abandon my activities."[30]

Despite the fact that Enwall turned down the appointment to be the medical officer on the *Vega*, he was nonetheless involved in the expedition's planning concerning provisions. He probably was involved with this duty because of his successful background in dealing with the outbreak of scurvy on the 1872–1873 expedition. There were equally neither reported cases of scurvy nor any death incidences on the *Vega* expedition, and the overall health condition was good during the entire wintering period. After the return home of the *Vega* expedition, Enwall was awarded Knight of The Royal Order of Vasa for his services rendered to the expedition.[31]

After just over 10 years in Växjö, new assignments presented themselves to Enwall. In the spring of 1887, he was appointed as head of the newly opened psychiatric hospital at Marieberg in Kristinehamn. During his time in Växjö, Enwall had acquired experience in psychiatric care and during his appointment as a doctor at Kristinehamn, and became a pioneer in introducing modern treatment methods within psychiatry. In *Sveriges läkarhistoria* (Swedish medical history), one can read the following "*E. was among the first, if not the first, who initiated the modern bed position treatment of which it provided asylums to convey the hallmark of a hospital more so than before. He was an ardent advocate as to the elimination of coercive treatments, such as isolation, straitjacket,* etc."[32]

Axel Enwall engaged in most of the running and the up keeping of the hospital, and his activities successively turned it into what was considered a paragon. Under Enwall's leadership, the Marieberg establishment was expanded to even comprise agriculture facilities, domestic animals, an apple orchard, a lido with its own jetty, and its own cemetery. The long-term ambitions for the institution envisaged by Enwall were recorded in a contribution made in *Bidrag till Sveriges officiella statistik* (Contributions to the Official Statistics of Sweden) of 1892. In this, the hospital and the care it provides are described in detail, starting from the wards for patients with different care needs, facilities for gardening and handicraft, kitchen area, washroom and premises for personnel. The whole establishment resembled a miniature community, and Axel Enwall's interest in photography is reflected in the pictorial studio within the hospital area, which was prepared and assembled by the patients' themselves.[33]

Over and above the detailed description in the Official Statistics of Sweden in 1892, Enwall in the same year conducted 6-month long international journey. He visited a number of renowned psychiatric clinics in, among other places, Austria, Switzerland, Germany, and Denmark. Upon his homecoming, the popular doctor was hailed with a triumphal arch erected by the patients and personnel at the hospital.[34]

In his dealings, Axel Enwall was known to be stringent but just and demanded high standards of performance from his assistants, but above all from himself. His activities as a doctor and administrator were characterized by very high standards of organizational capabilities, and as witnessed by his contemporaries he had an excellent caring hand with his patients. He was an advocate for a humane treatment of psychiatric patients and according to sources, he could "*burst out in anger - - - when someone talked about maniacs and lunatics.*"[35]

On several occasions, Enwall was even entrusted as an expert in the planning of other hospitals, for example at Vänersborg and Åbo. An example of homage attributed to Enwall for his care and consideration for his patients is concisely referred in the newspaper *Göteborgs Aftonblad* of 1894. Under the heading "*Lunatics (sic) on a pleasure trip*," the article goes on to relate that patients from Kristinehamn psychiatric hospital were on a boat outing on Lake Vänern on board the steamship *Våhlön*, that was hired with an extra barge for this purpose. The journey steered about 1½ miles out to sea under favorable weather conditions. "*Although the number of lunatic asylum patients were about 225, the atmosphere on board was far calmer and more peaceful than the usual trips of 'Våhlön' - - - for there were no inebriated persons among the passengers, who all showed much appreciation and happiness over this pleasure trip.*"[36] One can only speculate what Enwall would have thought of the heading.

Up to his retirement, Axel Enwall continued his occupation as a doctor in Kristinehamn, and to the end of his life resided there. By this time, his contacts with the polar research community were limited, albeit correspondence between himself, Nordenskiöld and Wijkander indicate that some contact was maintained, at least during the 1870s.[37] As mentioned above, a working cooperation with Nordenskiöld continued, as was evident in the case with the *Vega* expedition.

From these circles, even in later years, Enwall was not completely cut off. In 1896, chief engineer Salomon August Andrée (1854–1897) was preparing for his grand balloon expedition to the North Pole. This was at the time a most spectacular event and the project attracted a huge media coverage in both Sweden and the rest of the world. Even Enwall was affected, though at a more distant medical level, by this expedition. In the newspaper *Kristinehamns-Tidningen* of 1896, one can read: "*Andrée's North Pole expedition indirectly claims a victim. A civil servant, acting city solicitor and vice governor, by the name of A.W. Bergström from Stockholm, became unhinged after deeply pondering over the expedition. Escorted by a caretaker and a brother, he travelled on Sunday to Kristinehamn, where at the hospital the chief physician, doctor. Enwall was consulted. Nevertheless, he could not give any hope, as 5–6 cases of insanity in Bergström's family indicated that there were hereditary predispositions. The patient is calm but speaks continuously of the North Pole and is convinced of his discovery of it. He thereby harbours fears of being stripped of the honour by Andrée!.*"[38]

As a doctor, Enwall was more of a clinical practitioner rather than a theoretician and his publications were rather few. He did, however, translate from German, a few works of German psychiatrists such as among others Friedrich Scholz' publication *Om vården af sinnessjuka. Föredrag för skötare och sköterskor såväl som för bildade af alla klasser* (*On the caring of mental patients. A lecture for carers and nurses as well as for education of all classes*), in 1895, and Dr Ludwig Scholz' *Handbok för skötare af sinnessjuke* (*A handbook for caretakers of the mentally ill*), in 1899. The impact of the later work is reflected through the fact that the publication was purchased by the National Swedish Board of Health for distribution to the personnel of among other institutions St. Lars hospital in

Lund for educational purposes in 1905.[39] This pedagogical activity and of course, combined with his long medical service at the psychiatric hospital in Kristinehamn, led to his appointment as honorary doctor at Uppsala in 1907. Likewise, as in Växjö, he was active in local politics in Kristinehamn.

Last years

The last years of his life Axel Enwall suffered health deterioration and he also became a widower. Less than a year following the death of his wife, Enwall passed away due to angina pectoris, i.e. angina of the heart, and was buried at the cemetery in Kristinehamn. A significant medical career was now at an end, and the funeral turned out to be a major manifestation for the innovator of psychiatric care in Sweden.[40] Renowned psychiatrists such as Bror Gadelius and Gösta Harding later on acknowledged and brought to attention Enwall´s work.[41]

At his demise, Enwall´s estate left a substantial sum for those days, worth just above 90.000 Swedish kronor. In order to administer the capital, the Axel and Agnes Enwall Donation Foundation was established. The return on the capital was to be put aside in a foundation to cover the costs of a general hospital in Kristinehamn. Furthermore, every 5 years, eight-tenths of the return was to be distributed to the directorate of the Kristinehamn psychiatric hospital to be utilized for welfare ends. Finally, Enwall in his testament endowed his medical book collection to the hospital.[42]

Conclusions

Already in 1861, the Swedish endeavor of polar photography had its beginnings during the interdisciplinary Swedish expedition to Svalbard under the leadership Otto Torell. It then was Axel Goës, also a doctor, who stood behind the camera. There are but a few pictures by Goës preserved. The technique at the time did not have the capacity of photographic reproduction in print, one had to transfer the motif through sketching or xylograph to include it in a book format.

Axel Enwall's contribution as a photographer is worthwhile to draw to attention, and according to Tyrone Martinsson and Per Holmlund, it relates to the development of landscape photography and camera fieldwork, which by then was ongoing in other areas of the world.[43] They refer to the American historian of photography Weston J. Naef, who named the period between 1860 and 1885 as "the golden era of landscape photography," and that "this period of development of the aesthetics and techniques of landscape photography is a result of an increasing interest in nature and man's place in nature in art, literature, philosophy and science."[44]

Unfortunately we know very little about Enwall´s field working methods nor how he chose his motifs. It is highly probable, as Holmlund and Martinsson suggest, that certain motifs are chosen because of their accessibility and others chosen based on cultural traditions.[45] Even Herbert C. Chermside worked under similar circumstances during the Leigh Smith expedition, and certainly one can here discern similarities.

Despite all, today Enwall is mostly remembered for his contributions to psychiatric care in Sweden toward the end of 1800s. One could say that his contributions to photography were less acknowledged during his lifetime, but with his demise, a pioneer in Swedish polar photography also passed away. His memory lives on in the polar regions with the naming of a small island in the vicinity of Hinlopen Strait: Enwallöya (Enwall Island).

Notes

1. Nordenskiöld, A.E. *Redogörelse för den svenska polar-expeditionen 1872–1873.* (Bihang till Kongl. Vetenskapsakademiens handlingar, II:18, Stockholm, 1875, p. 35.
2. *Ibid*., p. 35.
3. On the hunters death and their possible causes, see also Kjaer, Kjell-G, Aasebø, Ulf, *Tragedien i Svenskhuset*, 2012.
4. Från Spetsbergs-expeditionen. Bref från Axel Enwall. *Hygiea. Medicinsk och farmacevtisk månadsskrift*, 35, 1873 p. 408.

5. *Ibid.*, p. 408–409.
6. *Ibid.*, p. 410.
7. *Ibid.*, p. 411.
8. *Ibid.*, p. 411.
9. Kjellman, F.R. *Svenska polar-expeditionen år 1872–1873 under ledning af A.E. Nordenskiöld*, Stockholm, 1875, p. 268.
10. Envall, Axel, Official report to the (Swedish) Royal board of health on the hygiene and care of the sick during the Swedish polar expedition 1872–3. In Leslie, Alexander (ed.), *The Arctic Voyages of Adolf Erik Nordenskiöld 1858–1879*, London 1879, p. 406. In Swedish in: Envall, A., Rapport till Kongl. Sundhetskollegium öfver hygienen och sjukvården under den svenska polarexpeditionen 1872–1873. *Svenska läkaresällsk. Nya handl., Ser. 2, D. 5:3*, 1874, pp. 87–122.
11. Nordenskiöld, A.E. *Redogörelse för den svenska polar-expeditionen 1872–1873.* (Bihang till Kongl. Vetenskapsakademiens handlingar, II:18, Stockholm, 1875, p. 71
12. Envall, Axel, *Official report to the (Swedish) Royal board of health on the hygiene and care of the sick during the Swedish polar expedition 1872–3.* In Leslie, Alexander (ed.), *The Arctic Voyages of Adolf Erik Nordenskiöld 1858–1879*, London 1879, p. 409.
13. Letter from Axel Enwall to Sten Stenberg June 17th 1873, KVA.
14. Kjellman, F.R. *Svenska polar-expeditionen år 1872–1873 under ledning af A.E. Nordenskiöld*, Stockholm, 1875, p. 78.
15. Letter from Axel Enwall to A. E. Nordenskiöld June 13th 1872, KVA.
16. *The Photographic Journal*, December 11, 1873 p. 13.
17. *Ibid.*, p. 13.
18. *Ibid.*, p. 14.
19. *Ibid.*, p. 18.
20. *Ibid.*, p. 18.
21. List of photographs from the Swedish Polar Expedition 1872–1873. Hand written list of Enwall's photographs in the August Wijkander archive, GLA.
22. Letter from Axel Enwall to August Wijkander March 1st 1874, GLA.
23. On Jaeger, see also Söderberg, Rolf & Rittsel, Pär, *Den svenska fotografins historia 1840–1940*, Stockholm 1983.
24. *Blekingeposten* July 17th 1874.
25. Hallbäck, Dan-Axel, Läkaren på Nordenskiölds skepp tog även alla bilderna, *Läkartidningen* nr 35 2009, vol. 106, p. 2156. Reprinted in Hallbäck, Dan-Axel, *3 fjärilar kring Vänern & 33 andra berättelser*, 2017 pp. 166–171.
26. Axel Enwall to August Wijkander September 13th 1876, GLA.
27. *Ibid.*
28. Axel Enwall to A. E. Nordenskiöld April 29th 1875, KVA.
29. *Ibid.*
30. Axel Enwall till A. E. Nordenskiöld March 16th 1878, KVA.
31. Hallbäck, Dan-Axel, Läkaren på Nordenskiölds skepp tog även alla bilderna, *Läkartidningen* nr 35 2009, vol. 106, p. 2157.
32. *Sveriges läkarhistoria. Fjärde följden. Första delen (A - E)*, ed. A. Widstrand, Stockholm 1930, p. 557.
33. Enwall, A. Beskrifning öfver Kristinehamns Hospital. *Bidrag till Sveriges officiella statistik 1892, K. Helso- och sjukvården II. Öfverstyrelsens öfver hospitalen underdåniga berättelse. Bil. C*, pp 32–35.
34. *Kristinehamnstidningen* August 11th 1892.
35. *Nya Kristinehamns-Posten* July 17th 1912.
36. *Göteborgs Aftonblad* September 20th 1894.
37. E.g. see letter from Axel Enwall to A. E. Nordenskiöld December 10th 1874, and to August Wijkander August 29th 1874 and December 10th 1874 where Enwall among other things informs Wijkander about the addresses of Chermside and Leigh Smith; and also September 13th 1876 where Enwall refers to correspondence with F.R.

Kjellman and a meeting with the captain on *Gladan* Gerhard von Krusenstierna.
38. *Kristinehamns-Tidningen* June 11th 1896.
39. https://carl.kulturen.com/.
40. *Nya Kristinehamns-Posten* July 22nd 1912.
41. Gadelius, Bror, *Sinnessjukdomar och deras behandling förr och nu*, 1913 pp. 117–118, Harding, Gösta, *Tidig svensk psykiatri*, Lund 1975, p. 269.
42. *Nya Kristinehamns-Posten* July 22nd 1912.
43. Holmlund, Per och Martinsson, Tyrone, *Frusna ögonblick. Svensk polarfotografi 1861–1980*, Stockholm 2016, p. 60.
44. *Ibid*., p. 60.
45. *Ibid*., p. 60.

Unprinted sources

Göteborg. Riksarkivet/Landsarkivet (GLA) [Regional State Archives in Gothenburg]. Letters from Axel Enwall to August Wijkander

List of photographs from the Swedish Polar Expedition 1872–1873. (Placed with Enwall's photographs in the August Wijkander archive)

Stockholm. Centrum för vetenskapshistoria, Kungliga Vetenskapsakademien (KVA) [Centre for History of Science at the Royal Swedish Academy of Sciences]. Letters from Axel Enwall to A. E. Nordenskiöld and Sten Stenberg

Printed sources

Chermside HC (1873) Notes on photography in the arctic regions. Photogr J (245):13–18

Clase BL (1936) Kapten Clases dagbok från svenska polarexpeditionens övervintring 1872–1873. Utdrag av C. B. Erikson. Ymer (1):81–92

Enwall A (1892) Beskrifning öfver Kristinehamns Hospital. Bidrag till Sveriges officiella statistik 1892, K. Helso-och sjukvården II. Öfverstyrelsens öfver hospitalen underdåniga berättelse. Bil. C, pp 32–35

Enwall A (1873) Från Spetsbergs-expeditionen. Bref från Axel Enwall. Hygiea. Medicinsk och farmacevtisk månadsskrift 35

Envall A (1874) Rapport till Kongl. Sundhetskollegium öfver hygienen och sjukvården under den svenska polarexpeditionen 1872–1873. Svenska läkaresällsk. Nya handl., Ser. 2, D. 5:3

Gadelius B (1913) Sinnessjukdomar och deras behandling förr och nu

Hallbäck D-A (2009) Läkaren på Nordenskiölds skepp tog även alla bilderna, Läkartidningen 106(35), pp 2155–2157

Hallbäck D-A (2017) 3 fjärilar kring Vänern & 33 andra berättelser

Harding G (1975) Tidig svensk psykiatri, Lund

Holmlund P, Martinsson T (2016) Frusna ögonblick. Svensk polarfotografi 1861–1980, Stockholm

Kjaer K-G (2012) Aasebø. Ulf, Tragedien i Svenskhuset

Leslie A (ed) (1879) The Arctic Voyages of Adolf Erik Nordenskiöld 1858–1879, London

Nordenskiöld AE (1875) Redogörelse för den svenska polar-expeditionen 1872–1873. (Bihang till Kongl. Vetenskapsakademiens handlingar, II:18, Stockholm, 1875

Nyström C (1868/69) Om den sista svenska Spetsbergs-expeditionens utrustning och hygien. Upsala Läkareförenings Förhandlingar, Bd. 4 1868/69, pp 419–439

Sveriges läkarhistoria. Fjärde följden. Första delen (A - E), ed. by A. Widstrand, Stockholm 1930

Söderberg, Rolf & Rittsel, Pär, *Den svenska fotografins historia 1840–1940*, Stockholm 1983.

Blekingeposten

Göteborgs Aftonblad

Kristinehams-Tidningen

Nya Kristinehamns-Posten

https://carl.kulturen.com/

Postscript

This chapter on the doctor and photographer Axel Enwall had not been possible without help. First and foremost, my thanks go to my friend and former colleague Gunnar Holmlund. Without his tracking scent and judicious comments, the result had been something completely different. Photographer Hans Jonsson in Landskrona has supplied valuable information on photo technique during Enwall's time, and Eskil Malmberg has contributed with viewpoints and hints within the field of psychiatrics. Many thanks also to archivist Anne Miche de Malleray at the Centre for History of Science at the Royal Swedish Academy of Sciences, whose fast and efficient help with scanning of documents has been of great use for this work and to Mohammed Zahran for translation of the text to English language.

4 Alfred Eaton and the Biological Collections from Leigh Smith's 1873 Expedition

C. L. Devlin

Introduction

From the diaries of Herbert C. Chermside emerge a colourful picture of the shipboard biology: the collection and observations of rare Arctic flora and fauna, and dredging for deep-sea specimens. These scientific endeavours were largely overseen by the expedition's naturalist and surgeon Alfred Eaton (1844–1929), whose success on this cruise cemented his reputation amongst his peers and led to ventures farther afield of greater scope and fame.

A Cambridge reverend goes to sea

By the summer of 1873, twenty-nine-year-old Alfred Eaton had completed B.A. (1868) and M. A. degrees (1871) at Cambridge University and served as a clergyman in both Derbyshire and Bedfordshire. Eaton was a naturalist on the move: by age twenty-one, he was already a regular contributor to the *Entomologist's Monthly Magazine*[1] and an active member in the Entomological Society London.[2] How he was selected to join the crew of *Diana* is unknown; however, both he and Leigh Smith as Cambridge graduates moved in the same circle of gentlemen scholars. Eaton, therefore, would have heard of Leigh Smith's 1871 and 1872 expeditions to Svalbard. It could also have been by way of an introduction by Leigh Smith's half-sister Barbara Bodichon, with whom Eaton had been corresponding on the flora and fauna of Svalbard in the spring of 1873.[3]

Eaton's role on the 1873 expedition was twofold, serving both as naturalist and the "Ship's Surgeon and General Practitioner." Indeed, the expedition chronicler, Chermside, referred to Eaton as the ship's "sawbones." Eaton appears and reappears in Chermside's narrative like a favourite character. Of all the crew, only Eaton is described in such detail: "…Eaton, equipped with an Alpen's Pack, a straw knapsack like a carpenters tool bag, numerous vials, corks, instruments &c, an *enormous* pair of boots, a glass bottle with a hole in the bottom of it to carry his shot & a rusty single barrel, & followed by a little "Jack," who is apparently an aspiring entomologist, takes the hillside & gets to work as usual, like a sandpiper turning up every likely looking stone for worms & insects, & I expect like them congratulating himself in tones as expressive, ere he consigns each treasure trove to a fitting receptacle."[4]

Entomological researches

To the layperson, the activities of a scientist often appear baffling, and to the expedition narrator, Eaton's work was no different. On the first leg of the voyage north, at a stopover in the Shetlands, Chermside found Eaton "… near a [Lerwick] bridge about halfway home, on his hands &

C. L. Devlin (✉)
Penn State University, Abington College, Abington, PA, USA
e-mail: cld5@psu.edu

P. J. Capelotti (ed.), *The Coldest Coast*, Historical Geography and Geosciences,
https://doi.org/10.1007/978-3-030-67880-7_4

knees, apparently *mudlarking* in a peach-coloured burn, but in reality carefully capturing & bottling those specimens of entomology, which are to immortalise him & his labours (Fig. 4.1)."[5]

Chermside's words would prove prophetic; of all of Eaton's realms of specialization, his knowledge of insects would in fact "immortalise him." Eaton's shipboard entomological work gripped Chermside from the start. "Eaton [was] very busy poisoning flies & other insects. We all are very glad of this, as some of them bear so strong a resemblance, to the "domestic *bonser*" that we are a little nervous about their being in the cabin."[6] Weeks later, Chermside was still engrossed by Eaton's activities and wrote that the naturalist was "in great hopes of finding some sort of "bonser" in the dead specimens [of Malley birds], as none is known by naturalists to inhabit the covers of this bird. For this night at least no parasite is forthcoming."[7] One can imagine Chermside's concern: Eaton with a magnifying glass, sifting through bird feathers, in a determined search for some undiscovered mite or flea, whilst the young military officer dreaded an invasion of the parasites into the sleeping quarters. In the same journal entry, Chermside reiterated his concerns, "Eaton was much elated at having during his researches discovered a larva or parasite from some animal: forbid such in the cabin."

Fig. 4.1 Alfred Eaton (courtesy of the Royal Entomological Society)

Eaton's field research on the snow-speckled islands and peninsulas of Svalbard were conducted with equal zeal and energy as his shipboard studies. "Eaton had been to Walden Island in a whaleboat collecting, & had been away some time; his proposed half-hour on shore becoming three or four, with satisfactory results, however."[8] Eaton is oft described in solitary, single-minded pursuit of rare polar specimens; this, despite the presence of polar bears. In that same May entry, Chermside remarked that "I … find that Brunie … tracked my footsteps across from where I killed a snowbird this morning. They have a nasty habit of doing this." Undaunted by the threat of bears, Eaton overnighted alone on the shore, as when Chermside reported that "Eaton did not come on board 'til 8 a.m. this morning. He had been onshore collecting last night & when he came opposite the ship soon after our return was unable to make any signal, so lay down on the beach …".[9]

Chermside occasionally joined Eaton's forays over the undulating rocky terrain, thereby documenting the naturalist's collecting activities and accomplishments: "… I have not touched on entomology, knowing nix on the subject. Moths have, however, been captures [sic], both the small kind, known to inhabit Spitzbergen, & a larger & I believe entirely new sort …"[10] His speculation about Eaton's new discoveries would prove correct. Two male moths found in Wijdefjorden wound their way into auspicious hands—first those of the German entomologist Phillip Christoph Zeller (1808–1883), who analyzed them in 1874. Zeller then forwarded them to Britain's preeminent entomologist Henry Tibbats Stainton (1822–1892), who due to the "pressure of business of various sorts," and opening his

London mansion Mountfield to the public for numerous entomological events, did not get to his analysis of the moths until 1880, when he wrote that Eaton's two specimens "must certainly be a good new species, for which I should propose the name of *Plutella Polaris*."[11] The well-camouflaged *Plutella polaris* would not be seen for another one hundred and forty years until the moth's re-discovery in Wijdefjord by Norwegian scientists in the summer of 2015.[12]

Eaton had equal success with the Hymenoptera (the group comprising bees, ants and wasps) collected at Wijdefjorden, Lomme Bay and Hecla Cove, which were analyzed by the London entomologist Reverend T. A. Marshall. Of the five species studied, "four are new."[13]

Botanical researches

Eaton's other realm of expertise was plants. The flora of Svalbard was a source of delight for Chermside as well because lovely descriptions of the plant life abound throughout his journals, such as: "The slopes up here were really, for this land, covered with luxuriant vegetation, endless cushions of a little smooth leaved green plant like a shamrock, scurvy grass with its heads of countless white flowers, yellow flowers the size of the common buttercup in flower & many others, all of which with their proper Latin names (far too long to be inscribed here) written down, with all care & precision, on the lids of the nice little wooden coffins prepared for any rarity vain enough to bloom, & pining for sepulchre in a museum, by the indefatigable Eaton."[14]

Eaton's "botanizing" excursions appeared throughout the journals: "Eaton has been travelling daily, right up to the glacier feet, & on the whole has done excessively well in the way of plants in this bay."[15]

Perhaps by association with and tutelage in microscopy by Eaton, Chermside developed his own eye for scientific observation. Below he described the algae and unicellular phytoplankton, diatoms, responsible for red and brown snow. "The ice where washed by the water is much distorted. This has been observed by Dr. Hooker in a voyage to the Antarctic Regions, & is due to vegetable growth. Where the discolouration is red & of a colour like iron rust, it is due to *Protoeous nivalis* & where of a light brownish colour due to some of the diatoms specimens, specimens of which were observed this morning under the microscope."[16] The "Dr. Hooker" in the reference above was the famed botanist Sir Joseph Dalton Hooker (1817–1911) who joined James Clark Ross' voyage to the Antarctic during 1839–43, and wrote *The botany of the Antarctic voyage of H. M. discovery ships Erebus and Terror in the Years 1839–1843 under the command of Captain Sir James Clark Ross*.[17] The Scottish botanist Robert Brown (1773–1858), who upon examining samples from Ross' earlier 1818 expedition to Baffin Bay, tentatively classified the microorganisms in red snow as "Algum genus?… Minute globules, the colouring matter of the Red Snow, of extensive patches were seen at lat. 76°25′N., and long. 65° W."[18] Recent studies have shown that Brown's 'Algum genus' [currently *Chlamydomonas nivalis*] belongs to the green algae group, its photoprotective carotenoid pigment, astaxanthin, causing the red coloration responsible for red snow.[19] The "light brownish colour" of diatoms is due to the presence of the photopigment fucoxanthin.

In addition to the collection of vascular plants, algae and diatoms, "Eaton took to the rooks for lichens [symbiotic fungi and algae/cyanobacteria] … in parts there is a laminated black rock in layers amongst the other which is, I should expect, metamorphous. There were large patches of mosses & lichens of most beautiful colours, green & orange mosses & lichens through every shade from white to black," wrote Chermside.[20]

Upon *Diana*'s return to home waters, the plants of Svalbard were divided and delivered to experts at different corners of the British Isles. The phanerogams and higher cryptogams were sent south to Spencer Le Marchant Moore (1850–1931) at the Royal Botanic Gardens at Kew, and the algal samples were sent north to Scottish George Dickie (1812–1882), the Regius Professor of Botany at Aberdeen University. The diatoms were analyzed in Dublin by the minister-naturalist Eugene O'Meara (1815–1880), who remarked that of the one hundred and eighty-one species previously known to the Arctic, Eaton had single-handedly collected "no less than 92."[21] Two new diatom species from the expedition would later be named *Amphora Eatonia* and *Amphora Leighsmithiana*. Eaton's botanical work during the summer of 1873 culminated in his publication from the expedition, "A list of plants collected in Spitzbergen in the summer of 1873, with their locations."[22]

Marine invertebrates from the deep

As a Royal Engineer, Chermside was inherently interested in the technologies associated with *Diana*'s dredging operations as reflected by numerous references and an entire entry of July 28 dedicated to dredging: the "pros and cons" of dredge types and creative, make-shift contraptions of the crew. One swab, for example, was improvized from "… the long bundles or brooms of oakum used to clean the decks."[23] That same entry also documented the hazards and challenges of obtaining living, intact specimens: "… by the time he [a marine organism] is with infinite difficulty hauled to the surface, he resembles a bag of ballast, & the aforesaid rarity's most intimate friends would have the greatest difficulty in recognizing him. In other words, the bag gets full of stones, & all the animals are pounded to fragments or reduced to become amateur jellyfish." Despite the hardships of the work, dredging was a necessary means for Eaton and the crew to collect deep-sea specimens.

Dredging in the summer of 1873 was by no means a new technology. The 1830s and 40s dredging experiments in the Aegean by Edward Forbes (1815–1854) were well documented. Forbes had published on deep-water mollusks, echinoderms, and cnidarians, and proposed his Eight Regions for the ocean depths and his Azoic Hypothesis for an absence of marine life lower than three hundred fathoms at the 1843 meeting of the British Association in Cork.[24] Likewise, Eaton was very familiar with the 1860s dredging operations in the Shetlands and Hebrides by renowned conchologist J. Gwyn Jeffreys (1809–1885), onboard his own private research vessel *Osprey*.[25] By 1873, when Eaton was untangling invertebrates and seaweeds from the nets on *Diana*'s slick deck, Forbes' Azoic Hypothesis had been debated and disproved seven years earlier by the superintendent of Norwegian fisheries Georg Ossian Sars and his father Michael Sars, a zoology professor at the Royal Frederick University (now the University of Oslo).[26] *Diana*'s dredging activities provided Eaton with a unique opportunity to contribute to the nascent discipline of marine biology, specifically in elucidating organisms from different zones of the arctic seas.

Amidst *Diana*'s exasperating and back-breaking dredging efforts came hard-won successes like when they discovered "… a sort of anemone [*anthozoan cnidarian*] beast I believe never before known to exist in these seas …"[27] Their luck continued a few days later when "… the swab … had been dragging along, with its floating hairlike fibres distended like seaweed or feelers, & will have picked up, doubtless, starfish, five fingers, alectos [crinoid echinoderms], echini, stones, seaweed, *et hoc genus omne*."[28] A month later the invertebrate collection expanded with the capture of "… some corallines, an alecto, sea spiders, starfish of two or three sorts, some with 12 & 13 points, any amount of worms…"[29]

The hunt for marine specimens continued at the intertidal zone: "Eaton had been *mudlarking* along the shores & was rewarded with anemones, & other fishy substances, enough to afford him employment for days to come."[30] The hunt even led to unsavoury places like the innards of marine mammals: "[Eaton] is further much interested in the intestines of the seal [bladder-nosed seal *Cystophora cristata*], & fills buckets &c with detached portions of his carcass including head, flippers &c, a new regime has now to commerce, as we may at any moment be summoned to shoot seals…"[31] Edward Miers (1851–1930), the curator of the crustacean collection at the Natural History Museum in London, to whom Eaton's crustaceans were sent, corroborated Chermside's remark: "Some of the specimens in the collection were found in a Saddleback's [harp seal *Pagophilus groenlandicus*] stomach killed off the Western Ice in green water."[32] Miers wrote that

Eaton's "species are most of them well-known Arctic forms; but the specimens generally are of a large size and in an excellent state of preservation."[33] The value of the collection was in the precise locations and fathoms in which the specimens were collected. Eaton's mollusks were studied by J. Gwyn Jeffreys who, as mentioned above, was no stranger to the rigours and rewards of deep-sea dredging. He found amongst Eaton's collection a range of pteropods, gastropods, conchifera and brachiopods. Two of Eaton's gastropods, one *Eolis* from Fair Haven and *Natica* from Lomme Bay, had ambiguous species designations, as indicated by Jeffrey's question marks, so may have been new species.[34]

Multinational scientific endeavours in Svalbard

The crew of *Diana* was not alone in conducting science in Svalbard; the Chermside journals describe periodic encounters with vessels fishing and hunting in the archipelago whilst conducting varying degrees of science, from rudimentary to the sophisticated. *Diana*'s first encounter was with the crew of *Eclipse*, a steam whaler out of Peterhead and captained by Scottish David Gray. Gray, a dilettante collector of marine specimens, showed Chermside and others "... several curios in the way of natural history, such as the parasites off whales (a cross between a crab & a scorpion), those of narwhals, also a large mass of them like a sponge, from the stomach of a "bearded seal" [*Erignathus barbatus*], "whales food" [krill] &c. We are now no longer in blue but green water, the colour I believe being caused by the medusa &c it contains, whilst at dinner a small net was put overboard & some collected for Eaton ... The water here is no longer of the same specific gravity, as before. This "whale's food" consists of numerous small shrimps, transparent, with red markings & long red antennae. There are also fry shrimps, very like the domestic shrimp of commerce."[35]

Later that month *Diana* searched out a Swedish scientific expedition under the leadership of Adolf Erik Nordenskiöld, whose crew had been forced to overwinter at Mosselbukta. Chermside wrote that "The Swedes have of late years been extremely zealous in the exploration of Spitzbergen & the Polar Seas around it... It is principally from the first of these expeditions [in 1861] that the recent & best chart of Spitzbergen has been prepared, & they have done more than any others towards a complete arrangement & catalogue of the Fauna & Flora of that country, as well as determining the geological & physical features of the land & sea, having done a great deal of dredging in many of the western & northern ways, besides tidal & other observations."[36]

The industrious Swedes had even gone so far as to build outbuildings onshore dedicated solely to scientific experimentation. "Then there is a little wooden observatory, painted red, a small octagonal erection amongst the rocks 150 yds seaward of the house. In this are their astronomical instruments &c. They have a small, roofed house for their thermometers, with double walls, the walls & floor being louvred, & another wooden observatory for their magnetic instruments. A tide pole fixed in a hole in the ice in the bay, observable through a telescope from the house completes their outdoor arrangements... [The observatory] contained a small passage for transit instruments, sextants, horizons &c, they also had apparatus for observing atmospheric electricity, a Thomson's electrometer &c. To the E was another wooden observatory, containing three magnetic instruments, magnetometers I suppose, to measure electricity, the horizontal & vertical forest of intensity of the earth's magnetism."[37] One can imagine Eaton's delight at having fellow scientists to converse with. "... Eaton, who had gone to visit the Swedes & inspect their ménage, returned with some of them, to lunch. It seems that they have been making a very complete set of observations of all sorts, & Eaton was much pleased with their collections of seaweeds &c."[38]

A month later, *Diana* encountered Norwegians whose "skippers ... are encouraged to make observations by their home authorities, who seem really "right down go ahead," in pursuit of scientific inquiry, & are, I believe, in cases furnished with chronometers, barometers & all other meters but a gaso."[39]

For this international group of mariners, Eaton served as an ad hoc medical officer. His collecting expeditions were often interrupted to attend to an ailment or injury as when "Eaton who as amateur sawbones … was soon however brought up, & the boats rowed off…" to attend to a shooting onboard *Diana*, but luckily it was: "Nothing serious, the third mate had pointed a rifle at the dog & accidently shot the steward through the arm & the dog through the jaw…"[40] A month later: "The captain of one of the [Norwegian] smacks comes on board for a sawbones, one of his men having shot himself in the arm, & Eaton goes off to look after him."[41] When Eaton was unavailable, the crew were on their own for their medical treatment: "The little *Diana's* medicine chest has certainly been of great use, & we are a sort of itinerant dispensary. Eaton is of course Surgeon & General Practitioner, but we all have a turn, according to the nature of the ailment."[42]

Aftermath

Upon *Diana*'s return to Britain in the fall of 1873, "Rev. A. E. Eaton presented a collection of Crustaceans, Annelids, Echinoderms, Tunicates, Coelenterates and Sponges, which he made during a visit to Spitsbergen" reported Dr. Albert Günther, the Keeper of Zoology at the Natural History Museum to the Board of Trustees.[43] As with the Svalbard plants, Eaton's invertebrates were distributed amongst Britain's experts for analysis, thus cementing his reputation as a hardy and versatile naturalist-collector.

After this, invitations to expeditions farther afield were fast in coming. The following June, 1874, amidst Royal astronomers and their clutter of telescopic equipment, the Cambridge naturalist lugged his sea bag aboard *Windsor Castle*, destination South Africa, and then on to Kerguelen Island on *H. M. S. Volage*.[44] The then-secretaries of the Royal Society, T. H. Huxley and G. G. Stokes, described the rationale for Eaton's nomination for the 1874 British Transit of Venus Expedition: "The naturalist attached to the Kerguelen's Land expedition was the Rev. A. E. Eaton, who was well qualified to bring to bear his experience of the Arctic fauna and flora upon those existing in the southern hemisphere under parallel physical conditions."[45] Upon return from the Transit Expedition, the organisms collected from the southern islands were "entrusted to competent persons for examination and descriptions."[46] Eaton analyzed the insects. In a special 1897 volume of the *Philosophical Transactions of the Royal Society* dedicated to the findings of the Transit Expedition, Eaton had an impressive four publications. Two newly discovered fungi collected for Sir J. D. Hooker during Eaton's stopover in Cape Town were named *Daedalea eatonii* and *Galera eatonii* in honour of his achievements during the Transit Expedition.[47]

Still, it was the 1873 Leigh Smith expedition to Svalbard that proved the critical testing ground for Eaton and prepared him for later successes on Kerguelen Island and elsewhere. Herbert Chermside's astute observations and vivid writing style created a picture of the shipboard activities and field excursions of Eaton during a formative phase in his scientific development and provided the modern reader with a context for understanding the challenges and rewards of conducting science in the harsh environment of nineteenth century Svalbard.

Through Chermside's eyes, we witness Eaton's energetic pursuit of arctic science, his "botanizing" and "mudlarking," searching pinniped intestinal tracts and sleeping ashore despite menacing polar bears. We also glimpse Eaton's elation at a rare discovery amongst bird feathers and his delight when lunching with Scandinavian colleagues, but little else is gleaned about the personality of this most prolific naturalist.

That is, until some clues appeared over twenty years later in a volume of the *Entomologist's Monthly Magazine* from 1896. In it is a brief vignette about a collecting trip gone terribly wrong: an English collector had been arrested in Algeria for collecting bugs without a permit. The Englishman is described as: "Dressed like a dirty tramp, without a collar, wearing a hat that had seen better days … Six feet of "sugar cane" (bamboo) with brass fittings, a sun-umbrella, square straw basket, muslin nets not strong enough for fish, bottles full of dragon-flies, and *Neuroptera* amongst bits of paper, a walking stick and pocket compass; no resources, and no paper of identification—clearly a suspicious character, or wandering lunatic."

For most individuals, detainment by authorities would be cause for alarm. This prisoner, however, was one unflappable, world-wise Alfred Eaton, and he wrote of his African misadventure with amusement and good cheer. The naturalist was at once polite and helpful in explaining his predicament to his captors: "I … amuse myself with herborizations and entomology (bottle of dragon-flies produced for inspection). Prompt discharge of prisoner, who returned to the Douar, and dined after all at Le Tarf. And so was lost a fine opportunity of sampling *Pediculi* [human lice] that night in an Algerian lock-up."[48] Perhaps as Eaton penned that last humorous line, a distant memory emerged … a memory of a ship off a misty polar shoreline where his companion, a young military officer, watched as he sifted through bird feathers for fleas and mites. "God forbid they invade the sleeping quarters."

Notes

1. See, for example: Eaton. 1864–65. *Entomologist's Monthly Magazine*.
2. Peters, J., L. Arvy, and W.L. Peters. 1980. *Pictet and Eaton: The first mayfly specialists*. In *Advances in Ephemeroptera Biology*. Flannagan, J. F. and K.E. Marshall, eds. Boston: Springer, 534.
3. Capelotti, P.J. 2013. *Shipwreck at Cape Flora: The expeditions of Benjamin Leigh Smith England's Forgotten Arctic Explorer*. Calgary: University of Calgary Press, 107.
4. Chermside, H.C. *Journal on board of the SS Diana*. SPRI, Vol. II, June 10, 1873. All quotes from this unnumbered journal are hereafter cited by their volume and entry date. The journals themselves are part of the general reference at SPRI: GB 15 Herbert Chermside, with the specific reference: MS 300/4/1–2-3.
5. Chermside 1873/1: May 13.
6. *Ibid*., May 14.
7. *Ibid*., May 25.
8. Chermside 1873/2: June 29.
9. *Ibid*., July 22.
10. *Ibid*., July 25.
11. Stainton, H. T. 1880. On the singular new species of *Plutella* (allied to *P. Cruciferarum*) collected in Spitzbergen, in 1873, by the Rev. A. E. Eaton. *Entomologist's Monthly Magazine*. 17: 109.
12. Nordahl, Marianne. 2017. "Arctic moth turns up after 140 years." *Science Nordic*, February 7, 2017. On-line at: https://sciencenordic.com/arctic-moth-turns-after-140-years (accessed August 13, 2018).
13. Marshall, T.A. 1877. "Descriptions of Hymenoptera from Spitzbergen, collected by the Rev. A.E. Eaton." *Entomologist's Monthly Magazine* 13: 241.
14. Chermside 1873/2: June 28.
15. *Ibid*., July 24.
16. Chermside 1873/1, May 26.
17. Hooker, J. D. 1844. *The botany of the Antarctic voyage of H. M. discovery ships Erebus and Terror in the Years 1839–1843: under the command of Captain Sir James Clark Ross*. London: Reeve Brothers.
18. Brown, R. 1819. Appendix, No. VI. "List of plants." In: Ross, J. *A voyage of discovery, made under the orders of the Admiralty, in His Majesty's ships Isabella and Alexander, for the purposes of exploring Baffin's Bay, and inquiring into the probability of a north-west passage*. London: John Murray, 195.
19. Müller, T., W. Bleiβ, C-D Martin, S. Rogaschewski, and G. Fuhr. 1998. "Snow algae from northwest Svalbard: their identification, distribution, pigment and nutrient content." *Polar Biology* 20: 14–32.
20. Chermside 1873/2: June 28.
21. O'Meara, E. 1874. "On Diatomaceae from Spitzbergen." *Quarterly Journal of Microscopical Science* 14: 254.
22. Eaton, A. E. 1876. "A list of plants collected in Spitzbergen in the summer of 1873, with their locations." *Journal of Botany: British and Foreign* 14: 41–44.
23. Chermside 1873/2: July 28.
24. Forbes, E. 1844. "Report on the Mollusca and Radiata of the Aegean Sea." In: *Report of the 13th Meeting of the British*

Association for the Advancement of Science. London: John Murray, 130–193.
25. Anonymous. 1885. "Obituary of Dr. Gwyn Jeffreys." *Athenaeum* 2988: 154.
26. Rozwadowski, H. M. 2005. "Dredging the moon." In: *Fathoming the Ocean: the Discovery and Exploration of the Deep Sea*. Cambridge: Harvard University Press, 148.
27. Chermside 1873/2: July 25.
28. *Ibid.*, July 28.
29. *Ibid.*, Aug 7.
30. Chermside 1873/1: May 16.
31. *Ibid.*, May 25.
32. Miers, E. 1877. "List of the Species of Crustacea collected by the Rev. A. E. Eaton at Spitzbergen in the Summer of 1873, with their Localities and Notes." *The Annals and Magazine of Natural History: Zoology, Botany and Geology* 19, 139.
33. *ibid.*, 131.
34. Jeffreys, J. G. 1876. "List of Mollusca collected by the Rev. A. E. Eaton at Spitsbergen during the third Voyage of B. Leigh Smith, Eq. Stel. Pol., in the Greenland Sea." *The Annals and Magazine of Natural History* 18, 499–500.
35. Chermside 1873/1: June 1.
36. Chermside 1873/2: June 13.
37. *Ibid.*
38. *Ibid.*, June 13.
39. *Ibid.*, July 14.
40. *Ibid.*, June 10.
41. *Ibid.*, July 11.
42. *Ibid.*, July 30.
43. Günther, A. 1912. *The History of the Collections Contained in the Natural History Departments of the British Museum: Appendix. General History of the Department of Zoology from 1856 to 1895*. London: British Museum, 29.
44. Airy, G. B. (ed.) 1881. *Account of observations of the transit of Venus, 1874, December 8, made under the authority of the British Government: and of the reduction of the observations.* Her Majesty's Stationery Office, 405.
45. Stokes, G. G. and T.H. Huxley. 1879. "Preface." *Philosophical Transactions of the Royal Society* 168, v–vi.
46. *ibid.*
47. Plug, C. n.d. "Eaton, Rev Alfred Edwin." *A Biographical Database of Southern African Science.* On-line at: https://www.s2a3.org.za/bio/Biograph_final.php?serial=830 (accessed August 13, 2018).
48. Eaton, A. E. 1896. "An English entomologist arrested in Algeria." *Entomologist's Monthly Magazine* 32, 186.

5 Chermside's Observations of Marine Mammals During a Svalbard Hunt

C. L. Devlin

Introduction

As a top student at Eton and the Woolwich Military Academy, Herbert Chermside climbed onboard *Diana* well-prepared for Leigh Smith's expedition to Svalbard. Amidst his preparations, Chermside had studied the university-educated whaler William Scoresby, Jr. (1789–1857), who in 1820 wrote perhaps the seminal book in arctic science, *An Account of the Arctic Regions, with a History and Description of the Northern Whale-Fishery, Vol. 1*.[1] Scoresby's impact on Chermside is seen in several mentions in a variety of contexts throughout the Chermside diaries, in particular within Chermside's descriptions of the marine mammals encountered by the expedition.

Scoresby's mode of conducting arctic science served as a template for other explorer-scientists thereafter: the exorbitant cost of the polar adventure underwritten by a large-scale hunt. For Scoresby, it was the hunt of whales, as described in *Arctic Regions*. For James Lamont, the owner of *Diana*, it was the hunt of "seahorses" (walrus) and seals. These he wrote of in his own Svalbard travelogues *Seasons with the sea-horses*[2] and *Yachting in the Arctic seas, or notes of the five voyages of sport and discovery*.[3] In the 1870s, Leigh Smith continued the tradition of his predecessors in the collection of blubber, meat, feathers, hides, and furs. It is no surprise, therefore, that Chermside, a man handy with both a rifle and a pen, was selected for the 1873 expedition. Like Scoresby and Lamont before him, Chermside copiously documented his thoughts on the pinnipeds, bears, and whales of Svalbard. Often, these thoughts occurred amidst the savagery of an arctic hunt.

Pinnipeds

It was early on the voyage, on May 23, when the first pinnipeds were spotted. Chermside wrote that "We saw one "school" or lot together, probably young seals. It was very exciting to see them rolling playing & turning." Under the tutelage of *Diana*'s sailors, he was soon able to identify the various types of true (earless) seals native to the Arctic: "I saw a very small seal before turning in. The mate called it a "floe rat" [the ringed seal *Pusa hispida* or *Phoca hispida*]."[4]

On Kong Karls Land, he observed harp seals (*Pagophilus groenlandicus*) that "number from ten to 20 or 30 as a rule, & seem all females or the young of the year, none of the males with their black faces & grey hoods. They go at a great pace, coming up like porpoises, & at about the same intervals, & frequently leaping right out of the water like a salmon."[5]

The bearded seal (*Erignathus barbatus*) he called: "Grim beasts … as they look at you from the water."[6] The seals had a reason to be grim. "One of them had a narrow escape from a brunie

C. L. Devlin (✉)
Penn State University, Abington College, Abington, PA, USA
e-mail: cld5@psu.edu

P. J. Capelotti (ed.), *The Coldest Coast*, Historical Geography and Geosciences,
https://doi.org/10.1007/978-3-030-67880-7_5

[polar bear], losing a hind flipper & carrying a very clear impression of his enemy's claws on his coat as well."[7]

Another time, a "fat young bladdernose" [*Cystophora cristata*] was shot and dissected by the naturalist Alfred Eaton for scientific study. "[Eaton] is further much interested in the intestines of the seal, & fills buckets &c with detached portions of his carcass including head, flippers, &c. A new regime has now to commence, as we may at any moment be summoned to shoot seals …"[8] The practice of searching for parasites in the gastrointestinal tracts of marine mammals was not unique to Eaton, but had been conducted by the naturalists onboard the ships of both Scoresby and Lamont.[9]

Seal social behaviour and modes of communication held Chermside's interest. "We were much excited at seeing a whole school of seals, playing about in the water, at no great distance … The sailors call it a "seals wedding."[10] And a month later: "It was very interesting, at the end of our stalk, to watch the seals, even at the distance we were. Some were asleep, others rolling & scratching themselves with their flippers, to say nothing of a tremendous parliament, continually going on. It was rather like the grunting of the hip [hippos], but rather a more whistling noise."[11]

Dangerous elements of the seal hunt were also described: "…I made towards his [a bladdernose seal] position. The travelling was to me of a very novel kind, jumping from block to block of ice, all said blocks being snow covered into which covering you often sank over your sea boots. These blocks were dancing & shifting, opening & closing with the heavy swell & it was ticklish work jumping from one to the other…"[12] But most frequently, horrific elements of the hunt—when animals, often mortally injured, slipped away into the depths—were described with indifference: "We stalk him, by getting the boat behind hummocks, & when I think I am near enough I fire, & put a bullet clean through his neck… He floats for a few seconds, but as we get there, he is sunk & only the discolored water & bubbles of blood show where he has gone down."[13]

The crew of *Diana* was not alone in the Svalbard seal shoot. When they stopped to converse with the crew of *Vega* on 26 May, the Swedes reported slaughtering three hundred seals. Later that day, they came upon *Active*, a Scottish vessel also holding three hundred dead seals. Two days later, they encountered the Scottish vessel *Jan Mayen* with 3,500 seals onboard. But nothing compared to the Peterhead vessel *Eclipse*: "She has 45,000 old seals on board & 3,000 young ones, & 2 whales, in all 250 tonnes of oil. She has had to throw overboard the greater part of her coal to make way for blubber. She is going S to try & get whales as he has room for 50 tonnes more, & wants to fill her."[14]

As *Diana* sailed north, as the landscape became increasingly devoid of seals, the sites of butchery became more frequent and grisly: "… point after point passed not a living seal in sight, only a few "*krengs*" or "flensed carcasses," a prey to ivory gulls & stately "Burgo Masters" or full joyed "Mallies," the relics of the spoil, scanty ones, too, of the patient seal fishers we have met."[15]

Chermside also conceded that: "These Norwegians seem much better seal fishers than our men. They go with less noise & more care, & stalk the seals, & generally instead of long shots get shots close enough to make sure. They also don't get to windward & perform various other absurdities, as seems the habit, of the British sailor…"[16]

So successful were the hunting practices of the Scandinavians that the extinction of another pinniped, the walrus (*Odobenus rosmarus rosmarus*), had occurred in Norway and Iceland, forcing fringe populations into polar extremes like Svalbard. Therefore, it was not until *Diana* entered the northern parts of the archipelago that the crew finally spotted the largest of the extant pinnipeds. The walrus responded to the ship's presence with a well-justified wariness: "Going down Smeerenburg, they viewed two seahorses on a piece of ice to the westward of them, & sailed down on the them but did not get near enough for a shot. The brutes, as happened with me, suddenly hurling themselves back with a crash in the water."[17]

Chermside's journal entry of June 9th detailed the violence of a walrus hunt:

> … we had previously seen two very fine "bulls" in the water. A boat is lowered at once luckily I am ready with a rifle & off we go… As we near them & row quite noiselessly, I get a good view of the immense brutes. They are lying all huddled together, & as the outside one puts up his head for a moment, I see the dark mask & long ivories of an old bull… It is now intensely exciting and I can hear my own heavy breathing as we are within the reach … we are almost on them. Bang – off they all go. But we are "fast" and as soon as enough line has been taken out, a turn is taken round a bollard in the bows & the walrus allowed to amuse himself by towing us. Generally, when one is fast, the others stick to him, attacking the boats. They are kept off with lances & rifles &, if possible, hand harpoons stuck in so as to get another fast… … suddenly up he comes with a rush but dives at once. I fired as I was told at his neck … an "express" hollow-headed bullet driven by 4½-drs of powder has a wonderfully quieting effect. As soon as he is finished, we haul him to the ice & try to get him out… [Another walrus] now puts up his great head & looks at us … within 5 yards of the boat, rearing his great bewhiskered head out of the water, & looking suspiciously at us with fierce little red eyes… He is looking straight at us, his head alone above water, not quite to the tips of his tusks. I am at the water line, *crack* … shot under the chin, through the windpipe, & he blows great gulps of blood every breath… He turns round & round & seems very bad, and at last disappeared for good & all …[18]

Earlier, Scoresby had documented the defensive behaviours of walruses when a member of the herd was threatened: "… an attack made upon one individual, draws all its companions to its defence."[19] So well-camouflaged were the walruses against their rocky shoreline environment that at times the crew nearly missed them: "… we … were electrified by the intelligence that there were seahorses on the ice, these we had seen some time before, but had not made out to be such, in fact I was told they were stones."[20]

It is the pinnipeds' amorphous 'stone-like' form that makes them perfectly adapted to both their arctic and aquatic habitats, for their reduced surface area minimizes heat loss, the streamlining facilitates ease of movement through the water, and grey colouration conceal them from apex predators like polar bears—and humans.

Polar Bears

Of all the marine mammals observed, the polar bear (*Ursus maritimus*) most captivated the imagination of the hunter-explorers and was considered the ultimate Arctic prize. By summer 1873, there was a long-standing tradition of acquiring bears for the growing zoological gardens of Europe. In 1812, Scoresby captured a live bear for his professor-mentor Robert Jameson, who exhibited it at Edinburgh's Natural History Museum and fed it a diet of liver and horsemeat.[21] After capturing two polar bear cubs in 1859, Lamont was able to observe bear behaviour close up, albeit in the unnatural confinement of pens constructed of driftwood on the ship's deck. These observations led him to correspond with Charles Darwin and offer the naturalist his own brash ideas on the evolution of the polar bear.[22] Lamont's two bears had a tragic fate when, during the siege of Paris, in the Franco-Prussian War of 1870–71, hungry Parisians seized upon the zoological gardens and ate them.[23] Leigh Smith also acquired a live polar bear during his expedition of 1871, when he purchased a cub from a ship returning from Novaya Zemlya. This bear eventually came to reside at the Zoological Gardens in Regents Park, London.[24]

Chermside's first sighting of a polar bear occurred at Walden Island:

> In the foreground lay the huge bloody *kreng*, with a circle of busy & hungry Burgo's & Snowbirds sharing the feast with the "Brunie." This beast, tho' small (about the size of those in the Zoo), looked very handsome as he shook the spray from his yellow coat, & picked & ate like a dog, apparently not over hungry too, or about "through" for from time to time he played with pieces tossing them about. The whole thing made an Arctic picture … not to be surpassed in beauty… Brunie soon sees us, & leaving the *kreng*, walks along the edge of the ice to the point nearest to the large floe piece… But he has his doubts about leaving the "*kreng*" & returns to it again, then he looks at us & sniffs & sniffs, "hang it, discretion you know" &c & biting at his paw in which, I suppose, some snow had balled he makes again to the extreme point. But he has been wasting golden moments, & he goes on doing so, as instead of taking a plucky

> header, he sways his head & body about over the water, & finally ignominiously turns round, & lets himself down hind legs first, & turning heads for the large piece—too late, *mon ami* ... we ... have cut you off... We are now taking across straight at him & the only one that occurs seems to stand with all four feet together like a goat & his head toward us ... the rest was very tame & rowing to within 20 yards a couple of bullets (head and shoulder) & all is over. He is soon in the boat ...[25]

On July 8th, "an old & a young bear appear on the floe," certainly a female and her cub, for adult males are solitary, but the pair smartly "refuse to be beguiled towards the ship, not even by burnt offerings of loons & eider ducks." The next day another bear is spotted: "He is a very deliberate brunie, but for all that his soft flexible limbs & broad feet get him along a great pace..."[26]

After a number of effortless kills, Chermside was smugly unimpressed by the polar bear's fearsome reputation, albeit from the safe shelter behind his rifle: "We certainly have lost all faith in the ferocity & terrors of the Polar Bear, the "lion of the north" & all that twaddle not only from our own slight experience but from all we hear... In fact, as an object of terror, he ranks, I think, lower than a hungry wolf..."[27]

Polar bears, like the pinnipeds, are well adapted to both the aquatic and arctic environments, and thus exhibit marked anatomical differences from their southern cousin, the grizzly bear (*Ursus arctos*). Chermside's remark above about the polar bear's broad feet was astute. When on land, the broad feet produce a snowshoe effect for effective locomotion on the snow whereas, in the water, they serve as paddles for swimming. Webbing between their toes further enhances the paddle-like function of the foot. Compared to their terrestrial cousins, the polar bear's narrow, pointy head, and reduced ears, streamline their bodies for sleek movement through an aqueous medium. Larger lungs and hence greater lung capacities are another adaptation for their semi-aquatic lifestyle. These are combined with features of short limbs that minimize surface area and reduce heat loss, black skin to absorb and retain heat, and a unique fur coat to keep water out and heat in. All act in concert for thermoregulation in the frigid climate.

Whales

Of the arctic cetaceans, the most iconic are the white whales: the narwhals (*Monodon monoceros*), and belugas (*Delphinapterus leucas*). Like the pinnipeds and polar bears, the evolutionary adaptations of these odontoceti whales make them perfectly suited for arctic waters. Their light coloration offers camouflage amidst the ice floes so as to avoid detection by their natural predators, killer whales (*Orcinus orca*) and polar bears. Another unique feature, their lack of a dorsal fin, allows for unimpeded navigation amidst the maze of ice. A bulbous forehead housing the melon, an organ of echolocation, permits communication by a complex language that cements their social groups, pods.

Arguably the most curious of the white whales is the narwhal—due to the prominent coiled tusk of the males. The modified canine tooth evokes a fearful image—a spear for impaling an enemy—but instead it is a highly innervated sense organ for determining the chemical properties of water.[28] Narwhal tusks washing onto the beaches of northern Europe reputedly formed the basis of the unicorn myth so, not surprisingly, mariners referred to them as "Unis."

On May 30, Chermside documented his first narwhal hunt: "...we went in chase of a Uni that was lying near the pack edge. I tumbled into the boat, which was rather slack at getting away owing to its being the first time. We got within about 40 yds of the beast when it started. After a longish chase, during which he several times went down just as we were going to fire, we lost him & returned."[29]

The "smiling" belugas were the whale most often encountered by *Diana*'s crew. "A large school of white whales then came across us, rolling & snorting, appearing & disappearing with great rapidity, but all making the same way, some black & some spotted, these latter being the young." Chermside was correct in identifying the juveniles by their dark, mottled coloration, a protective adaptation that provides camouflage in the shallow river mouths and bays where they migrate in the summer months. Unfortunately, arctic hunters were also aware of this migratory pattern: "This [Loome] bay … is a great resort for white whales, & the low sandy beach a very convenient point for hauling through on; their abundance is testified by the endless skeletons all along the shore for some miles."[30]

Such bone-strewn beaches were evidence of centuries of industrial-scale whaling operations in Svalbard. Of Smeerenburg, Chermside wrote that: "The name is Dutch & means simply "Blubber-town," and from 1614 to 1660 the Dutch had a factory, warehouses &c here, boilers, coppers &c, & all the means for converting blubber into oil. This was, of course, in the days when the fishery was carried on close to Spitzbergen itself. The ship's mooring in the bays, which abounded in whales for the summer, sending out the boats to fish. The whales, at one time so abundant, have long since left the bays, fiords & harbours, & are now only to be found in the open sea near the ice edge, or among the fields & floes of the western ice. I find from reading Scoresby's book, that frequent & serious quarrels having arisen among the ships' crews, belonging to the various nations, who frequented the fishery …"[31] The belugas were prized because they produced a superior-quality oil that burned without producing smoke, wrote Lamont.[32]

Though *Diana* was equipped with harpoons and whaleboats, whaling is a highly specialized form of open-water hunting so her crew largely avoided the dangerous endeavour. Besides, more often than not, the baleen whales spotted in summer 1873, were not the slow, docile right whales (*Eubalaena glacialis*) that produced volumes of blubber and whale oil, but swift fin whales (*Balaenoptera physalus*). On June 5, Chermside wrote: "… they saw a "blast" [of water] & pulled for it… They got quite close up to the fish, but it proved to be a "finner." These fish are not attacked in the same way as the right whale as they are a more powerful fish. They go at a much greater pace … [and] … are easily distinguished as their section is different, being almost cylindrical in the middle of the body, instead of the broad flat back of the right whale. Towards the tail they have ridges …"[33] These ridges observed by Chermside are throat pleats that expand like an accordion to take in voluminous amounts of water and planktonic prey during the process of filter-feeding. Like the crew of *Diana*, Lamont's men made no attempt to haul in fin whales as they deemed it too difficult and deadly an enterprise.[34] With summer profits in mind, Chermside said that "… [the fin whales] are not nearly so valuable as the right whale, as they produce no "baleine" [baleen], or whalebone, & the blubber is of a different kind, I believe tougher, courser, & thicker in fibre…[35] In short, fin whales lacked the right stuff.

Conclusions

Like his predecessors William Scoresby, Jr. and James Lamont, Herbert Chermside had a complex, paradoxical relationship with the pinnipeds, bears, and whales of Svalbard. He was at once eloquent in describing the beauty and behaviours of the marine mammals, then indifferent or even gleeful during their hunt for sport and profit. In the end, Chermside's diaries expose the appalling magnitude of destruction of mammal populations in Svalbard: multinational hunting frenzies consisting of often indiscriminate shootings that resulted in mortally wounded animals sinking to the bottom, and centuries of slaughter as evidence by beach after beach of bones. By summer 1873, populations of marine mammals were already severely depleted, forcing remaining individuals to ever more remote northern extremes. Most of the marine mammals described in this essay have been assessed as critically endangered by the *International Union for Conservation of Nature's Red List of Threatened Species*. In some cases, their numbers are so low that they are considered functionally extinct, even as global climate change adds another ominous stressor to the dwindling numbers of these species that are so beautifully adapted to the unique polar environment.

Notes

1. Scoresby, W. 1820. *An Account of the Arctic Regions, with a History and Description of the Northern Whale-Fishery. Vol. 1*. Edinburgh: Archibald Constable and Co. https://dx.doi.org/10.5962/bhl.title.49370.
2. Lamont, J. 1861. *Seasons with the sea-horses.* London: Hurst and Blackett Publishers.
3. Lamont, J. 1876. *Yachting in the Arctic seas, or notes of the five voyages of sport and discovery in the neighbourhood of Spitzbergen and Novaya Zemlya*. London: Chatto and Windus, Piccadilly.
4. Chermside, H.C. *Journal on board of the SS Diana*. SPRI, Vol. II, May 26, 1873. All quotes from this unnumbered journal are hereafter cited by their volume and entry date. The journals themselves are part of the general reference at SPRI: GB 15 Herbert Chermside, with the specific reference: MS 300/4/1-2-3.
5. Chermside 1873/3: August 19.
6. Chermside 1873/3: August 24.
7. Chermside 1873/3: August 28.
8. Chermside 1873/1: May 25.
9. Devlin, C. L. 2015. The influence of whaler William Scoresby, Jr. on the Arctic observations of Sir James Lamont. *Arctic* 68(3): 317–330. https://dx.doi.org/10.14430/arctic4510.
10. Chermside 1873/1: May 26.
11. Chermside 1873/1: June 2.
12. Chermside 1873/1: May 30.
13. Chermside 1873/1: May 31.
14. Chermside 1873/1: May 31.
15. Chermside 1873/1: May 28.
16. Chermside 1873/1: June 3.
17. Chermside 1873/2: June 11.
18. Chermside 1873/2: June 9.
19. Scoresby 1820: 504.
20. Chermside 1873/2: June 9.
21. Stamp, T. and Stamp, C. 1975. *William Scoresby, Arctic Scientist.* Whitby: Caedmon of Whitby Press: 47.
22. Devlin, C. L. 2015. "The letters between James Lamont and Charles Darwin on Arctic fauna." *Polar Record* 51 (5): 492–500.
23. Lamont 1876: 330.
24. Capelotti, P. J. 2013. *Shipwreck at Cape Flora: The expeditions of Benjamin Leigh Smith England's Forgotten Arctic Explorer*. Calgary: University of Calgary Press: 74.
25. Chermside 1873/2: June 29.
26. Chermside 1873/2: July 9.
27. Chermside 1873/3: August 30.
28. Nweeia, M. T. et al. 2014. Sensory ability in the narwhal tooth organ system. *The Anatomical Record*. https://doi.org/10.1002/ar.22886.
29. Chermside 1873/1: May 30.
30. Chermside 1873/2: July 11.
31. Chermside 1873/2: June 10.
32. Lamont 1876: 194.
33. Chermside 1873/1: June 5.
34. Lamont 1876: 7.
35. Chermside 1873/1: June 5.

6 The Discovery of Chermside's Photographs

Håkan Jorikson

In 2014, the late Doctor Dan-Axel Hallbäck, Sweden, donated two photo albums to Grenna Museum. The smaller one was an album with Doctor and photographer Axel Enwalls images from A. E. Nordenskiöld's expedition to Spitsbergen (1872–1873). The other album was supposed to be from the same expedition, though with captions in English.

Due to his profession and interest in the polar region and history, Hallbäck had previously, in 2005, bought the albums from the council in the Värmland region in Sweden, as they were to close down a minor hospital museum in Kristinehamn—obviously also containing artefacts from Axel Enwall. Hallbäck realized that the albums were to be thrown away if he did not act, since no one else was aware of their content and importance.

Some days after the donation, I more thoroughly examined the albums and found in the English one, the ships *Diana* and *Samson*. As a coincidence, P. J. Capelotti had then recently published a biography of Benjamin Leigh Smith (*Shipwreck at Cape Flora)*. It was from there that I recalled *Diana* and *Samson*. Capelotti confirmed the album's origin: Herbert C. Chermside and Benjamin Leigh Smith's third expedition in 1873. Dan-Axel Hallbäck was happy to know about it and the rareness of its existence.

One may wonder how the album of Chermside ended up in a tiny hospital museum in Sweden. As Anders Larsson tells in a previous chapter, Axel Enwall had a relation to the hospital and city, where he also ended his life. Probably, the two photographers changed albums after coming home. Hence, an album of Enwall's images, from 1872–1873, still might be somewhere in England. -Håkan Jorikson, Director, Grenna Museum.

H. Jorikson (✉)
Grenna Museum, Grenna, Sweden
e-mail: hakan.jorikson@grennamuseum.se

P. J. Capelotti (ed.), *The Coldest Coast*, Historical Geography and Geosciences,
https://doi.org/10.1007/978-3-030-67880-7_6

7 Magdalena Bay: Chermside's Observations in the Present Day

Tyrone Martinsson

Magdalenefjorden, on the northwest coast of Spitsbergen, is one of the most historic places in Svalbard. On 4 September 1873, the steamer *Diana,* carrying Benjamin Leigh Smith's third expedition to Svalbard, anchored at Trinityhamna in Magdalenefjorden. The expedition was on its way south and home after enduring an Arctic gale and sought relief in the famous bay. In his diary, the expedition's surveyor and photographer, Herbert Chermside, was struck by the spectrum of colours in the bay and the gorges filled with ice between the mountains. As he tried to put words on the scenery of the bay, he wrote himself into a long line of observers of the area. More importantly, his six surviving photographs from Magdalenefjorden are the first surviving prints we have from the area.

Chermside used Captain Philip Broke's 1807 name, *English Cove*, for *Diana's* anchoring site. He envisioned the name as given to the protected safe harbour by English whalers centuries earlier. He was partly right, as the name was given to the cove by Robert Fotherby in 1614 when claiming Magdalenefjorden for King James I:

> I went to the Harbour and there caused a Crosse to be set up, and the Kings Armes to bee nayled thereon; under which also I nayled a piece of sheet Lead, whereon I set the Moscovie Companies Marke, with the day of the moneth and yeere of our Lord. Then cutting up a piece of Earth, which afterward I carried aboard our ship, I tooke it into my hand and said, in the hearing of the men there present to this effect.
>
> I take this piece of Earth, as a signe of lawfull possession (of this Countrey of King James his New-land, and of this particular place, which I name Trinitie Harbour) taken on the behalfe of the Company of Merchants, called the Merchants of New Trades and Discoveries, for the use of our Sovereigne Lord James by the grace of God, King of great Brittaine, France, and Ireland, whose Royal Armes are here set up, to the end that all people who shall here arrive may take notice of his Majesties Right and Title to this Countrey, and to every part thereof. God save King James.
>
> This is a good safe harbour, and is under the latitude of 79. degrees 34. minutes, as I have found by good observation, and have of Westerly variation 25. degrees.[1]

Even though the fjord was claimed by the English in 1614, it had been visited and noted by the Dutch on their discovery journey in 1596. In particular, the safe harbour of the fjord had been pointed out by Willem Barentsz:

> This bay, where we were, went right in with a cove at the end. On the S. coast there was a low point, behind which one could navigate, entering near the N. coast, and lie there behind the point, sheltered from all winds. Our men found there tusks of morses, or sea-cows, which gave the bay its name of Tusk Bay [Magdalenefjorden]. We also found there a quantity of head of deer, and some wool like sheep's wool. Just S. of the point there was a little cove, like a harbour.[2]

T. Martinsson (✉)
Valand Academy, University of Gothenburg, Gothenburg, Sweden
e-mail: tyrone.martinsson@gu.se

P. J. Capelotti (ed.), *The Coldest Coast*, Historical Geography and Geosciences,
https://doi.org/10.1007/978-3-030-67880-7_7

The site where *Diana* was anchored is close to the peninsula (Gravneset) in the centre of the south coast of the bay. As the snow melted in the changing weather, Chermside became aware of the number of graves located on the hill of sand and rocks surrounded by rugged and jagged peaks of the mountains.

> Lit by the evening sunlight, those inaccessible needles standing clearly out against the skyline form a fitting tiara to the amphitheatre of hills that encircle the bay, the end isolated as is almost always the case by the icy cliffs of a glacier, the remains of that, that scooped out the bay. Heavy snow is over everything, but from the force of the wind it was in drifts on the mountains, many dark portions of the scars being this bare, so that the whole has the effect of a drawing in sepia & white, except indeed the edges of the glaciers showing in striking enough blue & green tints now, as owing to the season fractures are continually being made.[3]
>
> In 1671, Friedrich Martens noted on the fjord:
>
> At the Magdalen-Haven the Rocks lye in a round or semi-circle, at each side, by one another, stand two high Mountains that are hollow within, as if they were dug out: after the fashion of a Breast-work, with points and cracks at the top, like Battlements; at the bottom, within the Hill, stands a Snow-hill, that doth reach to the very top of the Mountain, like a Tree with branches and twigs; the other Rocks look rudely.[4]

In 1839, Léonie d'Aunet described the fjord with its sounds of calving glaciers and cracking mountains breaking the Arctic silence:

> One imagines always Spitsbergen as a barren and silent frozen world, but that is thoroughly wrong—nothing can be compared with the noise at spring thawing up here. Waves hitting the ice blocks and rocks, glaciers calves with deafening crashes, the mountains crack, ice floes are pressed against each other and triggers sounds like musket volleys, the wind howls and swirls the snow. It is magnificent but frightening as a premonition of the world's ending, never have I seen or heard anything the likes of what I experienced here, it passes one's wildest fantasies! It is like opening the door to an unreal confusing world that at the same time gives a sensation of fear and ecstasy.[5]

Léonie d'Aunet came with the French *La Recherche* Expedition led by Joseph Paul Gaimard. French artist Barthélemy Lauvergne made a series of watercolour drawings in the area, working alongside the artists August Mayer and François-Auguste Biard who stayed in the area for several weeks. Léonie d'Aunet made a remark on the nature of the land quite contrasting to our own times:

> Glaciers are located between the mountains and becomes thicker and thicker every year, when a few short weeks in the summer is not able to overcome the large masses of snow that covers the entire Spitsbergen for months. It is anticipated that all the mountain tops, over time, are to disappear under the ice.

Chermside refers to a survey of'Magdalena Bay' as made by *HMS Trent* in 1823. Here he is most likely referring to *HMS Trent* visiting the fjord 1818 with *HSM Dorothea* on David Buchan's expedition. The Captain of *HMS Trent* was then John Franklin. Having encountered severe weather and a bad storm, they undertook a survey during their lengthy stay at the fjord. In this way, they were continuing the work that had been conducted by Captain Philip Broke on *HMS Shannon* 1807. Broke made the first detailed mapping of the fjord with depth markings and created charts where glaciers and surrounding mountains are named. Buchan revised and followed Broke's work and made some further mapping in the area around the fjord where also William Beechey and Charles C. Palmer made a series of drawings showing glaciers and the fjord's character. It was Beechey, who in 1843, published the official account of the expedition where he described the hanging glacier:

> The bay is rendered conspicuous by four glaciers, of which the most remarkable, though the smallest in size, is situated, two hundred feet above the sea, on the slope of a mountain. This glacier, from its peculiar appearance, has been appropriately named the Hanging Iceberg. Its position is such, that it seems as if a very small matter would detach it from the mountain, and precipitate it into the sea. And, indeed, large portions of its front do occasionally break away, and fall with headlong impetuosity upon the beach, to the great hazard of any boat that may chance to be near.

Beechey and Palmer had come to Magdalenefjorden twenty years before photography and have achieved a visual description of the landscape while avoiding overly wordy aesthetic essays on the views they observed. Palmer, who was onboard Buchan's ship *HMS Dorothea*, made a panoramic view of the south coast of the fjord from the ship.

Chermside, as the photographer of Leigh Smith's expedition, does continuously throughout his diary notes make remarks related to his search for picture opportunities. In Magdalenefjorden, he gets a brief window of two or three hours in the weather on 5th September, when he obtains, as he writes,'a fair series.'

Chermside was not the first photographer in the bay though. In 1861, the young medical student Axel Goës came to the fjord on a Swedish expedition led by Otto Torell. Goës carried a camera and created a panorama of the fjord's east and south coast. The images today exists only as a lithographic illustration for the 1865 publication from that expedition.

Technically, Chermside worked in a time of continuous developments in photography as regards the collodion process he used. According to his diaries, he tries both wet and dry techniques. It was a quite bulky process of having to prepare plates before being able to take a photograph. The dry option required longer exposure times but was a bit easier than the wet plates. The technique further required a quite rapid process of development. Chermside bears witness to the photographic process having to be carried out in darkness under a canvas on a rolling ship. As they try to develop the glass plates, they sometimes slip through their hands and are damaged.[6]

Chermside photographed four glaciers in the fjord: Waggonwaybreen, Buchanbreen, Gullybreen, and the 'hanging iceberg.' The images are made from different view points on Gravneset. Waggonwaybreen in Chermside's images is connected to what today is Miethebreen, but in 1873 was a glacial branch following the mountain valley east of Alkekongen. That is the valley followed by Chermside as he went on an excursion on 5th September to seek a way across the mountains towards Bjørnfjorden and Sørgattet. He tried the gully at the intersection between the mountain and the glacier. This was a daring solo hike in terrain that is precarious and often dangerous. He was eventually forced by glaciers and fading September light to return.

The 'hanging iceberg' was possibly the most remarkable glacier in the area even at Chermside's visit and is marked on both Broke's and Buchan's maps. Beechey held the glacier as the most spectacular in the fjord, and Chermside describes this glacier as the largest and most noticeable of several hanging glaciers in the fjord.[7] It no longer exists and Chermside's two photographs of it are valuable records of its features on the slopes of the Trankollane mountains above Gravneset. These photographs are not only comparative records of changes in the landscape, but also represent cultural heritage artefacts with potential qualities of immaterial heritage value. Photographic images are not only silent witnesses, but are also charged by the stories they are connected to. Chermside's images are part of a long story of Magdalenefjorden where relationships between different documentations and representations of the fjord are critical to understand transformations over time.

Today, we use these photographs in time series and through the technique and methods of rephotography to visualize changes to the glaciers over time. Within rephotography, historical documents meet physical locations creating comparisons where contemporary made views can be linked to stories of geographical locations found in archives, libraries, and museums. We use multiple photographic documents from a specific site to try to find a sequence forming a timeline enabling us to observe differences over time. Through this, we try to establish when transformations in the landscape become apparent. Chermside's images form a baseline for such observations.

Nineteenth century photographers came to Magdalenefjorden between 1861 and 1896. This thirty-five-year time span shows no apparent changes to the glaciers in the fjord. The nineteenth century saw increasing numbers of tourist ships in the fjord leaving a record of images. Ever so slowly, beginning in the 1920s, a process of change in the fjord's glaciers starts and, since the 1960s, it has been accelerating exponentially. Rephotography and the processing of historical material create a visual narrative that, among other things, indicate that particular shift when the changes in glaciers start to become very evident.

Through series of photographs, we get clear evidence of what once was in front of the camera. The recordings of the camera, in their capacities as instruments, differ from pre-photographic recordings relying on eye/hand relations and renderings often based on artistic interpretations. Comparing an 1873 Chermside photograph with a 2016 view of the same vantage point gives us dramatic evidence of the profound changes of glaciers in the area. This type of evidence is both convincing and powerful. Hence, Chermside's photographs today are invaluable for the continuous process of increasing awareness of a changing Svalbard in a climate in rapid transition.

Notes

1. Purchas, Samuel. 1905. *Purchas His Pilgrimes*, Vol. XIV, London: Hakluyt Society: 65–66.
2. Barentsz, Willem, in Hessel Gerritsz. 1613. "Historie Du Pays Nommé Spitsberghe." Translated into English by Basil H. Soulsby, in Martin Conway. 1902. *Early Dutch and English Voyages to Spitsbergen in the Seventeenth Century*: 17–18.
3. Chermside, 1873/3: 4 September.
4. Martens, Friedrich. 1694. 'Voyage into Spitzbergen and Greenland', in John Narborough, Jasmen Tasman, John Wood, and Frederick Marten (sic) *An Account of several late Voyages & Discoveries to the South and North. Towards The Streights of Magelan, the South Seas, the vast Tracts of Land beyond Hollandia Nova, &c. Also Towards Nova Zembla, Greenland or Spitsberg, Groynland or Engrondland, &c.* London: Royal Society: 23.
5. Aunet, Léonie d'. 1968. *En pariserinnes reise gjennom Norge til Spitsbergen anno 1838*, translation to Norwegian by Generieve Jul-Larsen. Oslo: Aschenhoug. Originally published in 1854 as *Voyage d'une femme au Spitzberg*.
6. Chermside, 1873/1: 7 June; 1873/2: 8 June.
7. *Ibid*., 4 September.

8 Revisiting the Locations of Chermside's Arctic Imagery

Andreas Umbreit

When asked to assist with identifying the locations of the newly rediscovered 1873 photographs of Herbert C. Chermside, I saw this as an exciting challenge to combine my 35 years of experience in Svalbard with my fascination for the history of the archipelago. Some locations were quite obvious, but I soon realized that others were a true challenge.

Introduction and challenges

Chermside's images were not necessarily in chronological order so, in principle, they could have been taken anywhere along the route of the *Diana*. What to do with pictures that only showed fragments of landscapes—and in some cases in poor quality with limited sceneries such as a small flat rocky island (there are hundreds of these in Svalbard), or just a long glacier front emerging from a fog-covered background, or a picture of pressed-up ice masses, with just a few mountain tops sticking up behind the ice?

Then there was the issue of dating: the 1873 expedition of Benjamin Leigh-Smith visited some sites repeatedly during their voyage. Even if a location could be identified, during which of these calls was a picture taken? The Chermside diaries are not very precise on dates and locations of his photographic work in Svalbard. He mentions photography repeatedly—but also that some of it failed due to the many technical problems associated with the complicated procedures of early photography (bad quality of plates, problems with chemicals, and so forth). Also, long exposure times were required and these must also have been an issue—but to what extent? Seemingly, Chermside did take some shots from the ship, but as a potentially unstable platform for long exposures, where was it calm enough to make these attempts a success? Which of the exposed plates were discarded later due to too poor results? Where, for instance, did dense drift ice have a sufficient calming effect on the waves, and thereby on the ship movements even under windy or even stormy conditions to allow successful photography with long exposure times?

We can also look at this from the other way round: as his notes on photography in the diaries do not seem to be especially precise, there is the other possibility that he also took pictures in locations without noting it down, at least not in the available diaries (though it can be assumed that he had some other notes on this work, given the difficulties of each shot at that time, giving each image quite another value compared to the ease of digital photography nowadays). Even where Chermside mentions landings in which he participated (and he seems to have been an enthusiastic hiker), this does not mean that he had his most likely quite heavy and clumsy camera gear always with him. Another possibility is also that parts of his photography during the expedition is lost (or not found, yet).

A. Umbreit (✉)
Munich, Germany
e-mail: terrapolaris@gmx.net

P. J. Capelotti (ed.), *The Coldest Coast*, Historical Geography and Geosciences,
https://doi.org/10.1007/978-3-030-67880-7_8

A further uncertainty were changes within Svalbard over time: Chermside took these pictures 147 years ago. Scenery changes due to human activities (buildings, etc.) are not the biggest issue in a mostly uninhabiteted wilderness like Svalbard—though in some very interesting cases, this is an issue on the Chermside pictures.

More interesting was the question: to what extent did the natural scenery change over that long period? The expedition took place in 1873, which is some two decades after what is regarded as the approximate end of the so-called Little Ice Age—most likely the coldest period over the last 8,000 years. In the Arctic, a cold period, lasting for several hundred years, means quite an increase of both glaciers on land, covering more of the land mass and reaching out also into the coastal waters, and also a wider distribution and greater thickness of sea ice. The warming following the Little Ice Age was only in its beginning at the time of Benjamin Leigh-Smith's 1873 expedition, but had already effects also mentioned in Chermside's notes, where he refers to successful recent voyages of others, especially hunters, who managed to reach remote eastern parts of the archipelago (like Kong Karls Land and the more remote parts of Nordaustlandet) just a few years before—areas which were inaccessible for hundreds of years though vaguely known (like Giles Land, mentioned also by Chermside) from reports of the first human activities in seventeenth century, before the maximum of the Little Ice Age. So, in 1873, the retreat of the ice had started already, but obviously, many parts of the archipelago must still have looked different, due to still more glaciation, than they do today. This could also present an issue in identifying locations on the Chermside images and, at the same time, as Tyrone Martinsson points out in this volume, a potentially interesting feature of them, allowing comparisons between then and now.

So despite having the images and the diaries, the identification and dating of the photographs was more of a challenge than expected. At least, the diaries allowed the exclusion of parts of the archipelago, as they were seemingly too far away from the route taken by *Diana*. On the other hand, it proved fascinating to do such work on some of the very earliest photographic images ever taken of Svalbard.

Method (Locations)

Chermside's images were first sorted by landscape types, including those which I recognized immediately, as others might show the same area from different positions or angles. Luckily, Svalbard offers quite a variety of different environments, for example, the pointy alpine west coast, the flatter and much more glaciated landscapes around Hinlopen Straight, and the structures based on hard basaltic rocks. I am intimately familiar with this varied scenery from cruising along them year after year. Grouping the images in this way facilitated narrowing in the areas where they may have been taken.

This first step led to major clusters of pictures seemingly from these areas: Magdalenefjord (7, most already identified by Tyrone Martinsson in this volume); Sjuøyane (9 images); Lomfjord (6 images); and Sorgfjord (old name: Treurenberg Bay, 4 images); and Hinlopen Strait (7 images). In addition, Mosselbukta, with Nordenskiöld's wintering base Polhem was fairly simple, as this is the only picture showing an intact building not looking like a trapper cabin—and surely there were not many of this kind in Svalbard in 1873. Moreover, there were seven images showing just drift ice, or the ship in drift ice, or hunting on ice. With these, I did not deal any further because without land visible, they could be taken anywhere where the *Diana* moved into thicker drift ice (especially around the Seven Islands (Sjuøyane)) and in southern Hinlopen Strait area.

In addition, there were five more landscape images left, which fit into none of the above clusters. Identifying these took quite a search through my own personal collection of Svalbard imagery. These were eventually pinned down to Grønfjord (Green Harbour, showing partly the area of todays Barentsburg); Austfjorden (innermost Wijdefjord); and in northern Wijdefjord (showing the east side of northern Andrée Land). Similarly, identifying the pictures, which looked like somewhere in Hinlopen Strait, was also a challenge because of the uniformity of much of the scenery there—and here, Chermside depicted some relatively small scenes, like minor barren flat islands. These turned out to be Sværingen (the main island of the Foster Islands), Karl Alexander Island, Franz Island near Kapp Torell, and Kapp Oetker.

Method (dates)

After having located all images with scenery visible on them, the second issue was to date them as closely as possible. With areas visited only once, this was of course easy, but for places visited more often, tiny clues had to be found, like different degree of snow cover on the mountain slopes (Lomfjord (Lomme Bay) with Kapp Fanshawe and the views into the inner fjord). This allowed the determination of whether pictures showing the same place were taken earlier or later in the season.

In two other cases, depicted ships were a good help, allowing an image taken in Heclahamna and another one at Valhallfonna in the entrance of Lomfjorden to be dated to the second visits to each of these areas. Mosselbukta, with Nordenskiöld's wintering base Polhem, was also visited repeatedly. Here, the lack of sea ice and the blurred poor quality suggest that the image was taken during the visit there in mid-July with the ice already gone—possibly on 16 July when the ship voyaged into Wijdefjord, where movements of the ship could have caused the reduced quality.

Sea ice is also a critical dating hint for most of the photos showing the Seven Islands (Sjuøyane). Here Chermside describes heavy drift ice blocking the west side, with the *Diana* partly even fixed to ice floes by cables (visible in some photos) during the first visit, while the west side of these isles was clear of ice by the time of the second visit in late August (one image from the west, without sea ice). When *Diana* entered the waters between Sjuøyane to the north and Kapp Platen and Chermsideøya to the south, this allowed Chermside to take a shot of Sjuøyane from the south from a hunting party on drift ice.

As mentioned previously, Chermside's available notes on his photography are not terribly precise, either in time or space. Therefore, it is virtually impossible to date more of his pictures more precisely than to a first or second visit to an area, but not to a specific day within that period.

Revisiting the locations

Once the locations of the Chermside images had been fixed, the obvious challenge was to match these historic photographs with new ones showing the same locations as they look in the present-day. In a number of cases, my personal photo archive (and, in Magdalenefjord, also Tyrone Martinsson's) provided fitting images. Moreover, as this book project progressed on over several years, I had had the opportunity to intentionally visit some of the Chermside sites during cruises on which I was expedition leader, in an attempt to create modern visions of Chermside's Svalbard landscapes, *from as near as possible to the location where Chermside stood*. In this, especially, the pictures showing present-day Lomfjord (Cape Fanshawe) were taken with the intention to match the respective Chermside images as closely as possible. The same intention was behind visits to Mosselbukta and Grønfjord, but there, time pressures and other logistical problems did not allow us to get equally close to the perspectives chosen by Chermside 147 years ago.

Climate change

As mentioned, Chermside composed his 1873 photography in the early years after the approximate end of the so-called Little Ice Age, during which Arctic glaciers and sea ice had a massive growth in mass and extension. As for sea ice, a single image of a certain area is of little help, because fluctuations in ice cover of these waters were and are considerable. A number of Chermside's photos show heavy drift ice or even fast ice in Lomfjord, around Sjuøyane or in Hinlopenstretet in June or mid-July, but also today, these waters can be ice-covered in the same periods, either by winter ice from that area, or by ice fields being pushed by wind into these places (though on average not as often as in late nineteenth century). For judging the bigger picture, such coincidental single photos are therefore of little help.

On land, this is a bit different. Glaciers (and also lasting snow fields) do not change as much from year to year (and even less within a summer season). So here, old photos can be a good help for seeing developments in comparison with today. Unfortunately, only a minor part of the Chermside images shows glaciers or lasting snow fields. Most striking in this aspect are his photos taken in Magdalenefjord, where one shows a huge united glacier front of Miethebreen and Waggonwaybreen and even Brokebreen, only 2.5–3 km inwards (east) of Gravneset. This is 3 km further out than the glacier front of Waggonwaybreen today, while Miethebreen nowadays does not even reach the sea.

The situation is similar with Gullybreen just west of Gravneset, where Chermside depicted a huge calving front stretching even a bit out into the main Magdalenefjord, while Gullybreen nowadays has retreated by about 2 km in comparison to 1873, with a shallow side bay in place of the absent glacier masses. Chermside´s images are the earliest surviving photography of these glaciers, though in Magdalenefjord, changes of the glacier fronts have been documented fairly well also by earlier drawings and various mappings (see Martinsson, this volume).

Another Chermside image of a glacier front (most likely Valhallfonna on the west side of the entrance of Lomfjord) is unfortunately of little glaciological use, as it includes no reference points, which would allow to locate the front line back then. Furthermore, he took a picture of inner Sorgfjord including Dunérbreen at its end. Though details are limited due to the long distance, it is obvious that this glacier at that time still extended all the way to the sea, with a big ice-core moraine on the beach and the open ice continuing inwards right from that moraine. Today, there are about 2 km of low land between the shoreline and the current end of the glacier with much lower moraines and an obvious loss of thickness of its ice volume at least in its lower part. Some of the lowland in front of the current end moraine might be due to sedimentation, though, since Chermside was in the area. Similarly, Schrödingerbreen, in Austfjord, has lost volume visibly, when looking at the Chermside images and the recent comparison.

Some of his images also show minor ice and snow fields, for instance some detail shots of the shore east of Kapp Oetker, where massive lasting snow accumulations covered parts of the slope at his time, which have melted away to quite an extent, since.

Unfortunately, these realities are even a bit more complicated in Svalbard. It is assumed that most (if not all) bigger glaciers there are of the surge type, meaning that their flow downwards is too slow in comparison to the accumulation in the upper part, which leads to retreat in the front and growth in the upper areas, until the glacier gets out of balance and makes a rapid advance (surge), then retreats again, etc. Each surge-type glacier has its own rhythm—the fastest ones surging about every 30 years, while the slowest need hundreds of years between two surges. When now comparing old and new pictures of the same glacier, it needs to be known whether its visible front position is due to a surge/retreat of its surge rhythm. Ongoing climate changes, then, come on top of this complicated process.

Human presence

A special aspect of some of the Chermside images is the documentation of human presence in this arctic wilderness, and with it the transience of human presence.

Chermside takes us 147 years back in time, into a period where Svalbard was far more remote than nowadays and where the only human traces were the relatively small artefacts by hunters (whalers and trappers) and the first scientific activities. No relevant mines existed (except of minor attempts to replenish coal provisions from openly accessible coal deposits by some ships for their own use), and certainly no more permanent settlements.

Apart from the activities of the Leigh-Smith expedition itself and other vessels they met, Chermside depicted human remnants on land only in two places, and then only as tiny elements in the landscape. First, in Mosselbukta, where the *Diana* spent a number of days on two occasions, assisting the Nordenskiöld wintering expedition in their base Polhem with urgently needed provisions. Here he shows the area from on board the *Diana*, with the main building of Polhem and faintly also some more installations visible on land in their wider surroundings. Second, in Heclahamna, a tiny dot on top of the hill north of it (probably the Parry Hill in his diary) is most likely the flagstaff base of Parry, when he used that bay in Sorgfjord some 45 years before the visit of the *Diana*.

The Chermside images of Hecla Cove are touching also because of what they do *not* show: 25 years later, this place became the main base of the Swedish group of the Swedish–Russian Arc-of-Meridiane expedition, which set up a number of buildings and other installations here and used them for several years. While Chermside shows the site before, the current images show the same place with the Swedish constructions already fallen in ruins, slowly disappearing back into the endless arctic nature. Time and transience become quite apparent here.

Chermside depicted also an area in Grönfjord under the mountain Heerfjellet to the estuary of Gröndalen, showing pristine tundra landscape there. This place nowadays is occupied by the Russian (former Dutch) coal mining town of Barentsburg. In Chermside's time, the first mining attempts in that place were still more than 30 years in future, while now, some of the installations in the modern matching photo are falling into ruins again, already. In these poignant ways, Chermside's photography is also a strong reminder of human transience in the high Arctic.

9 Chermside's Diary of Leigh Smith's 1873 Expedition

P. J. Capelotti

Expedition to the Polar Seas conducted by B. Leigh Smith in the spring and summer of 1873

Volume 1

H. Chermside, Lt, R.E.

May 1873

Journal on board of the *Diana*, s. s. 250 tons.

Owner: B. Leigh Smith Master: Capt Fairweather

Passengers	Rev. A. E. Eaton	1st Mate	James Cumming
	R. E. Potter, Esq.	2d Mate	William Poldie
	H. C. Chermside RE	Skisman/Harpooneer	Jeffery Phillips

Crew		Crew	
1st Engineer	William S. Forbes	Boatswain	David Cobb
Asst "	James Kidd	A B	Allan Gillis [sp.]
Fireman	Nicholas White	"	Henry Mitchell
Steward & Seaman	William Hayward	"	Thomas Fenton
		"	Alexander Scott
Cook & Seaman	Joseph Shelton	"	Thomas Edler
Carpenter	Charles Harvey	O S	George Cartright
Cooper & Seaman	Robert Greene	O S	Owen Dolan
		A B	E. Th. Modsen
Powder monkey	Jack Eddie.		

P. J. Capelotti (✉)
Penn State University, Abington College, Abington, PA, USA
e-mail: pjc12@psu.edu

P. J. Capelotti (ed.), *The Coldest Coast*, Historical Geography and Geosciences,
https://doi.org/10.1007/978-3-030-67880-7_9

This is the third of a series of scientific expeditions made by Mr. Benjamin Leigh Smith for the exploration of the seas around Spitzbergen [Svalbard], of the North and Eastern coast of that country and of any land that may be N and E of it should it be possible to reach it, as well above all to penetrate as far Northward as possible. In 1871 and 1872, the former expeditions started from Hull: in the first expedition a Norwegian crew being got from Tromsö [Tromsø]; in the second the crew was composed of English sailors and Shetlanders from Lerwick. The first expedition was unusually successful, penetrating as far N as 81° 24′, and discovering a long tract of land on the N coast of Spitzbergen as far E as Cape Smyth [Kapp Leigh Smith]. The position of many of the islands, formerly inaccurate and in some cases showing islands that did not exist, was also laid down; the former supposed to be eastern boundary of Spitzbergen was shown to be imaginary. These alterations have been now included in the new Admiralty chart of the Polar Seas, drawn up, in accession to a petition from the Dundee and Peterhead whaling captains, and presented by Mr. Smith. During the second expedition, the pack ice from the N prevented penetrating beyond 80° 30′ and the little *Sampson* (150 tons) schooner (85 register) got seriously damaged by the ice. In both these expeditions, very interesting observations of currents temperatures (on one occasion 60°F was registered by the deep-sea thermometer) and soundings were made. Captain Wells, RN, was Mr. Smith's guest in the second. This year, he takes up the S.S. *Diana* (103 tons register, 250 builders) and it is on board her that I am now writing. I can never feel too grateful for the friendship and hospitality of BLS, which has given me an opportunity of taking part in an expedition in these regions, the exploration of which has always had a strange fascination for my mind. The *Diana* is provisioned and equipped for a year, except of course in the matter of coal. The *Sampson* under Captain Walker (13 hands) also takes part in the expedition, as a reserve. She is to rendezvous with us in Kobbe's Bay [Kobbefjorden], Spitzbergen, on July 1. We then hope to be able to penetrate eastward along the North coast as far as Gillie's Land [Kvitøya], should such exist and, if so, keeping up its western coast, having of course from the revolution of the earth, a lee pack, and as in the case of Greenland a chance of more open water than on the east coast of land, to penetrate as far north as circumstances will allow.

Diary

May 10, Saturday.

After a very busy morning in Dundee, I found myself about 1.30 p.m. on board of the *Diana*. The Dock gates were open, as it was high tide and all were ready for a start with one—to me most important—exception, namely the absence of my luggage. About this, however, I was by no means uneasy, as I had heard through the railway authorities that it was at Perth, that morning, the luggage would come by the midday train. Sure enough, it did come to my intense relief, and was safely shipped aboard just before we left the quay. My tobacco, however, had not arrived, and I left the ship in search of it, only to be informed that it was on its way, and just as we were getting underway an excited small boy appeared making all sail to catch the retreating vessel. It was thrown on board, caught and disappeared down the companionway, to be stowed with the rest of my belongings in a spacious chamber 5′6″ by 3′ 6″ inclusive of cupboards &c, which the steward was pleased to call my "state room." We are now out in the Tay, the engines working capitally, and as "the landsmen's cheers come fainter o'er the water," I retire to write my last batch of letters, which will go ashore at Broughty Ferry by the party landing on board. Next comes the lunch, and lots of fizz, during which repast I am busy writing in the commander's room but emerge in time to get a bite and come in for the speechifying and good wishes. As we came through the Dock gates, two photographs were taken of the

Diana. After several turns up and down the river, we land the shore-goers at Broughty Ferry and proceed down the Tay, which is certainly far inferior in scenery to the Firth of Forth. At 5 p. m., we are off the Fairway buoy, and the pilot leaves us. It is a fine evening and as the *Diana* is set on her course, N.N.E. ½ E, all looks prosperous. We have a fair breeze, coming from aft of the quarter and set our sails. Towards night the breeze fell, and we steam up the coast on a course of N.E. by E about eight miles from the land, sighting first the Aberdeen light, Girdle Ness, then the Buchan light. It however comes on foggy before we sight the light at Kinnaird's Head, but the wind getting up the sails are again set, though before morning the wind falls, and what there is of it coming from the N, sails are useless.

May 11, Sunday.

A good deal of sea on, and the wind getting up, we have long since lost sight of land, and steam on the same course at half speed to economise fuel. The motion begins to tell, and the Sunday service is postponed. Towards evening the wind increases, still coming from the North. We see a good many birds, guillemots, gulls, solan [barnacle] geese, and two or three divers, which we are not near enough to recognise for certain. At dinner parade I am for the first time a defaulter.

May 12, Monday.

This same wind but a good deal more of it. The sea looks very fine and there is a long roll on, which looks splendid after the choppy sea one is accustomed to in the channel. The roll, however, decidedly affects our appetites and we get through little breakfast, less lunch, and only manage some hot mulligatawny in the evening. It is decided to put into Lerwick to wait out what is evidently going to be a hard blow from the N, and to fill up our stock of coal, as steaming dead against wind considerably runs into that valuable commodity. Towards evening, we sight Fair Island on our port hand before we turn in to descry Mousa, Sambro [Sumburgh] Head, Noss Head, and other of the southern headlands of the Shetland Isles. I got up about midnight and realised we are past the Sambro light and soon came opposite Mousa, an island lying on the left as you come in. It is low lying and rockbound, with curious dark upright jagged rocks, standing up from the sea in isolated pillars just beyond the extremity of the island. There is an old ruined castle on the island, said to be the oldest in the three kingdoms, built by the Picts. It is a fine though cloudy night, and as we enter the sound between Bressa [Bressay] and the mainland, the effect of light and shadow is very fine. The moon is up and he's almost dead astern of us and shows through a break in the dark thick upper bank of clouds, casting a shimmering patch of yellow light on the long, white crested waves. The light is fitful as at times, the driving scud of the lower tier of clouds obscures the rays northward above the islands and the grey bank of clouds surmounts them, the outline of which seems almost like a range of hills. There is a bright white glare in the sky, which casts a long trail of light towards us. On the right rises abruptly the high dark cliff called Bressa Head, a little below which is the lighthouse. There is not much sea on round the cliffs as this is the leeside of the island. We steam slowly up the sound, the quiet waters of which form a strange contrast to the commotion outside, and hug the right side until nearly opposite Lerwick, which lies on the left, behind a jutting promontory, which completely conceals its position. As we come nearly opposite Lerwick —the position of which behind the promontory aforesaid is alone betrayed by a church and school built on it—we shape a course obliquely across the sound, which is here barely a mile in width. The moon is still shining astern, but the light to the N is paling, and away over the hills of Bressa, to the eastward, are seen the long pale red streaks of early dawn. *Crrr*—down goes the anchor and about 2 a.m. I turned in, as we are safe in Lerwick harbour and safe out of the game of buffeting between the wind and waves, which is commencing in the Northern ocean.

May 13, Tuesday: Lerwick, Shetland.

I cannot say that I was disappointed with the appearance of Lerwick, as many people seem to be. Perhaps the chief reason that it pleased me was the charm of novelty; it being utterly unlike anything I remember to have seen before, although its quaint irregular arrangement of houses in some degree resembled the fishing villages on the more desolate lands of the coast of the Papal States. We went on shore between 10 and 11 a.m., and met young Leask who belongs to the shipping agency at Lerwick and seems to conduct a general store and a good deal of business through all the islands. He was, however, first starting for the Northern Isles in the *Crab* and left us in the hands of his foreman, Mr. Roberts. He also kindly volunteered the loan of his trap [light one-horse carriage with springs]. We however preferred our legs to a "machine." After being conducted about by Roberts in search of a dinghy or a light Shetland "casey," we at length found a boat that BLS purchased, although I was of the opinion that it was rather too large and heavy for our purpose, namely to be managed by two hands and be easily lowered away at any and all [times] in pursuit of seals. It has a 10′6″ keel and like all Shetland boats, the keel has a considerable dip in the centre. The bow and stern are almost alike and resemble those of a whaleboat, but clinker not carvel built. It was a fine bright morning with a strong wind blowing from the N, and Potter and I started to walk to Scalloway, taking across the moor in spite of all the assurances to the effect that we should get hopelessly bogged as well as pounded. We made a pretty straight line across the moors, which are very bleak and bare of heather, and abound in peat bogs. All the natives we came across were women, who were either occupied in cutting peat or carrying it to their homes. The peat is carried in a large upright basket strapped round the shoulders, which much resembles those baskets that carry logs of wood in France and Belgium. The people look very healthy and hardy, and one is astonished at the absence of beggars. In fact, I think Shetland is the only place in the world where I have not found them. The peasant girls are a fine grown and decidedly bonny lot. The women wear curious shoes, formed of pieces of cowhide first laced round their stockings. They mostly do not wear hats but a shawl of Shetland wool is wound round their necks, which when occasion requires, they draw over their head. In crossing the moors, we came across numerous curlew, golden plover, snipe &c, and disturbed some of the latter species from their nests. We also found a lark's nest. We came across a lake lying very hidden in the hills beyond the Scalloway road and on this were a good many ducks, mallard and widgeon and a larger kind of duck that I was not near enough to recognise but two of them appeared to be of the same size as, and much resembled, female sheldrakes [shelduck]. Taking round the foot of this lake, we climbed the opposite hills and then came in view of the sea on the other side of the island. This view was very fine, as a large bold-featured panorama, and gives one a very good idea of the various scenery in different parts of the island. Immediately below us lay the two lakes of Asker and Tingwall, and slightly beyond the extremity of the former was the sound of Scalloway. The valley in which these lakes lie was well cultivated and contained numerous farms on the southern and southwestern slopes of the opposite hills. The pasture in the bottoms also seemed good and there were here to be seen a good many cattle and ponies besides the sheep on the hillsides. Away to the right the valley gradually rises, and is closed in by the hamlet of Tingwall, with a manse and old church. Behind us lay the bleak moorland and the lake we had just left, over and around which the ducks and other wild birds were flying, and from time to time patches of sunlight shone down on its troubled waters or flitted along the dun hillsides, lighting up small portions which made the dull contrast of the rest only the more striking. At times, the whole scene brightened until the driving scud from the N again hid the sun from our sight. In front and beyond the valley to the Northward rise ridge after ridge of hills, the heights of which are I believe Westdale Hill (842′) and the hills rising over Bixetter Voe, or sound. To the left we can see the old castle of Scalloway, and the sea beyond, showing here and there in the voes or sounds of the islands of Trondra and W and E Burra. The view in this direction being closed by the bold outline of Fitful Head, a rocky headland of striking profile rising to a height of 929′ above the sea. Away across the sea in our right front, we could see many headlands and islands. Those that were most striking we took to be Skelda Ness, Papa Stour Island, and Vaila Ness, and away to sea rose an island with three hilly points, two of which appeared very steep, which I suppose is Foula Island—a wild rugged little island, the highest point of which is over 1300′ above the sea. After resting for some time, we took along the hills to the left and down into Scalloway; here our first object was to get some food, for want of which we felt quite faint, as we had had an early breakfast, and had been rather off our feed the day before on board ship. Now came the part of the proceeding that I hate in all countries—most of all in the British Isles: making inquiries of a native. In fact, as long as I have either a map or a guide book, I make a rule of never doing so. On inquiring your road to anywhere one is usually met by a stolid stare of stupidity, and in answer to the question "does this road lead to," "oh aye: ye can go this way," and it is probably not till you meet a gentleman two or three miles further on that you discover that the turning you had passed within 50 yards of the native was the shortest road by two or three miles, besides which case the characteristic answer I have quoted would be only elicited after repeating your question three or four times. People who most pride themselves on their comprehension of provincial dialogue are apt to forget that although you can understand them, they cannot understand you; in fact the only way is patience and slow speaking. Then again, they never know the distances between places even approximately, and are only too ready with an answer certain to mislead you, if by any chance they catch the name of the place you are going to. I don't allude to the famous "mile and a bittock" ["a mile and a bit," later to be fashioned into a poem of the same name by Robert Louis Stevenson] but the [ease] with which they call seven miles four or vice versa. Should a bold but inexperienced traveller who has taken his own line manfully across heath, moor, or plain so as to enjoy the true features and natural beauty of the country together with the charm of watching the birds and beasts, for want of further landmarks, trust to the intelligence of the native, he is done for, the latter's only idea being that he has lost his road, and he will accordingly direct him to the nearest point on the road. The only way in which I have ever found it of any use to put the query being: "If you were going from here to A _____ in a hurry, how should you go?" He will then scratch his head, and altho' possibly his own dwelling is at A _____ he will if he have one within a reasonable distance, summon his mate to a consultation. This consultation always begins by the convener acquainting the remaining member or members, with all the details of your personal history, plans, and doings that he knows. This is more especially the case in France, and the excited gestures and even if you are not as is likely trespassing, induce a belief, that you are either very much out of your way or in some dilemma that requires aid. In Belgium, if you are lucky enough to have addressed a Flem with some slight knowledge of French, this consultation is carried on in Flemish or Walloon and generally ends in an offer of the president conducting you as guide for perhaps 50 or a 100 yds at the end of which distance your route, perhaps now not so easily discernible as on the side of the Parliament, is pointed out to you and baksheesh [a small bribe] must of course be proffered and accepted. Another awful query to make is for a place of refreshment. If you ask for the "Inn" you find on leaving the pothouse at which you have been forced to regale yourself, a comfortable hotel a few doors further on. But using the word

"Inn," you are told, "Oh, yer honor, I didn't know it was the "hotel" yer honor, wanted." If you ask for the hotel, you only get a blank stare of astonishment, or are informed there is none. "The best public, where I can get something to eat," only elicits from the rustic connoisseur of malt, the name of that special house of call, where the four [coin] small beer is most to his place. In fact, this is a difficulty that no ingenuity can ever overcome. Perhaps the best method of proceeding being to first explain that you are in search of a place of refreshment, and then to mildly inquire "which is the *first* house" in the place.

In this case, all inquiries were of no avail; there was emphatically no Inn, no hotel. But we could be provided with food at Mr. Roberts. This seemed to be the universal opinion of all the inhabitants with one important exception, i.e., that of the family Roberts, who politely but firmly refused our demands for food. Our next shot was for the place where they sold "beer or whisky"; after some conflicting directions we found ourselves at the Post Office. Here our troubles were at an end, for although a retail establishment, with liquor not to be drunk on the premises, they took us to upstairs, and we got "lashings and lavins" of eggs, bacon and new bread, beer &c, for which entertainment we paid only 2 s 8p. Having examined the castle, which is an old, square and not over-imposing building constructed of very durable stone, but apparently not of the sort now in use, we started home. After about a mile the road branches into two, both branches leading to Lerwick. Here we were overtaken by a native, very respectably dressed, and bestriding a sheltie barely the size of life. Potter rashly requested information with regard to which branch of the road to take, and the reply was worthy of a Greek Chorus. We, however, selected the left-hand road that follows the course of a valley we had crossed on our way out. We found Eaton, near a bridge about halfway home, on his hands and knees, apparently *mudlarking* [searching in river mud for items of value—a form of subsistence along the Thames and other British rivers in the eighteenth and nineteenth centuries and continued into the twenty-first century as a form of tidal archaeological exploration] in a peach-coloured burn, but in reality carefully capturing and bottling those specimens of entomology, which are to immortalise him and his labours. We took home across the moor and had great fun with the birds. I succeeded in calling a golden plover to within a few paces of where we lay. I never came across a more natural actor, than that bird. His mate had a nest close-by and accordingly, after a stone had been heft at him, he thought it a point of honor as a true plover to improve his opportunity. Accordingly down he came apparently winged. He then allowed us to approach quite close and just managed to flutter away and so on, 'til far enough off. The only mistake that he made is to let us get so close that we might easily have "arranged his affair" with a stone. We got on board in time for dinner and afterwards watched a successful herring fishery going on at a short distance from the ship.

May 14, Wednesday: Lerwick, Shetland.

Stopped on board, all the morning, reading &c. It was awfully cold on deck, with showers of snow and sleet. Eaton very busy poisoning flies and other insects. We are all very glad of this, as some of them bear so strong resemblance to the "domestic *bonser*" that we are a little nervous about their being in the cabin. Went on shore at midday and met BLS. He had bought another boat. This time I am pleased to approve the purchase. She is a handy little dinghy, square-sterned, not old, and copper fastened. Both the boats can be stowed abaft the wheel. We buy some Scotch Wool blankets, and then proceed to the Queen's Hotel for lunch. Here we stop some time, as the weather is squally with a cutting wind and storms of rain and snow. We are then again obliged to make inquiries on fishing tackle and, of course, are directed to the wrong place. We fail to get made up sets of hooks or flies for the necessary rigging of a "dandy line" for herrings, and so have to buy the materials. These materials we proceed to make up on board after dinner; after a long time spent in this occupation, just as all is ready for a start we find that the gut is hopelessly rotten, and *la pêche aux herrings* [fishing for herring] is given up for this night only.

May 15, Thursday: Lerwick, Shetland.

Colder than ever, still blowing hard from the N, makes us rather down on our luck. We had some intention of some fishing in lochs Tingwall and Asker but thought that the day was too cold and very uncertain information about the season of the fish was forthcoming. We decided to telegraph an inquiry on whether this Northerly gale had obliged the *Sampson* to put in, as she had only left Lerwick the Friday (9th) before our arrival. We did not go ashore until after lunch, and we then proceeded to Leask's to buy fishing tackle. We also invested in "fearnought clothes" [cold weather outer garments worn by sailors] of a whitish colour as we expected them to prove useful in seal hunting. We then adjourned to the Queen's and read the Scotch papers, as the mail had come in the night before. We borrowed some ready-made "flies," as they call them for the herring fishery, from some of Leask's people. We proceeded to walk 'round the town, which is built in a very irregular fashion: the shops are all down in the streets near the sea. These streets are very narrow and crooked with corners of houses &c projecting them in the most promiscuous way, and have flat paving, but are nevertheless used for carts &c. The upper town, which overlooks the sea the other side of the promontory, is connected with the lower by alleys (very steep, with steps in many places) and tortuous passages. It is built in a more detached style but is rather exposed. We also visited the photographic shop, as the proprietor had taken some photos of the *Diana*. After dinner, Potter and I and two volunteers proceeded to the herring fishery. The position for fishing and in which all the boats lay at anchor in close proximity, was 500 to 600 yards from the ship, up the sound. Jigging was the method of fishing and is occasionally very successful. I have practiced it in far deeper water on the Irish Coast, with a somewhat similar apparatus. In this case, we fished with about seven to eight fathoms of line out. At the end of the line came about a fathom of gut, with bright white metal hooks, either bare or with white quill attached, tied on at every 10 or 12 inches. Below this is a thin line four or five feet long with a three- or four-ounce lead sinker. The hooks are bare, or with quill attached as above mentioned, and are kept moving up and down, whence the name. This was a bad cold night with showers of hail &c. We, however, drank luck to our first fish and caught over a score, which was I believe more than any other boat, many of which caught nothing. We got back by 10 p.m.

May 16, Friday: Lerwick, Shetland.

We were engaged to go and fish at the lochs before mentioned today. Young Leask was going to drive us over in his trap, so Potter, Wallace (black retriever), and I went on shore about 10:30. We went with Leask to visit his uncle, a fine old boy, who remembered "Parry" [Sir William Edward Parry (1790–1855), British naval officer and Arctic explorer] coming to Lerwick in 1827, and then started. Leask has a very handsome and strongly built little pony phaeton and the whole turnout was very smart. His two ponies are white and stand 13.2 and three, the off one is a very handsome beast. It was a fine sunny morning though very cold and we enjoyed our drive very much. We passed several fine views, one especially up the Voe towards the Skerries. The moorland is all apparently of the same character, though some of the pastures we saw this day were apparently very good. The heather is short and does not give cover enough for grouse. I hear they and partridge are being tried with some success in a neighbouring parish. Arrived at Tingwall, Leask who was busy there with a school board procured us the loan of the "meenisters" boat and introduced us to that worthy, with an injunction to get him to play if we could. Accordingly, when conversation flagged, I asked him to give us something on the piano and was much astonished at the piece played—the music of the "Sindu [Sindhu] dance." After this we proceeded to launch our boat, which entailed a good deal of wading. Luckily, however, we had long boots on. It was a bright day and very windy on this secluded loch, but in spite of our trying several very likely flies as well as the "phantom minnow," we had no success, tho' I struck one fish and got back in time to meet Leask for lunch. Potter succeeded, whilst I was rowing, in losing four good flies, his knife, and our only cast. After lunch we drove home, disturbing a snipe off her nest on the way; these nests seem very plentiful here. Leask asked us to dinner, pleading that the others would be waiting on board. Eaton had been *mudlarking* along the shore and was rewarded with anemones, and other fishy substances, which were enough to afford him employment for days to come. [?] is determined to start early tomorrow if the wind will allow. Potter and I again go to the fishery and capture about 30 beautiful silvery herrings; a fresh herring is certainly the best one for breakfast.

May 17, Saturday.

I go on deck at 2 a.m. There is little wind and smoke coming from the funnel showing that they are stoking up. A boat, which has been sent on shore, for a Shetlander who has been engaged as "cooper," returns with said Shetlander, and we were now only awaiting the arrival of the pilot. The sun has risen over the hill to the E (the Ward of Bressa), and after waiting in vain for the pilot we decided to go out towards the Southern, instead of the Northern Channel. It is very annoying this brute not putting in an appearance, as he was to have come off at 11 p.m. the night before and I had been trusting him to take off a note of apology to Leask, whom we had asked to lunch as well as the lines we had borrowed. However, the capstan's at work, round she comes, the anchor is weighed, catted, and fished, and in the glory of the golden sunrise, we steam down the silent sound. Lerwick is asleep and the only sign of life are the few boats at the early herring fishery. We couldn't have got off more quietly with the "swag" even if we had a spring on our cable. We decidedly "spoil the Shelties" for let alone the return of hospitality in the shape of a lunch, and seeing the vessel, the fishing lines &c, we have got "fearnought" clothes, tackle, curry powder, and miscellaneous articles of different sorts, besides about eight tons of coal, and to our credit be it said—though to our debit be it entered that nothing has been paid for. On our left as we steam down, the dark black rocks and the sides of the hills are all in deep shadow, which they cast on to the dark waters of the sound. To the right there is the glorious sunlight making Lerwick look more beautiful than I could have imagined it and throwing the hills beyond into bold reality and contrast every knoll, every fold of ground, and every tussock having a dark as well as a bright side. Faraway it lights up the stretch of mountains away to Mousa and [?] rocks all of which are wrapped in that soft delicious hazy halo of sunlight that reminds you of a summery sunset but differs on you can't say how. As we pass under Bard Head and turn eastward, we come upon the lighted side of the cliff and island and get a better view of it. It is a fine bold headland 400′ or 500′ above the sea, with the waves surfing and rushing in and out of the numerous caverns at its base: here are hollows, there caverns, pillars, channels, and complete passages through some of the lower portions. Above there are steep slopes of grass, coming to the sheer edge of the cliff on which sheep are feeding. On any and all sloping parts of the cliff, there is dark green vegetation, apparently much of it being samphire, the ledges, *ça va sans dire* [it goes without saying], are occupied by sea birds. The cliffs are made up of layers of rock (aqueous) originally horizontal but now with a strange "cant over to starboard"; in fact, as soon as you round it to the E, it looks as if the whole promontory were slipping into the water. The tilting of the layers of slabs varies greatly and in some places gives an absurd effect as if the part "b" were actually in motion and falling. The water here was not very cold (46°) but very clear and blue. The old Northerly breeze was getting up, and though still in calm water we can see a heavy sea on the horizon outside. Between 6 and 7 a.m. we come to the Noss Head (577′). This is a magnificent cliff, lighted up as it is by the morning sun, and the effect is grand: parts are in light and parts in shadow. The cliff is of varying colour, and there is a good deal of it covered with dark green vegetation. It contrasts wonderfully with the bright blue of the sea below, parts of which are in deep shade. There are countless fantastic arches and caverns at its foot, in and out of which great torrents of water come rushing and one longs to see it, in a Southerly Gale. At some distance, the profile shows a face almost exactly like the "Red Queen" in Alice [from Lewis Carroll's 1871 *Through the Looking-Glass, and What Alice Found There*]. The water is alive with countless birds and the air as well: on the topmost edge, on the water at its foot and for some miles near, on every nook and ledge, chink and cranny are countless sea birds. Those on the water are chiefly loons, guillemots, and divers; those in the air &c gulls. The sight of this crag brings back so forcibly all the stories so often dwelt on in boyhood of the hardy islanders descending the cliffs for the eggs, and one can picture the whole scene with vivid distinctness. We saw in all: cormorants and shags, razorbills, guillemots, loons, mackerel cock [in the Irish Sea, the name "mackerel cock" is used for Manx, as they predicted the arrival of mackerel. Since the whalers sailing north mainly came from the North Sea side of Scotland, gannets are common there and do follow mackerel. So here Chermside likely refers to gannets], all the common sorts of gulls, grebes and divers, red throated (?) and some other species, and I think a Manx shearwater. We also saw some Peregrines—that is one pair—pretty close so that you could tell the hen and cock birds apart; the other pair were farther off: they were hawks but of what genus I am not sure. When we got outside it was roughish, we were glad to be able to lay a course a little E of N.N.E. We, however, apparently made more easting than was allowed for. We had to put a reef first in each of our topsails, then in our fore and aft sails, and finally took in the jib. We knocked off steam at 2 p.m., but there was such a sea on from the late gale that we did not make such way. We passed in sight of Whalsey [Whalsay] Island and could still see the Skerries at dark, and these are a number of small rocky islands, and there is a light on in one of them called Gruna [Grunay]. It came on fresh at evening, but the wind dropped at night. Fetlar and Unst Islands were the last land we saw and they were, I believe, still in sight at one of the early watches next morning (Sketch 9.1).

May 18, Sunday.

A smartish sea on still, which made some of us rather uncomfortable, but little wind. We got up steam in the course of the morning. We did very little with our sails but, towards night, a fresh breeze came on and we let the engines run down. They stopped about 10 p.m. Soon after it became calm, or rather I should say the wind fell. A large number of birds joined company with us this day, consisting of a flock of kittiwakes of various ages, marks and [?], and birds the sailors call "mallemuks," which I recognised as fulmar petrels. These "mallies" are tremendously common in the regions we are bound for.

May 19, Monday.

Breeze fell, a good deal of roll on, which prevented us all shaking down into our usual ways, and made me too unsteady to get to work preparing photographic chemicals &c. The way the things steeplechase about, even on a properly "*fiddled*" table, is very unpleasant. We are certainly desperately unlucky in not getting a fair wind. Soon after midday, we got up steam, as we were only doing 1½ knots. We somehow steamed into a breeze, which just allowed us to

Sketch 9.1 Fault lines in the Shetland islands

lay our course. We knocked off steam towards evening. Dinner was worse than ever, everything most awfully jumpy. Potter and I amused ourselves with the "*Péche aux mallies*" today, however, but with little success, and the final loss of line hook &c. He is an awful lad that for losing fishing tackle; we hadn't got over much of it. The process was simple: a line with a hook at the end baited with a piece of fat. I think if it had been calm, we should have caught some as when we were going slowly just before getting up steam, they settled on the water and had a most animated discussion of the bait. But the hook was too large, or they were too cunning as they only bit the bait off. As I had my guns upon deck to clean them, I shot some and some kittiwakes, but regretted it afterwards although it certainly was pretty shooting and good practice. The "malley" is a bird of graceful flight and at first reminded me of an owl. He is nearly of the same size as a barn owl but has a more graceful and sweeping flight. In fact, it is between that of an owl and a swift, the way that they can move up, down, or across wind with perfectly extended wings and without a single flap is most surprising. There was one bird especially who was watched after dinner, and certainly was the "tallest" flier I ever came across. Talk of Mark Twain's jumping frog [Twain's first book, published in 1865, was a collection of short stories entitled *The Celebrated Jumping Frog of Calaveras County*], but he was nothing compared to our "malley." He made a point of honour: never to give a flap, and the way he sailed and used the action of the wind was one of the most beautiful sights in nature I ever remember to have seen. It was indeed the poetry of motion and the grace with which he turned was beyond words. Some of these birds—I suppose the one-year-old or the females—are a dirty brownish white all over. The old birds are grey on the back and wings like a seagull, white head and breast, and yellow bill. The bill has a sort of curved knob at the end. They are covered with a very thick layer of fat under the skin. We saw a "boatswain" gull, as the sailors call them, which is apparently their sobriquet for the Arctic gull [Arctic skua]. Yesterday we were accompanied nearly all day by a wheatear, who must have been rather out of his reckonings. He lighted several times on the ship, but finally disappeared and I fear perished. After dinner, some fish [whales] appeared a short distance off on the starboard quarter. They came up like porpoises but had a different shaped nose and were far bigger. They moved in single file and always showed a large triangular fin, which seemed to project at least three feet from their back. They were first going south and came up to blow very frequently. They blew like whales. The sailors called them "swordfish," but whatever they were [killer whales], they weren't that, as they had no "swords" unless indeed they can "return swords" at will. I think they were sort of dolphin whale, of the genus *Delphinapterus*, as they had the dorsal fin. Shortly afterwards a "bottlenosed" whale came up, close to the ship, but went down and was no more seen.

May 20, Tuesday.

Something like a fair wind, at last more westerly and on the quarter made a course a little west of N to recover some of the ground. We have got to the East of the Greenwich Meridian, on which we are, by way of travelling. We are going too fast (about 5 knots) for a *pêche aux mallies*, and I have registered a vow not to shoot any more, unless as specimens, after the way that bird flew last night. More of gun and rifle cleaning. Towards evening, the breeze increased, and when we went on deck after dinner under tremendous difficulties, it was blowing very fresh. It came on murky and squally and we shortened sail so that by 12 p.m. it was blowing a gale and we were flying along under only three sails: foretopsail, square foresail, and main topsail. Before eight bells, we reefed the foretopsail—the wind since midday had been dead aft.

May 21, Wednesday.

The gale continues and is still pretty well aft. I have been in several gales before and harder ones, so it is hardly worthwhile describing one's sensations (altho' I have been out in a stiff one in such a sized boat as the *Diana*, but once and then it was anything but comfortable). The sea did not run the conventional mountains high, tho' it was a good sized one, and owing to our having stores of clothing &c, aboard, sea boots and other necessaries, we did not suffer much. As for our mishaps, these commenced yesterday: we were rolling a good deal, hatches fastened down &c, and I was standing on the forecastle talking to some of the watch, the others had turned in, when about 10.30 p.m. a sea struck her. Bang on board it came but that was nothing, as she had already dipped frequently both sails under. This time, however, she heeled more than usual and two of the pork casks aft, which were lashed near the mizzen, bounded fairly out of their lashings like a horse backing out of its girths. One rolled to the port and was seized by the Captain and held firm against whaleboat; the other rolled to starboard and was just prevented by the mate from the stoving the bulwarks. We found, all of a heap over one another, the baskets out of the rack, tins, soap, the rations of salt meat, which were soaking everything adrift. Below in the forecastle, chests were careening about and from the galley came a rattling like the opening chorus of the "Christie's." As soon as possible we got to the assistance of the Captain and Mate, and in due time the casks were secured. Other seas followed and by the time I went below the cabin was in a pretty state and the other lads glad of my assistance. The medicine chest was adrift, charts and books strewed the floor, rifles, cushions &c; in fact, there was general hay. I left my door open that night as it was very close and the hatches were down; however, I was nicely roused before morning, as the sea forced open the little doors above and coming down the companion, wetted a good many of my effects, self and bedding included. I managed next day to find some dry things. Things continued much as they were, but the sea was certainly higher, owing to the continued action of the wind. Towards evening it was very bad and came clean down one's neck in spite of waterproofs &c—in fact nearly knocked one bang over. The decks were very full of water, too. She takes very long to clear herself when the sea is on board. Endless seas came down the crevices of the companion: there is a sort of perpetual waterfall going on the stairs and when you go above it's a lottery whether as you emerge you don't get a facer. At dinner, a sea came through the chink of skylight we had open and this, once commenced, continued so steadily that soon sofas, books, everything was wet and I was driven on deck. Here the gale was at its height. It was now and had been since morning getting more easterly. We had reefed both topsails and the men go aloft to reef the foresail. I must say it made me giddy to look at these fearless lads out at the end of the swaying yards rolling from side to side, the wind whistling and howling thro' the cold rigging, and the sea seething below, being cruel and cold. It was with a feeling of relief that I saw them all at length safe on deck and warming their numbed fingers. Below things are much as they were, but I received a sea full length as I lie on a sofa, wetting me from neck to boots. Smashes are frequent: several glasses, plates, and tumblers as they say, "in charge," which either means adrift or a smash. I had the barometers and then go on deck again, but get such a souser that after one or two pipes you fain to come below and turn in.

May 22, Thursday.

Gale continues but is more from the E and has apparently moderated, but the sea is higher I think, more of a cross sea. Most of us are affected by it more or less. Little of interest happens today as it is mostly on deck and one cannot see far. Potter and I are there for some time. We feed promiscuously on cold meat at odd times and I make up for all shortcomings about 5.30 p.m. on cold meat and potatoes. I saw a seaman's sou'-wester float close by us this morning. Towards evening, the gale is over and we shake out the two reefs in the fore and aft sails that had been hoisted since the wind had been so much in the East. The thermometer is down to 40° and the water to 38° and then to 36°.

May 23, Friday.

About 4 a.m., we first come across ice, and by 8.30, there is a good deal of it. There is a heavy swell on and a nice breeze from the East that still holds. The ice floats about in white snow-crowned detached blocks, never much above five feet from the surface and of every shape and form, the pieces being eaten into most fantastic shapes by the action of the water. The sides are in beautiful green, and if only a bright sun were upon it, the effect would be lovely. Bump, there we go, as I write, bang against a block. As far as the eye can see the sea is covered with these floating blocks, which are the remains of the floes or field or pack ice and broken up by the late gale. Bump, bump, she shakes, so as she strikes and it is so thick ahead that about 10.30 a.m. we turn 'round. We are now fairly in the ice and in the long wished for Polar Regions. We are in about lat. 72°; after getting out of the ice we get up steam, and steam and sail a bit to the E when we again try to the North. We, however, have to turn 'round after some hours again. The ice is in larger blocks than at first and some of them a good deal higher—the pillars caverns are of most wonderful shapes. Now that the sun is shining on them the effect is magnificent. The whole sea is in motion as there is a long heavy swell on as a result of the late gale, and as far as the ice is concerned this adds an indescribable change to the same, although we would much sooner be without it, and it is this alone that prevents us penetrating the ice.... [A break in the journal occurs here.]

After turning the second time, we beat Eastward a bit under sail and steam. We are today whilst in the ice favoured with our first view of seals. They are all with one exception "on the passage," that is, not playing about or stationary. This lad, who was on the ice, was too far off for a shot and it was not worthwhile lowering away for the chance. We saw one "school" or lot together, probably young seals. It was very exciting to see them rolling playing and turning. All the rest of the day we kept getting to the E and to the N when possible. We saw a schooner who we at first took for the *Sampson*. She was probably a Norwegian and was beating when we first saw her. About 4 p.m., she was pretty close, bore up, and made apparently for home. The "crow's nest" was today hoisted on the mainmast. The skipper now spends a great portion of his time up there. I suppose I must go up someday if only for the view, but I cannot say that I ever fancied myself either as a "daring nor an experienced climber." It is still wet on deck, and we ship a casual sea at times, just I suppose to show there is no ill feeling. It is very cold here as it is always near the western ice (about 32°), and amongst other things the experienced navigators are much surprised at the amount of "foul or black ice," that is ice smudged from contact with the land. I have got my sea fingers by now and if I don't write more legibly than of late, at any rate do so with more comfort. We first come across a number of little birds called "rotches" today. These, I suppose, are alcanthe, the little auk, but 'til I see one closer I cannot tell for certain. They are jolly little birds and look like diminutive guillemots or loons. They fly like those birds or auks, if not faster, and are all these at diving. What amuses me the most is to see them get up, you see say five or six on the water, away goes one, two, three, four, five, six, and to your astonishment also seven, eight and perhaps nine, the latter emerging from the sea, in single file with about three feet interval, perhaps even from the centre of a wave. The manner of emerging, too, defies description. It is not as if they were diving below and by some signal or because it was time to return to the surface appeared swimming even for an instant, and then seeing how things lie, followed their leader. Not a bit of it: they apparently know what is going on as well as those above and as the last on the surface gets up, the next emerges, as I have said, at the proper interval, not swimming or even getting up from the surface however hurriedly, but absolutely "at the fly" and so on with the others. It has a most peculiar effect, to see three or four of them fly absolutely out of the heart of a big wave. They are nearly as clever at flying into the water, but there the eye can perceive a check as they enter the more resisting elements.

May 24, Saturday.

I did not turn in 'til late last night, and determined in anticipation of the hope for cry of "a fall a fall" when we get further N. To practice getting up, accordingly about 4 a.m. I turned out. It was a very cold raw damp morning, no ice to be seen. We were still on the starboard tack with the wind from the E, but not going very close. I sniffed about, pitied the man at the wheel, and having satisfied my conscience, turned in 'til waked at 8 a.m. by the steward… [A break in the journal occurs here.]

…our longitude, but it is too cloudy to shoot the sun. We did get a meridian altitude and find we are about 72° 22″. It being the Queen's birthday we drank Her Majesty's health in a bottle of [?] for 33. We saw a barque apparently bound southward. She, however, still had her crow's nest up aloft. She was out of sight by 8 p.m. We still continue tacking towards the N and E.

May 25, Sunday.

After failing once to get into the ice, we get up steam, and get along pretty well through the ice, in a N.N.E. (true) direction. The ice in some places is very thick but much broken, and now that the dangerous swell of the last two days has somewhat subsided, we get along well enough. There must evidently have been a gale here from the N.E. that we know nothing of. The swell is still considerable and very long. I fix the distance from crest to crest of the waves to be from 120 to 150 ft. We see a few seals on the "passage" and pass by two "flensed" carcasses, and a good many marks of seals. We have rather a successful "malley fishery" and capture several, and after submitting them to various indignities we let them go, either striped with ink or minus part of the end feathers of the wings &c. Like many seabirds, they can't fly unless the water be under them and are very helpless on the deck. We saw several "Burgo-masters," as glaucous gulls are called here invariably; also, for the first time, the "snowbirds" or "ivory gulls." These have a very peculiar harsh cry. And doveca's [dovekies, or black guillemot] which are birds of the guillemot or *Colymbus* genus, with red legs and a white patch across their wings, which has a curious effect when they are flying. About 4.30 p.m., we knocked off steam. It was now very cold and misty and a ship that we had seen for some time is out of sight to the Eastward. We are no longer in such thick ice, as sometimes today we have simply forced our way through it. It was a matter, however, of no great difficulty, as the "floe" has been broken into very small pieces. Whilst at dinner or rather just as a bottle of "Black Strap" [rum]—far superior to the "strong military ditto" as the fiery port of messes is called—was making its first round, we are electrified by hearing that a seal is on the ice dead ahead. Rifles are got out and we come noiselessly up to him. I have first shot but falling into the mistake of resting my sights on the gunwale behind which we were crouching, miss him clean and clever as there is a heavy roll on. The captain also misses, but BLS gives him his quietus in fine style. A boat is lowered and he is soon on board and proves to be a very fat young bladdernose. Crowds of "malleys" hurry up to the feast, and a most exciting fishery ensues. There is no longer any difficulty in catching: they crowd up to the stern and cackle and scream incessantly for who is to be caught. We get a great haul sometimes three at a time. They are then painted red green and generally fantastically adorned and turned adrift, with the exception of two or three, which Eaton required as specimens. Some of the old birds (as I suppose), which are much whiter than the others, are very handsome. Eaton is in great hopes of finding some sort of "bonser" in the dead specimens, as none is known by naturalists to inhabit the covers of this bird. For this night, at least, no parasite is forthcoming. He is further much interested in the intestines of the seal and fills buckets &c with detached portions of his carcass including head, flippers &c. A new regime has to commence now, as we may at any moment be summoned to shoot seals. One has to be ready to turn out sharp, and this would be effective, I fancy, by keeping a suit of clothes to sleep in at night and arranging boots, coat, cap, gloves, rifle &c, handy so as to be assumed at a moment's notice. I turn in between 1 and 2 a.m.

May 26, Monday.

About 7.30 a.m., I am roused and informed that we are going to speak a "Swedish Ship" and have the satisfaction of finding myself on deck in, I should say, less than a minute. In about ten minutes we are alongside the Swede, which is a fine steam vessel of, I should suppose, 6 to 700 tons (builder's measurement). She has a very large ensign at her mizzen. She is evidently surprised at our flying the "Blue Ensign" [of the United Kingdom] and at once she asks if we are an "expedition." Nothing of much interest transpired: we had no news to give from Europe. She assured us of open water (as it is called) to the E and N, gave us some information about the projecting points of ice, and told us that the *Active* was not far to the N. She had an immense number of boats at her davits and a very large crew. There were 43 on deck at the time. Her captain, who was in the chains [a small platform on the side of a ship where a sailor could stand when sounding with a lead], spoke remarkably good English. The name of the ship was the *Vega*. She had been N as far as 79° and had 300 seals on board, and was not equipped for whaling. It seems very strange that these Scandinavians have abandoned this fishery and that it should yet be well worth the while of other nations to prosecute it, especially as the provisions for the crews are less in quantity and inferior in quality, and their wages far cheaper. Though the vessel was "Swedish" her crew were, I think, "Norwegians." [The 150′ whaling vessel *Vega* was only a year in commission at the time of this meeting with *Diana*. Launched in Bremerhaven, Germany, in 1872, it would go on to fame in the history of Arctic exploration in 1878–1879, under the leadership of Adolf Erik Nordenskiöld, as the first vessel through the North East passage.] Another craft was in sight at the same time, on the port tack, in fact rather going from us. She was a very small smack (60 tons at most?) with the Dutch tricolour flying. She was a yawl or "dandy" and carried a crow's nest hoisted, on her little broomstick of a mizzen.

We went on steaming most of the day and in the afternoon came upon the *Active*, a Peterhead seal and whaling ship. She sent a boat on board and we subsequently sent one off to her. She gave us very little information but stated that, some way to the N, the ice was very fast. I forgot to say that the Captain of the *Vega* asked us on board. We, however, declined, nor would he come to us, of which I was rather sorry, as we were just going to breakfast and I should have talked to him. The *Active* had about 300 seals on board. She had last fallen in with them on Saturday. Whilst lying we conducted a malley fishery. We were much excited at seeing a whole school of seals, playing about in the water, at no great distance. It is, however, of little use going after these as you cannot get near them for a shot and even if you do shoot one it sinks. We, however, tried and failed. This is the second school that we have met the last two days disporting themselves in this way. The sailors call it a "seal's wedding." The swell was tremendous this afternoon and was unlike anything I have ever seen before. It is of itself alone wonderful to see a swell of very high waves, with their crests a prodigious way apart, when there is not a breath of wind and the sea is like oil, but it is easy to think how the wonder of this effect is enhanced by the fact of this sea, being an intense colour, apparently almost a turbid ultramarine; although in reality very clear, its surface covered with various sized blocks of ice, all snow-crowned, but the sides of almost every hue are of blue and green, and in places where the ice is fresh water, almost black. These blocks, of course, show a motion like our own so that at times many of the most conspicuous are out of sight, while at other times, standing in bold relief against the sky, but only for an instant disappearing to be replaced by others and so ad infinitum. It is to the eye far more wondrous than would be the same behaviour in a snow-clad moorland, for to the eye the distance is almost as solid looking as the moor. But, instead of the uniform shroud only relieved here and there by the dark overhanging side of some large rock, there are those endless shining blue surfaces, lifted ever and anon with a motion so slow, so stately, as to produce an effect hitherto unknown to me. This blue which so strikes me is, of course, due to the effect of the blue seawater on the snow and the ice. It does not, however, seem merely the reflection or reflected light from the depths below, but to be most intense where the floating mass has a white submerged base. This, of course, is covered with blue water and seems from its whiteness to reflect up very strongly any light on to the overhanging sides of the mass above, the blueness of these latter or intensity of colour depending on the depth to which the reflecting base is submerged.

There is something curious and unusual in the ice we have been amongst for the past few days. It is more to the Eastward than usual, and it is in a far more broken state and will not allow for any close approach to the fast pack, wherever that may be, as all endeavours yet to penetrate into it have failed. It is also, I think, very doubtful whether, as they seem to think here that, it ever was a close pack. It would seem more as if it had come from Spitzbergen or somewhere there, and not to be the Westward ice at all, but merely driven against it and stopped there in its progress. The fact that so much of it having been in contact with the land tends to support this idea. The ice washed by the water is much distorted. This has been observed by Dr. Hooker in his voyage to the Antarctic regions that it is due to vegetable growth. Whether the discolouration is red and of a colour like iron rust, it is due to *Protoeous nivalis* [*Chlamydomonas nivalis*] and if of a light brownish colour then due to some of the diatoms [plankton] specimens—the specimens which were observed this morning under the microscope. [Sir Joseph Dalton Hooker (1817–1911), famed botanist and close friend of Charles Darwin, made a voyage to Antarctica in 1839–1843 under James Clark Ross. Between 1844 and 1860, Hooker published *The botany of the Antarctic voyage of H.M. discovery ships Erebus and Terror in the Years 1839–1843 under the command of Captain Sir James Clark Ross*.]

Eaton was much elated at having during his researches discovered a l—-a or parasite from some animal: forbid such in the cabin. It was a charming evening and it was not 'til late that we knocked off steam. We now, however, keep the fires banked, as it is very convenient to be able to get up steam at short notice, and the water not to be warmed all afresh. Of course, as soon as we knock off, the breeze falls, and we are helpless and rolling about in a most unpleasant way. You would think it was a gale, as she rolls right down to the rail first, on one side and then on the other, and once or twice nearly brought the boats at the davits onto the water. The deck and vessel were in a tremendous mess today, but the men, under the influence of a double ration of grog, have done a great deal. An immense number of stores have been brought from the hold, the boxes broken up and made into firewood &c, and the Australian meats &c stowed forward at the bottom of an empty water tank. The whaleboats are at the davits &c, and already improvement in the new distribution of weight, lower down instead of so much between decks, is apparent. We see a few seals, one ahead on the ice, which of course disappears as soon as we get up steam to approach him. I saw a very small seal before turning in. The mate called it a "floe rat." Turned in about 2 p.m., three or four hours after the others.

May 27, Tuesday.

Felt awfully fit. Turned out soon after 8 a.m. and pulled the other lads out of bed. Whilst on deck waiting for breakfast a little land-bird came flying round the ship and finally on to it. It was apparently an ordinary red pole but rather larger with greyer and lighter tail coverts &c [Greenland hoary red pole]. It perched quite close to me and for some time I did not think of the molesting of it. Then, as it occurred to me that Eaton might want it as a specimen, as it is a disputed point, whether it occurs here or not, I went below for a gun. But luckily, he flew away just as I emerged from the companion. We got up steam in the morning as the ice ahead was very thick and in some places no way through was apparent, we then rammed at it, and gradually turned the pieces aside with the bow (Sketch 9.2).

One would think at first that these shocks must do the vessel a great deal of harm, but apparently, they don't, and one *accoutume bientot* [quickly accustoms]. I must say I would like to see the *Glatton* [a Victorian Royal Navy monitor launched in 1871] or the *Devastation* [a Royal Navy turret ship also launched in 1871] have a good wire in at a piece [of ice] with their rams full steam up, and going ten or 12 knots, just *exempli gratia* [for the sake of example].

When we left the *Active* we for some time steered a Northerly direction through tolerably open water and this morning came into thick ice. Finding large and heavy blocks ahead we turned a bit westward, and suddenly about 11 a.m. come into open water. To the right alone is ice to be

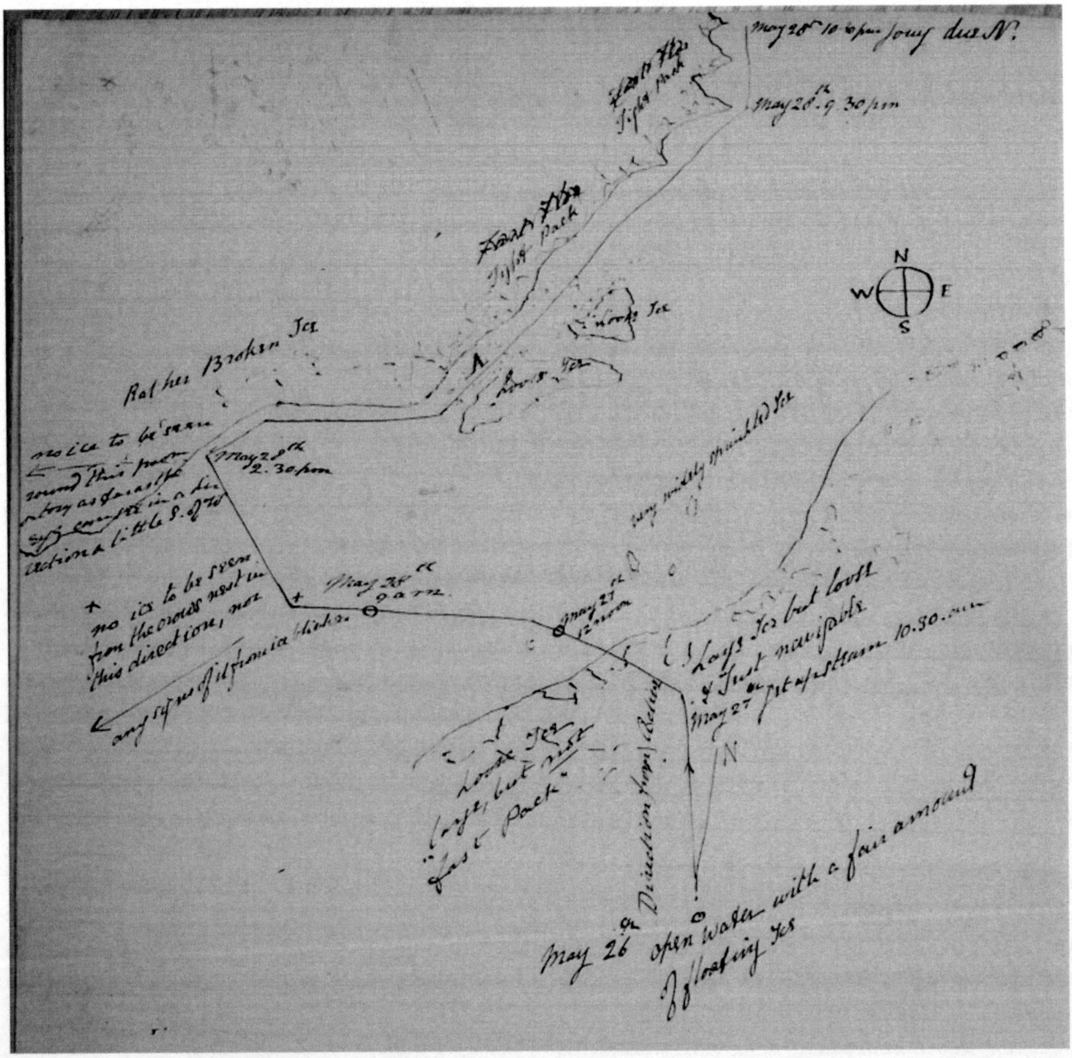

Sketch 9.2 Ice conditions, 26–27 May 1873

seen in front of us. The ice line through which we have broken trends away to the left rear. At noon, we get our position from a good sight of the sun. Steam is knocked off about 2.30 p.m. and it afterwards became so calm with the little wind that there was dead ahead and that the ice was visible astern from the crow's nest up to 10 or 11 p.m. We are now in what we took to be a bight of the ice at first, but if so, it is of enormous size as when I turned in at 12 p.m. there was no ice to be seen in any direction, nor was there much indication of it. By the ice blink in the sky except that astern over the ice that we came through, still there was indication of ice to the N and to the E tho' faint and distant, but none to the W or SW. It therefore seems that we are in a large island lake, so to speak, among the floes, with an arm stretching away southward and westward goodness only knows how far. How far this idea was correct will be seen from the diagram on the opposite page, which gives my subsequent general idea of our course, and the position of the ice. Today was a busy day for the crew. They were busy cleaning and fitting, harpoons, harpoon guns and lances. We took a sea temperature at 50 fathoms tonight. Before turning in I had a shot at a seal in the water, at between 250 and 300 yds and I aimed point blank in order to try my new rifle (Westley Richards Express). It was a goodish shot, right in line and apparently only a few feet short.

May 28, Wednesday.

Very cold in bed. My breath froze on the porthole window about a foot from my face, in spite of having my door shut for once and having the very modest dimensions of my stateroom. On deck at 8.30, it was 25 °F, but once up one doesn't seem to feel it, and we prefer not having a fire in the cabin. The Captain was anxious to get up steam and push on to the ice and accordingly, by 10.30, steam was up. We then took the course shown on the diagram or thereabouts and made the ice about 2.30 p.m. It was then that we turned along the ice with a fair wind. It was not without some reluctance that we turned Eastward, but alternatives are always turning up in life and resolutions have to be made. It seems well known to whalers that there is a bight, as it is called, in the ice, at about lat. 75 and 76° (our position) but it is seldom that anyone penetrates far into the W here. Here was our chance: open water as far as the eye could see, besides which where can the seals be? None of the vessels we have spoken have fallen in with them in numbers of late. They have come from the N along the edge of the very floe we take along, the appearance of which and its position, as well as its character of impenetrable fastness, we at once recognise from their description. All the other ships we have seen including the one that passed us this morning appear "light," and as the Captain expresses it, they float on the water like a cask. So much for the pros; now for the cons. The wind, which freshened (of course) from the moment we got steam up, is the most favourable one for these seas; it is from the SW and will carry us along the floe in search of seals in and out of each bight and away from northward and eastward. Along that long white line, hazier and hazier, with here and there, a sharp white point showing clearly out against a dull leaden blue sea and sharp steel grey sky loses itself in the far northeastern distance. Along this very line or thereabouts last year, the little *Sampson* was running the same course with very fair luck in the seal fishery and splendid *sport*. Certes it would seem folly to steam right in the teeth of that cold west wind, dead against us, using measure after measure of that precious coal, the want of which might mar our faintest prospects in the higher latitudes we hope so to reach, later on in the season. So the die is cast, "*noli respicere*" [do not look], for is not after all our object to get North, and we *know* that to get N we must follow that line of ice. As well, we know that the line which runs now so fairly N and promises us such an easy path will of a surety stretch out floating promontory after promontory and bay after bay gradually more and more easterly, and leading slowly and surely to the coast of Spitzbergen. However, we still hope that ours may be the luck to find a more westerly path through the floes, a more direct route, overseas unknown and unsounded (Sketch 9.3).

But if we choose to please our fantasy with such visions, at any rate goodbye to that of penetrating westward down the channel we see so fair, forget that lurking idea that it might take us straight to the E coast of Greenland with a straight run of open water, to teem of course near land with narwhals and walruses, and seaward with seals. I was not personally so impressed even in making a run of that sort, altho' one never likes to pass by what in these seas is a rare opportunity of prosecuting researches in waters not usually so open. But I was intensely eager to see whether as we ran W and S or W, we should not find some channel leading up N, either through the floes or near the land either as at B or C. Beyond this there is of course a chance, but less, of finding seals along the edges of the recently visited floes that we shall coast along and this unexplored bight was provokingly suggestive of their retired watering place, wherever it may be seal fashion to spend the spring. We others, of course it is nought to us, but I so long for a good cargo to repay some of BLS' immense outlay on this expedition. As we sail along now and I write (12 p.m.) point after point not a living seal passed in sight, only a few "*krengs*" or "flensed carcasses," a prey to ivory gulls and stately "Burgomasters" or full joyed "Mallies," the relics of the spoil, scanty ones, too, of the patient seal fishers

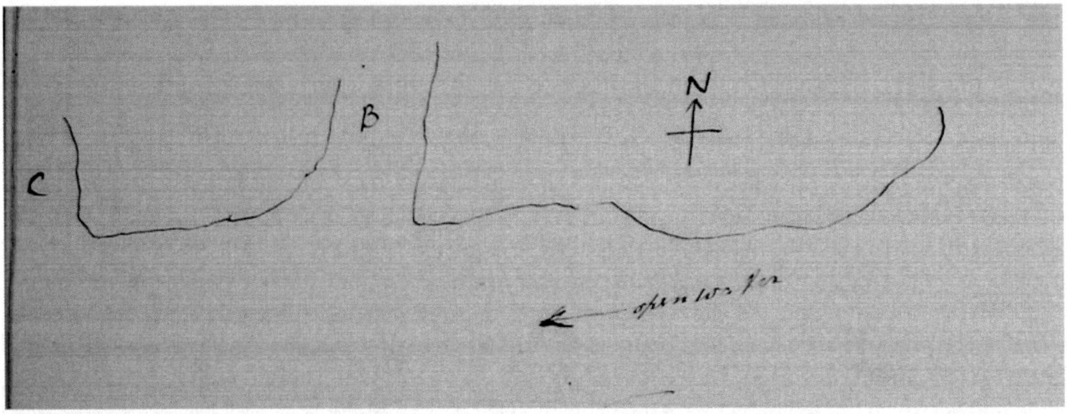

Sketch 9.3 Ice conditions on 28 May

we have met. Goodbye to the West. I leave it to be explored by the dear little travellers bound "that gate" that we saw today: snow buntings, first a single one, in spring plumage all but the bars on the wings, there a group of five, one white, two pied, and two brown, and then another group of five an hour or two ago, all bound westward, and some of them resting a short time on the ship. I suppose the Derby is run today. I wonder to whom the Blue Riband of the Turf has fallen. I have been very busy all day, ruling tables for the observations I take every day of barometers, chronometers, temperatures of air and water &c. We come across the ice again as I have said, about 2.30 p.m. It is here we found large but broken ice and one could not penetrate any distance. We take eastward and soon find ourselves easting along the edge of a tight pack we pass up a chance of "A," under the impression that we are at length leaving the inland lake (as we thought, of course, more correctly "bight"). However, the ice to our right ceases after a time, it being only a sort of island, and we are again, as will be seen from the diagram, in the bight. This time, however, we have reached its edge, and like hounds hitting off the scent after a check, or Theseus on finding *his* thin white line, we follow ours far away N and E. We have a fair wind, and the pack is on our weather hand, so we coast along quite close to its edge except where we take across the open bays and bights from point to point. This is quite a new experience to us, as hitherto we have never come across a well-defined edge of ice. Here, however, we go gaily sailing along (not overfast however, say five to six knots, for the *Diana* is certainly neither the hardiest nor fastest vessel under canvas of my acquaintance) with a westerly wind. On our right or starboard hand lies open sea, with a fair amount of swell for a time. That is, as long as the wind has been blowing *on* not *off* the ice. Hardly any but bits of stray ice to be seen here and there. On our left lies the "fast pack," a mass of floating blocks, floes &c, in such close order that one could almost apparently walk across it and with a wondrous resemblance to a low snow-clad coast as you see bay after bay and point after point. It looks so firm and land-like that the heavy swell is killed and disappears long before it has penetrated even 100 yds. It is only at the edge where you see (and hear) the difference. Here you can see not only the impossibility of a boat penetrating or even, I should say, landing while this swell is on it. Large masses are ever rising and falling, meeting and parting, cracking and splitting, and finding their sides one against the other, and shooting up long spouts of foam as they come suddenly together. Soon as this dies away, as you look inwards, yet when you are close to it keeps forcibly in your mind's eye that it is ice not land you see. Now for the first time you hear the mighty roar of the ice, as block after block, ever chafing to press forward farther and farther in, dashing against each other and those that are already firmly wedged into the pack, running before the wind and crowding into gaps and corners like sheep before, wear away their life and individuality, and dispersing into thousands of fragments, add their quota to those endless morsels, that form the fringe that decks the outskirts of this floating shore. As we go further along the edge and round a point into smoother water, the roar and its component sounds or notes—which are at times heard distinctly and are due to the different sounds produced by the pieces of different size that are grinding digs away —a new and stronger sound is heard, a weird unnatural whisper, all the more striking that it is heard in these Arctic…. [A break in the journal occurs here.]

Turned in at 2 a.m. While sailing a NE course in crossing a bight we see, or rather the skipper sees, a ship far away into the ice. As her movements seem suspiciously like that of a ship whose boats are engaged in fishing—as it is termed for whales and seals—we get up steam and go in towards her. In crossing a thin belt of ice into a long interior lagoon, we see the marks of "bar" [bear] on a piece of ice, but no other signs of bruin. When we come up from dinner, we are pretty close to the stranger, who hoists the British ensign, and is recognised as the *Jan Mayen* of Peterhead. Her skipper comes aboard and is very hospitably treated to fizz &c, and a great game of deep and dark purposes goes on between him and our captain, the stranger stating that there is nothing in the bight, but that having come in last night he was coiling his whale lines all day in expectation of coming across fish southward, whither he was bound. He was a wily Scotch skipper and it was certainly a game of Greek meet Greek, as the Dundee, and Peterhead skippers are rather jealous of one another. We wanted seals did we, quite so, for sport more than for profit. Ah well, he could tell us exactly where to find them: round the point of the bight and about 15 miles eastward along the ice. "Why had he left them?" "Oh, because they were "saddlebacks" in bad condition with very little blubber"; moreover so shy that it was very hard to get them. He was going south for the "bladdernosed seals" (which are always coming N about this time). As a return he was informed of the open water westward between 75 and 76°. How likely it was that he could there get into the floes and find fish? These statements, on both sides, were inwardly, I fancy, accepted—if not, as the journals say, with reserve, then certainly "*cum grano*" [with a grain of salt]. We had a letter for him and he was hardly over the gangway and rowing off when these statements were communicated by the skipper to "James," a very old salt and an ancient hand at the whale fishery, who at once gave free vent to his disbelief in undisguised laughter. I must say I am inclined to believe him for many reasons, chiefly because one never caught him tripping and next because his statements are really probable. He has 3,500 seals on board and getting a few "saddles" daily is very slow work when "fish" are to be won and lost. The bight that he described in 77°40" as being full of "unies," which is short for unicorn, the sailors name for that narwhal, may be so, as his reasons for not pursuing them are that his lines were not coiled and also that they are not quiet, being outside the ice and "on the passage." However, he is not off southward yet, nor are we making round the point for the seals, as we almost implied we would. It is the general opinion that at least he has seen a whale or something come in here, and that he wants us off. It may be so. Anyhow, although I stick up for him, if the seals are there they can wait, and I certainly think it the right course to have a look about here for ourselves. So up and down we dodge, and see him safe out and reaching away S. We wait long enough to be sure that this is not a ruse, and about 4 a.m. come out and again coast the ice from point to point.

I forgot to state that all day we have been sailing along pretty well clear of the pack as it is useless going into the shallow bays along this coast line of the ice. We have been going N and E all day, crossing from point to point, generally two to three miles off. It was not 'til about 5 p.m. that we took into the bight, I have spoken of, about 12.30 p.m. I took a deep-sea temperature at 300 fathoms. Turned in at 2 a.m. Men have been coiling whale lines all day.

May 30, Friday.

A cold raw foggy misty day. The fog freezes very quickly and coats everything with ice. There is a slight breeze, but we can barely make out the pack edge, as being a lee pack we give it a fair offing. We can hear the roar well enough. It is not very interesting all day and we see nothing, not even a "uni." It cleared up soon after dinner and we went in chase of a "uni" that was lying near the pack edge. I tumbled into the boat, which was rather slack at getting away owing to its being the first time. We got within about 40 yds of the beast when it started. After a longish chase, during which he several times went down just as we were going to fire, we lost him and returned. Later on, a "bladder" was seen lying some ways into the pack and Potter and I sent away after him. We failed to land owing to the swell, besides which we could not see the beast. We all at length landed and then began our travels. The Captain and I made towards his position. The travelling was to me of a very novel kind, jumping from block to block of ice, all said blocks being snow-covered into which covering you often sank over your sea boots. These blocks were dancing and shifting, opening and closing with the heavy swell and it was ticklish work jumping from one to the other heavy swell.

I soon, however, got to look where there was firm footing at the edges, not "cave ice" or "cave," as projecting snow ledges are called. The worst place to be obliged to jump is a submerged tongue as they are very slippery. The chief thing is to judge your time when the pieces come together. A fall is very dangerous as you may get squashed between two pieces. When the gaps are too wide to jump and not likely to come together you, if possible, secure with a "seal club" or an "ice pick," a small floating mass and use it as a raft. We penetrated about a quarter of a mile, but [the] seal was gone, and with some difficulty we got back to BLS and Potter and managed to retrieve them off an isolated piece. About 1.30 a. m., we came in sight of a point covered with seals, who were on a very narrow strip. This is evidently the bay the *Jan Mayen* skipper spoke of. We determined to leave them 'til about 10 a. m. tomorrow, as I always call it, as it is sunny, they are likely to lie well. Turned in at 2 a.m.

May 31, Saturday.

Roused at 7 a.m. I was told that the boats are to start at once, as a "Dutchman," as any foreign vessel is called, is bearing down on. Awful scuffle to dress, and stow one's pockets with "baccy," "pipes," lights, flask, biscuits, ammunition &c, as the boats are already being lowered. We bolt a few mouthfuls of breakfast and then away go three whaleboats, BLS and myself in the large "dinghy," with a crew consisting of the engineer and "steward." We are an oar short and also waited to see the others fairly off. The attack, however, is very badly managed, as the leading boats go straight at the seals, instead of leeward side of them. Round the point of ice, one boat getting "hung up" in the ice had to be dragged across a piece and the seals getting the alert saw what was up. We, however, make the best of our way in. It is painful to hear the fusillade going on all round and to make such a slow progress. The ice, however, is not favourable for our purpose—it is too open to enable one to cross it as we were doing last night on foot and too close to allow boats through with ease. By and by we see a single seal, though there turn out to be two there. We stalk him, by getting the boat behind hummocks and when I think I am near enough I fire and put a bullet clean through his neck. Thinking he is done for, BLS stops the boat to let him get his aim at the other, whilst I shout "give way" lads, as in a last throe he struggles off the floating piece of ice. He floats for a few seconds, but as we get there he is sunk and only the discoloured water and bubbles of blood show where he has gone down. It is very provoking, but a common enough misfortune, as it is rare except with a swift light boat to get seals out of the water, so on we go. It is, we find, no easy work getting any of these shy brutes and so you must fire longish shots. What is more, it is no use just hitting your seal—that with a good rifle becomes easy enough as soon as you get steady—you must kill him dead, as one or two convulsive kicks, roll him into the water, and then he probably sinks to rise no more. If slightly wounded, you sometimes get them in the water. The head is the place to fire at, because you are most likely to kill him, and even if you don't, the amazement caused by the whistle of the bullet, may get you a second shot, or give the other gun a shot if there is one. It is, however, a great effort of resolution, when you see a great plump brute lying on the ice, to neglect such a target as his body offers and aim at the head. But I must get on, all the other boats are farther into the pack and here we are without a seal. We get on but slowly as the long oars are greatly in the way in the ice, and we have no boathook, only "seal clubs" to punt ourselves along with. BLS, steering with an oar, keeps her off; then round with her, here we are stuck, no, I (who am in the bows) see a lead as it is called, there, to the left, and through or over that tongue get one leg out, on the ice; here goes. I am alongside on the ice: crack, grind; now she moves, hooray! She is through and again afloat in a little lake with small bits of ice floating all over it and the snow-crowned hummocks all round, with their blue glint, for as you look into the recesses of snow there is a stony misty blue colour, or "bloom" about them (like the blue of starch). The sides are of an almost emerald green in places, in others blue. *Sssshh*—as I hold up my hand and wave in which direction the boat is to be steered, round with her, round with her to the right, as is shown by the frantic waves of my hand, I see two seals, and now we are out of sight—I whisper that as we come round that block we shall see them. "Give way," "two good strokes," "way enough," "let her glide," the oar astern sending her head round, ah you brutes (Sketch 9.4).

Up go their heads, shuffle, shuffle, shuffle, as they wriggle across the ice. Plunge, there goes one. The other is at the edge; crack, crack; he never moves again as a couple of bullets settle his affair. Give way as his head overhangs the water; he may wriggle in. I seize a seal club and as the boat runs against the ice drive it into his head. We now disembark and the operation of "flensing," as taking off the skin and blubber is called, has to be gone through. The steward alone knows how to do this, and he is anything but an experienced hand. Alas! The knife we got from the cook initially did not cut. We used the sharpening steel never so wisely. So, my small knife (as it is for the purpose, though in reality a good big seaman's clasp knife) is in requisition. We are much delayed, the operation taking nearer 20 min, than two (the proper time). Meanwhile, I shoot at a seal who came up in shallowish water; that is, above a submerged tongue. He was not however killed dead, so we don't get him. I may as well say here that the seal club, which is used for killing young seals on the ice—early in the year I wounded seals at all times, on the ice or in the water—consisted of a wooden handle about 4′6″ long, rather thicker than a mop handle, and at the end is an iron head, something like a slender pick-head but with points at both ends. These clubs are very useful as one can easily imagine on making one's way across the ice, and pulling pieces towards you, or pushing them aside &c. We are now again afloat and as before following leads through the ice and making towards "patches of seals" which seem unattacked by the other boats. As soon as you have made out, seal or a patch on the ice, that you think you can approach, you look out leads and stalk them as quickly as you can by getting the boat as near as possible, and then if the ice will allow landing. There was great rivalry among the boats, enhanced by a prize of a pound of "baccy" to the boat with most seals and had they not raced jealously straight up to the seals, I am sure we should have got more. It was more especially annoying to us as we, of course, did our best to avoid the other boats, and not interfere with their seals. But as we were, of course, slower than the four-oared whaleboat, they several times raced up to isolated patches, which we had been for a long time stalking and selected because not near the other boats, so as to improve the general bag, went straight at them, and frightened them all away, perhaps getting one seal, or as it happened more than once missing the lot. I was in the bows and of course got all the shooting which BLS generously insisted on giving up to me. We were on one occasion, close to a patch in the boat and well concealed, I fired at the nearest two, at long shots, and hit both. One, however, was evidently not quite dead, and though I thought as he didn't move if left he would die as he was, I did as I was bid,

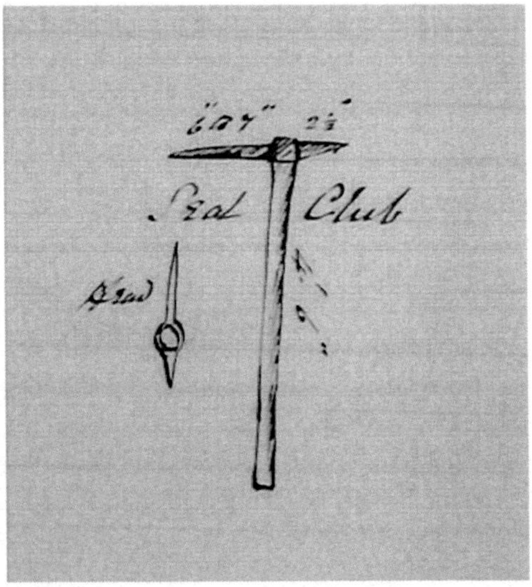

Sketch 9.4 Seal club

and fired again, luckily hitting him in the neck. We were then hidden from our dead seals by a hummock, and on getting up found to our intense surprise only one. The other had been near the edge and had got off; how it happened I can't imagine as the amount of blood was tremendous from each wound, and the ground covered with a dew of it, from his breathing as I hit him in the neck. We thought, however, that it is always dangerous to leave a man on the ice, or to leave dead seals, since it is difficult if you want to find them again. We had such a little distance to go (about 100 yds) to take up a nice position for polishing off several of our friends in the patch, and that we should not have been out of sight. We, however, decided not to think that the other two boats who were racing up would be ready to attack just as we should be done. Not a bit of it, up they come, right at them, bang, bang, bang, and not a seal. However, here we go, and later on come up with a patch at which another boat had opened fire. This is all right, as they are well hidden, and it brings a crossfire on them. Ours is, however, at a long range. But we kneel on the snow, in the proper military position, and crack—two… three… four…—we have got four. BLS, making an awfully tall shot at the farthest off of the lot, dead as a nail. Of the other seals who were all in motion from the fire of the other party, two were first over the water, and one unluckily floundered in. He floated for some seconds, but we could not get him. The other party, I am sorry to say, got none at this patch. The "Jack" is now up at the mizzen, so the boats collect and we have a long pull to the ship. One boat has six, another five, another two, and the dinghy (ourselves) five, in all 18. We get back to lunch, and I am of opinion that "sealing" is one of the best sports I ever experienced. I find myself a good deal better shot, with a sporting rifle, then I expected, as I seldom miss a seal, but I certainly have great room for improvement, as I ought to be able to kill them dead. In the evening we saw some more seals farther along the ice; we were aware how little use it was putting off to them, but the "Dutchman" who had got nearly as many as us today or at any rate nine or ten, and another Norwegian brig, were coming on along the ice, which is what we have been doing since one, and so we lowered away. I went in the dingy this time by myself. This time the attack was better arranged: two boats having a long pull round to leeward of the point. We, however, find the ice too open for stalking and the seals on the alert at the first shot. There is a heavy swell on and we soon have to give it up. I got two or three very long shots, and killed one in the water, but did not get him. Rowing back to the ship, one came up and showed his head. Considering the swell, which moved both boat and seal, I made a lucky shot and we were still luckier to get him, didn't I just drive in the "hack pick" as we came up with him. I never saw such an effect as those cartridges of mine have: there was nothing of his head left but bits of the jaws. One of the other boats got a seal. Almost 10.30 p.m. we saw a large steamer coming down on us. She was close to us by 11 p.m. and a boat came away. It was, as we had known some time, the *Eclipse*, a fine large steam whaler (Peterhead). David Gray, her captain, came on board, he being a friend of BLS' and a very intelligent clever man, who has been in these seas all his life. She has 45,000 old seals on board and 3,000 young ones, and 2 whales, in all 250 tons of oil. She has had to throw overboard the greater part of her coal to make way for blubber. She is going S to try and get whales as he has room for 50 tons more and wants to fill her. BLS was in bed and remained there. Gray and our skipper came down to the cabin and we did the usual "fizz." We had letters and papers as well as a certain amount of news for him, as he sailed in March. He agreed to keep his ship here all tomorrow, and he would have lunch with us, while we will dine with him. Turned in at 2.30 a.m.

June 1, Sunday.

It was cold, damp, I saw this morning, with the usual frost. This is the only time I feel the cold; this fog that chills you to the bone, as ordinarily I sleep with my door open, and of course all the air &c from the companion comes straight down. It does seem strange to have the temperature of one's bedroom perhaps as low as 28° at times and not to feel cold. I wrote letters today to go by the *Eclipse* on the chance of her arriving home before us. It is so foggy that we can hardly see her coming through the mist, and the ice pack is barely visible. We guess our latitude well enough from our reckoning, as "shooting the sun" is out of the question. We went on board the *Eclipse* to dine. She is a fine vessel and seems very well built for work amongst ice, being doubled all over and extra strengthened at the bows. Her decks seemed very large after ours. She was far cleaner than I expected any whaler at the fishery to be. She has got fine engines, though not compound ones. They work up to 360 indicated hp, but not being compound engines consume nine tons *per diem* under steam. She has an iron boss on her propeller shaft, into which moveable brass blades are fitted. It does not, however, like most of these that you see in these ships, lift up. I am inclined to like this myself though of course my experience is nil. But it enables you to strengthen the stern post, which is the weak point of these ships; that is, for the work they are intended for. The *Eclipse* burns seal and whale "krang," as the carcasses are called, what is used for fuel being the scrapings from the cleaning of the blubber. This is found to succeed very well. The captain told us he had steamed over 100 miles on the tail of a whale. The best plan is to get up a fire in your furnaces and throw "krang" mixed with small coal upon it. The young seals "krang" is the best as it contains more fat. The ship had no vestige of the smell so much talked about in whalers, as nothing putrefies in these latitudes; moreover, the tanks are sealed down and it is only when the cargo is being disembarked that there is much smell about the vessels. She has a large and well-lighted cabin and seems very comfortable. We met the captain and the doctor and had a capital dinner, and after some time a regular tea, we stopped on board 'til 12 midnight. We were shown several curios in the way of natural history, such as the parasites off whales (a cross between a crab and a scorpion), those of narwhals, also a large mass of them like a sponge, from the stomach of a "bearded seal," "whales' food" &c. We are now no longer in blue but green water, the colour I believe being caused by the medusae &c it contains, whilst at dinner a small net was put overboard and some collected for Eaton. He subsequently got some for himself from on board the *Diana*. The water here is no longer of the same specific gravity, as before. This "whales' food" consists of numerous small shrimps, transparent, with red markings, and long red antennae. There are also fry shrimps, very like the domestic shrimp of commerce. We were also shown a "Palinurus" or instrument for correcting the local deviation of iron ships compasses [from Roman mythology, Palinurus is the helmsman of Aeneas's ship]. We had a long conversation and numerous yarns, about the Polar seas, and the fishery. Captain Gray is very well informed and tolerably well known. He wrote a very good letter this year on the subject of a close time for the seals, whilst producing their young on the ice, during which time they are slaughtered by thousands. Also, letters on the subject of polar exploration by Smith's Sound: his idea seems to be to penetrate in westward in 73, 4, or 5° to the east coast of Greenland, near Shannon Island, and that one can thence work up N in open water along the coast. It would be, of course, a lee shore as regards the ice pack, from the revolution of the globe, but from his own experience and the nature of the prevailing winds, he thinks there is open water there. He is against the Spitzbergen route, and thinks that even suppose Gillie's Land exists—which would be a weather shore as regards the ice pack—the N coast of Spitzbergen is almost always unpassable, owing to the drift of the polar current. However, *nous verrons* [we will see]. We saw a good lot of "unies" today first and last. They were all, however, on the passage. Turned in about 2 a.m.

June 2, Monday.

Sailing along the pack towards a point where Gray says we shall meet with seals. The brig is gone, but the "smacker," as the skipper calls her, is following us in the distance. As usual the excitement comes in the evening. We sight seals and for fear of these vile "Dutchmen," both of whom it seems are following us, we lower away two boats going to windward of the point and two to leeward. I am in one of the latter. We are, however, furthest from the seals, who are very shy and leave the ice in "schools," coming out mostly our side. The ice is too close for boats and the windward boats cannot land owing to the swell. After shooting one or two seals in the water, which we could not get, the second mate and I determine to try and penetrate to where they are lying in great numbers. Two hands are left with the boat, and one is to come with us—unluckily as it afterwards proved only one hand remained with the boat, the other coming on with a seal club or something. For some reason they could not keep pace with us and were subsequently more harmful than good. Eddie (the mate) is a good man on the ice and we got on capitally. It was very bad travelling and need to be slow, as the snow was soft and the ice wide for foot travelling, with a good deal of swell on. We also had to stalk. Often, we missed the boathook men dreadfully, but in spite of their "hack picks" and hooks they were in rear hopelessly pounded. We frequently missed a boathook very much, especially when we wanted to "ferry" and pull in the pieces for so doing or push others aside. At length we got near the seals, but couldn't see them, and had to wait about 20 min before the ice shifted enough to let us get on to a large piece we had been waiting for. Off this we got after a hard climb, over hollow snow, on our bellies to the top of a hummock, whence we fired. We calculated that we could stalk to another hummock, whence we could command the thick of the flock in from 10 to 20 min, but the ship thinking all the seals were gone hoisted the recall. This was very provoking but perhaps we could not get the two seals we had killed; however, had our boathook men been up, we could have easily reached one of them. After some time, we came up with them and their hooks were very useful, though it was hard work to get them along. It is always more difficult to get a large party over moving ice, as you have to watch your moment when the pieces are coming together and jump. Well, of course, it often happens by the time one or two have jumped such that a yawning gulf is left. Eddie is first rate; he and I while going did a great deal of the stepping stone business, i.e., jumping quickly from small piece to piece, none of which perhaps will bear your weight, many turning over, but your foot of course is only on them for a second. The folly of having left only one hand in the boat was now apparent, as the edge was by no means distant, and our boat far away, and the wretched man in her could hardly move a whaleboat, with her lines coiled in her, against wind, by sculling over the stern. We made the edge as near him as we could and luckily another boat took us off. Going back to the ship, I made a good shot at a seal's head in the water, and we nearly got him. He was a fine big beast, and floated for some seconds, and I got so near the water before he sunk that I drove in the seal club but was only rewarded by a shower of blood and water. We hope, however, to get the seals tomorrow if it is fine. We took a temperature at 500 fathoms tonight. Cleaned my guns and turned in about 1.30 a.m. I saw marks of "bar" this evening. It was very interesting, at the end of our stalk, to watch the seals, even at the distance we were. Some were asleep, others rolling and scratching themselves with their flippers, to say nothing of a tremendous parliament, continually going on. It was rather like the grunting of hips [hippopotami], but rather a more whistling noise.

I saw some birds tonight, which I have not seen before, first a single one and then five. They were like swallows in appearance and flight, and they had a flight something between that of a swift and a "summer snipe." They flew about like and amongst the "mallies." They were chiefly black with forked tails and a white patch above them. They were I think Leach's [storm] petrel, and if so, I should say very rare here.

June 3, Tuesday.

Early this morning that vile "Dutchman" came up, and instead of waiting for the sunshine sent away his boats and frightened all the seals off. *Hang* him. If he does so again, we are going to wait 'til he has lowered away, and then if we can't drive the ship into the middle of them, fire shell or rockets to frighten them all, as this brute not only follows and sticks to us, but spoils both his own and our sport. Luckily, he got none this morning. These Norwegians seem much better seal fishers than our men. They go with less noise and more care, and stalk the seals; generally, instead of long shots, get shots close enough to make sure they also don't get to windward and perform various other absurdities, like the habit of the British sailor, and go to work in a more sportsmanlike manner. We stood on all day, and about midday took into the ice to cut off a point. Here it came rather calm. There was a beautiful sunlight out and it was smooth as an inland lake. The Dutchman close by us; I went away in the Shetland dinghy, in pursuit of a "bladder" (these do not sink). On returning, they told me to go and shoot some loons [thick-billed guillemot] &c as they are good to eat, and as we get nearer Spitzbergen these increased greatly in numbers. Potter and I went off, and had great fun; some of the wounded birds give us splendid chases. I had a shot at a very fine bladder (slipping a ball cartridge into my shotgun) but it glanced or something as though hit. He didn't after a short time seem much the worse and before we got to him dived. We only at first took away 16 cartridges, and had to return for more, bringing with us five trace of loons [guillemots], dovekies [black guillemots], and rotches [little auks], which considering the race they fly was not bad. We got two more afterwards and they make a fine show of game. I was very unlucky in falling through the snow today, just as I covered a seal. Luckily the rifle didn't go off, but he was gone before I got another chance. When we got near the ship, we perceived that two whale boats were away. They were after "unies," two or rather three of which were together, as the female had a young one on her back. They came up close to us, but of course we didn't fire as they wanted to strike them with harpoon guns. They would have got one at first when they were asleep, but on approaching them the single boat that had been lowered found they had no caps for their harpoon gun and had to return. By the time they had got them, the noise of the water pumped from the engines woke them and off they went. The boats returned after a long and unsuccessful chase. Potter and I got near one lying hidden from the ship. We signalled to them, and finding no notice were thinking of attacking him with a seal-harpoon and scatterguns with ball cartridges but decided not. We then got up steam for the object of cutting out the "smackie" before he can get to the point that Captain Gray told us. The usual evenings excitement is wanting as it comes on thick and misty, we knock off steam at 8.30 p.m.

It is my father's birthday today, always a sad day for my mother. I suppose she will be very busy, as she leaves Brussels tomorrow, I think. Everyone has turned in early at 9.30 a.m. I don't know how I am to get on by myself as I am tired of writing, and I find six to seven hours plenty of sleep for the exercise I take. I fail to get an altitude of the midnight sun as happens in about 11.40. I forgot to state that I saw him in all his glory two nights ago. Turned in about 2.30 a.m.

June 4, Wednesday.

I turn out early this morning—that is a little before 8 a.m.—and find it sunny but cold, and a sun out which I shoot before 9 a.m. We are in the blue water again, coasting eastward along the ice edge. We get a small seal this morning, a "whitey coat," about 10 a.m. We got up steam and towards 4 p.m. make the point, where we expected to find the seals. It was about Long 5° E and Lat 78° 40′ N. The seals, however, have gone and we could see only a casual one in the water here and there. The water here is of a curious brown colour, and soon after we come again it turned into very green water, the temperature and specific gravity of which both seem about the same as those of the blue. The sails are now useless as the wind is north-easterly, so we steam along. There is a heavy swell on and things roll about terribly. We now see signs in the sky of open water to the westward. This is charming as we want to penetrate westward about this latitude to look for bears and whales. The sky is in these latitudes invaluable to the whaler and ice navigator as it shows him the general lie of the ice and water far beyond the reach of the best glass, from the crow's nest. Above large masses of ice, there is always a glare in the sky, varying in degree and colour, being sometimes almost golden, but always clearly discernible (except in fogs and even there is at times partially so). This glare is called the "ice blink," and accordingly as it is present, to its extent right and left can the position of ice be judged. It forms on a dull day (or night) a strange contrast as it is either blended gradually or stands out abruptly from the dull leaden grey sky above the rest of the horizon. A whaling skipper says, oh there was a bad "ice sky" in that way. It was of no use in trying, and it is wonderful, how correctly by experience, they guess beforehand telling you where the patches, and "streams," as detached broken floes are called, lie, and how they spot a passage in what is to you an almost continuous "blink." There are crowds of dovekies, rotches, &c, flying about in all directions in a tremendous hurry. The next time when we hove, we must fill up the larder. Being the "glorious Fourth," Potter and I, as good Etonians, drink "*in piam memoriam*" ["in loving memory." The "Fourth of June" marks the birthday of King George III (1760–1820), a regular visitor to and patron of Eton College]. We steamed on 'til about 10.30 p.m., going across several sort of lagoons of ice and finally coming into a position S (Sketch 9.5).

It then appeared that by taking a tack out from "P" instead of getting up steam we might have reached the position "S," without the use of coal. While penetrating in, however, we were not aware of this and imagined we were getting well into the westward, when at "S" it came on to blow a bit, and we came out about 12 p.m. as shown in the diagram. BLS and I both took altitudes of the midnight sun, but the horizon was not good. I made the latitude to be 79° 8′ 40″ and about 79° 12′ or 13′. We were to dodge about outside for the night, and I turned in about 2 a.m.

June 5, Thursday.

Turned out about 4.30 a.m. and found that the "blast" or spouting of a "fish" was visible. It was very exciting to see it, being dead in the winds eye. I turned in again after waking BLS, and we have to make several tacks to get near it. As I was not called again as I asked to be, I was not ready, when between 6 and 7 a.m., the boats were lowered away. After pulling towards some ice over a mile off, they were to turn their boat's heads and row along it. Unfortunately, when they got to this point, they saw a "blast" and pulled for it, this being in the wrong direction. They got quite close to the fish, but it proved to be a "finner" (*Balaena Physalis* [*Balaenoptera physalus*]). These fish are not attacked in the same way as the right whale (*Balaena mysticetus* [*Eubalaena glacialis*]) as they are more powerful fish. They go at a much greater pace and in a direction parallel with the surface, very often, instead of sounding as the right whale almost invariably does. They are a larger and more powerful fish and are easily distinguished as their section is different, being almost cylindrical in the middle of the body, instead of the broad flat back of the right whale. Towards the tail they have ridges, above below and at the sides the section becoming a rhombus with rounded angles. They are not nearly so valuable as the right whale, as they produce no "baleine" [baleen], or whalebone, and the blubber is of a different kind, I believe tougher, courser, and thicker in fibre. When fished for, the boats and gear are differently arranged to those of the ordinary fishery: the line running out through a hole in the bottom of the boat (through a tub with a hole, or well, I suppose), as the danger of the boat being capsized is thus diminished. A "rocket" or "shell" is also, almost invariably, fired into them. When they are struck, in order to put a stop to their going powers, as if some

Sketch 9.5 Navigation, 4 June

method of this sort is not followed, they will tow boats clean out of sight of the ship, and frequently run out all the lines. It is for these reasons obviously unsafe to attack them in the way of the ordinary fishery, as for a fish of comparatively little value, the loss of almost the whole of the gear on which the success of the fishery is wholly dependent may be incurred. Another objection is, I believe, the fact that when killed they sink, and have to be hauled to the surface with great labours, or else you have to wait, until sufficient gas is decomposed to float them, buoying them meanwhile, with of course the risk of losing them in a storm. The boats having inspected the fish at very close quarters returned on board. The appearance of this "finner" was very inopportune, as we were in apparently beautiful feeding ground for whales and it is always supposed that these "finners" scare away the other fish. The skipper, who is an experienced whale fisher in Davis Straits, has many a yarn about the incidents, hardships, and adventures of the fishery, especially in "March Month" and "October Münth in Dāvēēs Strāyāit." He was positive that he had seen a "right" whale from the "nest" and considered it probable that the arrival of this "finner" had scared her. About 10.30 a.m., we came in sight of seals on a point. It was a dull cold day, but there being no prospect of sun, we lowered away. There was a good deal of swell on, among the ice, but otherwise, it was far more favourable for shooting, than I have yet seen it. The seals were, however, very shy, and all took off at once. Some of the boats got a good many long shots, but we only got three seals. I was in the "Shetland dinghy" with one hand and got one, far the longest shot I have made. I had two other shots in the water but didn't hit either. I shot a fine "Burgo-master" or glaucous gull, but he is useless for preserving as the bullet made an awful hole. A fine old bladder came up near the ship. BLS hit him but didn't get him. We then sailed along the ice a bit and got into a sort of lake. Outside of us and all round was a tolerably thick compact stream and we had plenty of room to ply about inside. Towards evening it came on to blow, harder and harder from the E, and at dinner we congratulated ourselves on our snug harbour, as away to the E, we could see the tall spouts and clouds of foam and spray as the waves break on the edge of the stream. Short lived was our cause for such rejoicing. Fast and sure the ice was closing in and though we did not use it we got up steam in case of accidents, just managing to beat out in time, before midnight. It was blowing at any rate what I call a gale, and all the boats had to be got in board. The topsails were reefed, and the foresail reefed, the mizzen double-reefed, and we proceeded to beat about. It was very clear, although we did not catch the sun at midnight and we could make out land to the E from the bridge, as we plied on the starboard tack, it must be the highland of West Spitzbergen to the N of Prince Charles Foreland [Prins Karls Forland]. After stowing all the litter, which in a short time strewed the cabin, and making fast guns &c, and all things likely to be "in charge," we turned in.

June 6, Friday.

Land was clear enough in the middle watch this morning, and I could see it clearly enough from 8.20 a.m. to about 10 a.m. since which time it is out of sight even from the masthead, I believe, as we are going a good deal westward and northward. A "rotche" came on board last night and has been captive all day in the cabin. He is bold and ferocious but will not eat and refuses all attempts at domestication. With regard to food, he is very greedy about bolting one's fingers, but refuses all other nourishment, and it is in vain that I beg Eaton to allow him to regale himself on his "whale food." Eaton doesn't seem to see it and suggests that his skin would form an excellent specimen. This cruel proposal is at once quashed. "Rotchie" is not musical and though intensely interested in the "Musique de Geneve" which plays all its tunes for his edification, he doesn't seem to catch the airs, as one would suppose he must, from the sapient angle of his head and twinkle of his eye during their performance. He is also much struck with BLS's chirrupings, but being, I suppose, from the season of the year lovesick, cannot find it in his heart to make any response. After a lesson of how to make the ascent of the "companion," he is about 11 p.m. placed on the binnacle, whence he shortly takes his triumphant departure and is last seen on the crest of a gigantic wave, busy in the mysteries of his toilette, and evidently, going to spare no pains to make himself irresistible to the next unengaged fair one he comes across, if not to plot against the domestic happiness of less artful of his kin. The gale is pretty well over. Moreover, we are in a bight, the celebrated "whalefishers bight," recurring year after year in about 79° 30′. This year we see nothing in it, though we sail up to the most north-western edge (79° 40′). The ice as might be expected with this wind is close-packed and we of course cannot penetrate. The water is rather colder, and a hopeless blue again—ugh! how unlucky we are in the wind. Now is our chance of fish, seals, and bears, if we can only get in a little way W, but we can't, although here we are in the best latitude for everything, except bladders. After some consultation between BLS and the Captain, it is determined to be across to Spitzbergen, a course ESE (true). This will take us rather to the N of Prince Charles Foreland, and it is about here that we hope to fall in with some bottlenose whales. We may after this, if the winds suit, run over back to 78°40′ where we saw the seals, unies, and whales, and then work up here again when perhaps the winds will be more favourable, and the ice a bit more open. There is a strong breeze on still, and a vile cross sea, but it is fine, which goes for something. Perhaps too, these winds may have a good effect on the ice to the N and NE of Spitzbergen. Land is plainly to be seen now between 12 midnight and 1 a.m. and the breeze is still fresh. Turned in at 2.30 a.m.

June 7, Saturday.

When we came up on deck at 9 a.m. this morning, land was clearly visible, so clearly that it appeared only ten or 15 miles off but, of course, one has always heard of this peculiarity in Arctic regions, owing chiefly to refraction and I suppose the effulgence of the land caused as well by the white reefs of the snow as by the contrast between it and the dark rocks of which the mountains are composed. These mountains are extremely sharp-edged and steep and run up into acute and jagged peaks apparently mostly inaccessible, and on approaching it, the aptness of the name given to this country by the old explorers of *Spitz-bergen,* or "sharp mountains," is strikingly apparent. We were under canvas nearly all day and Potter, myself, and BLS got to work unpacking the photographic apparatus, cleaning and stowing the beakers, dishes, &c, and mixing a few chemicals. All day we were in sight of the "Seven Ice Mountains [Dei Sju Isfjella]," as the glaciers on this coast are called. These seven mountains lie between the mountains along the coast from the mouth of King's [Kongsfjorden] and Cross Bay [Krossfjorden] to that of Magdalena Bay [Magdalenefjorden]. Three mountains are apparently from 2,000 to 2,500′ above the sea and rise very abruptly. They form the lateral branches of a chain of mountains running parallel to the shore at five to ten miles distance. The valleys between them are filled with immense glaciers, some of which wind down the valleys like huge serpents; their upper surfaces in many places concave, from most beautiful inclined planes of snow, terminating in a sharp fracture which forms a regular wall at the sea edge. This fracture is in places white and others green, according to the masses of ice that have been of late detached or not, and their innumerable fissures form strange caverns with sides of every hue from sapphire to emerald. Some of these walls are only a few feet raised above the level of the ocean, and others some hundreds of feet; this, of course, depends on the form of the hills and width of the valleys on which they lie. As the summer comes on, watercourses are formed on their surfaces as well as in their innermost recess by the melting snow, and the whole glacier, or "iceberg," as they were always called up here, becomes a porous or rather venous mass. It now only needs a wind and heavy swell from the seaward to undermine the outer wall for large masses to be detached by the action of the running water. They, of course, fall over into the sea, with a tremendous crash, and it is only owing to the foulness of the ground onshore, which is mostly shallow with numerous sunken rocks, that instead of forming icebergs, the detached masses are dashed to pieces. This at any rate is the way "Scoresby" [William Scoresby (1789–1857), English Arctic explorer and scientist and author of *The Northwern Whale-Fishery* (1820)] accounts for the scarcity of icebergs in these seas, and the theory appears correct. But today, however, we are lucky as we come across several green and venerable looking bergs.

Prince Charles Foreland, which is on our starboard bow coming in, is a long island, excessively mountainous, about 50 miles long. It forms a sort of "counterface" or counterguard to the land between Kings Bay and Ice Fiord [Isfjorden], from which it is separated by a wide sound, narrowing however at one place to two and a half or three miles, the sounding at this point being only 2½ fathoms. The chain of mountains which forms its backbone runs from Fair Foreland [Fuglehuken], its most N point, about 35 miles down its length and there ends in table land, after rising into five distinct summits with extremely acute peaks, forming the "Middle Hook [Kaldneset]" of the Foreland. At the southern extremity, there rise three more peaks. This chain has a very high peak in it, estimated—I believe—at 7,000 ft. Our observations which from our distance were not likely to be over-correct make it 4,000 to 4,500, which, I think, nearer to the mark [Chermside here appears to reference *Grampianfjella*, the highest mountain on the Foreland at some 3,000 ft (1,000 m) high]. The chain on the W side rises abruptly from the sea, and on the E there is a little low-lying flat between the hills and the sea.

We get up steam as we come in past the Foreland, and as we come abreast of Quoad Hook [*Kvadehuken*, a point on the western side of the entrance to Kings Bay], which has a long projecting promontory of low land, we are in for a very smooth water. The ground around Prince Charles Foreland is all very foul and there is not good anchorage. We intended to anchor in Cross Bay but owing to the apparent incorrectness of the chart we stood on into Kings Bay and anchored on the southern side not far from the edge of the fast ice which filled all the end of the bay. After tea BLS, the Captain, Eaton and others went ashore, and Potter and I went shooting in the dinghy. We soon made our way, through some broken ice into a sort of inland lake, full of birds. I think I have never enjoyed an evening more: there was a glorious Arctic sun still high in the heavens, that even at midnight shone in all its pride above the northern hills casting on their coverings of the purest snow a sheen of golden light so pure, so dazzling, as to seem unearthly. All around us, ice or the still quiet waters of the bay, rippled only by the splash of the numerous wildfowl and closed at the end by a gigantic ice wall, 5 or 600′ high winding away up the valleys, where those magnificent hills the "three crowns" [Tre Kroner] with their lofty crests, lifted to heaven, shut in the view. These are indeed wonderful mountains and form part of a chain of five if not more, but three alone are surmounted by perfect crowns. These crowns are pyramids of very regular shape, formed of horizontal layers of rock stone, and are certainly one of the sights of Spitzbergen, the rock of which forms such a glorious contrast with the snow, especially when the latter is sunlit and the darker rocks bring of an intense purple, sometimes almost red, and in the distance assuming a bronze tint.

The shooting was capital sport, the "dovekie" being far the most sporting bird in the country. Beautifully, there were a great many eider ducks about, but they are very shy, and one gets few but long shots. We rowed across the bay and had some exciting seal hunts, besides a hunt after white whales. The latter are numerous here but very shy. I shot two seals, one a very long shot, but as he was killed dead, I failed to get him, the brute sinking. He was the largest seal I have seen (*Phoea barbata*) [bearded seal]. We rowed across the bay, about four miles, and did not return, until 4 a.m., having then run short of ammunition. We had only taken 50 shot cartridges, and our bag was 3½ brace [pair] of eider duck, 6½ brace of loons, 5½ brace of dovekies, and a sandpiper. After some cocoa and bread and butter we turned in, delighted with our first night in Spitzbergen.

End of Volume 1.

Volume 2.

June 8, Sunday.

This was a splendid morning and felt like summer although the thermometer was only up to 34½°. It had, however, been up to 60° in the sun, before we turned in this morning, with the water at 32°. The day was spent in vain attempts at photography with wet plates and dry, the latter failing through age, and the former, through the refusal of the "developer" to work in this climate. We landed to try the dry plates on a sort of foreshore of stones and ice and got a sandpiper there. After dinner, Potter and I again started in the dinghy. The ice was now much changed, a good deal having broken off and floated past the ship. We wound across the bay, shooting on the way two pin-tailed ducks [long-tailed duck], the drake of which pair escaped by diving and holding on to the bottom. We passed one of the boats in which the captain was acting "harpooneer" in pursuit of white whales [beluga]. He only got one shot which he missed. We gave up loon and dovekie shooting as the larder was plentifully stocked. Having rowed across the bay and found a landing place, with some difficulty, we hauled the dinghy bodily on land or rather snow, which was no mean labour as the level of the snow was in no place within three feet of the water. We then commenced to climb the hill, which was one of those deceptive ones with no summit in view but stretching ever and ever up. The snow was soft, often over the tops of our long boots, and a weltering sun on us. Add to this thick winter clothes, mitts, wrappers and field glasses, a bag with 50-plus cartridges, a breechloader in one's hand, ammunition for a rifle, and said rifle (an "express," no mean weight) slung on ones back, and you may guess that travelling in Arctic Regions can be hot and heavy work. However, we pushed on and getting into the shade of a hill made better progress. Near the top of the hill, I missed a sandpiper of some sort as the dog chased him, and I didn't get over an easy shot. Soon after we knocked over a brace of fine ptarmigan, to our delight. We now reached the summit of this Alp, and pulling down a cairn of stones, which with a broken flagstaff, adorned its summit, discovered a morsel of papers in Swedish and French, the latter part informing us that the ascent of this mountain was made by the Swedes in 1861 and its position determined that during the ascent the barometer stood as such and, finally, invited the finder to dinner sometime in 1865 [the Swedish Spitsbergen Expedition under geologist Otto Torell was A.E. Nordenskiöld's first visit to the islands] (Fig. 9.1).

Having failed to get Eaton a specimen of the snow buntings which were flying about, we proceeded to select various vantage points to enjoy the view, an enjoyment we had well merited. Looking down on the bay, we saw what I

Figs. 9.1 and 9.1.1 Chermside's view of Gullybreen in Magdalenefjorden. His views of four of the Magdalenefjorden glaciers were made from different viewpoints on Gravneset. Modern view from Tyrone Martinsson in 2012

have already described with Prince Charlie's Foreland in the distance. We were, however, surprised to find of what extent the bay now appeared, and that quite half of it was filled with fast ice. The extent of the glacier to the east closing the bay was now also apparent as we were above it. It falls from the valley of the "Three Crowns" towards the sea in a gentle and regular slope, with an ice wall of startling height above the shore. We now see that the chain of hills running parallel with the shore, the branches of which contain in their valleys the glaciers called the "Seven Ice Mountains," is only an inferior chain. The main chain, with the summits from 3 to 4 and I think 4,500 ft running inland—this nearly due N and S from the head of Cross Bay and branching apparently into three ridges—one ridge forming the chain running S from the

W point of Red Bay [Raudfjorden], the other turning eastward and ending in the lowland at Red Beach [Raudstranda], and the third and I should think highest runs I believe uninterruptedly NE ending in the round bluff headland between Liefde [Liefdefjorden] and Wood [Woodfjorden] bays. The three crowns were now in their glory, the crowns almost dark against the sky, rising above a mantle of snow, like samite [a heavy silk fabric], mystic wonderful, with a curious soft blue halo upon it. They are, I take it, a part of another chain running more eastward with other hills in it, whose crowns are broken. What the said crowns are I cannot say. They must, I think, from their appearance be aqueous rock, and might be such rock, deposited above amorphous or volcanic rocks. Away a bit down the other side, we come upon a steep glacier separating us from the opposite hill. He is excessively smooth and steep and barely to be crossed anywhere as the slopes from two valleys meet in a very narrow ridge or "saddle." A little to our left, we come on the view of a glacier, simply stupendous, full of cracks and fissures, but of a vast extent. A little further, we find that our hill like the "Elle women" of Scandinavian mythology is hollow behind ["Elli" (an Old Norse term for "old age") is the old woman who wrestles and defeats Thor during a series of challenges Thor faces in the *Prose Edda*]. He ends in a precipice, and far below we can peep over to where at the foot of the glacier, there lies an intensely green lake into which large fragments of ice are continually falling. Yet a little further we find our lake the head of a fiord (Blomstrand Harbour [Blomstrandhamna]), an inlet from the bay. Around us fly myriads of screeching rotches and, though in need of them, as they are a delicacy second only to "cock," we shoot none as they would fall into the sea. Our hill and the sea and glacier below are in bright sunlight, except where the glacier lies under the shade of the huge hill opposite, along the top of which is a long white belt of snow cutting the sky against a dark cloud from which the sun has first emerged. On this belt, he settles and absolutely appears to roll along it for a few minutes but there is no rest for him; he is at his nadir and must rise and he is soon on his way eastward again. We have a long climb down the hill and perform some of the descents by lying down and sliding, much less dangerous than it looks, as the snow even where it is frozen can be broken through with one's heel which soon stops one. We get a brace more ptarmigan on our way down, and at last call a halt as, after a long descent, we are still apparently only halfway down. Having drunk some water from one of the rills of melted snow we proceed onwards and in 20 or 30 yds, to our astonishment, find that what looked hundreds of feet below us is at our feet. Another and yet stranger phenomenon, showing horizontal as well as vertical refraction—or rather I suppose "astigmatization"—was that having to make out our "lead," as it is called, home to the ship. We had two alternatives: to row our boat either landwards or seawards. As it was now—or rather, the point on which it was ashore, in the middle of the floating ice, which was now drifting westward having left the pack —by rowing seawards and that is forcing our way through this ice 'til we came to open water, we should perhaps have a little shorter row; that is, if the ice did not continue to push us out. But, as the floe edge offered the additional attraction of the birds, we determined to row the course CD [see illustration previous]. But to our surprise when we came to row it, we were never within a mile or a mile and a half of the floe edge, which as seen from the point E on the hill appeared almost in a direct line between ourselves and the *Diana* (Sketch 9.6 and Fig. 9.2).

However, I anticipated we had got to the ice edge and found our dinghy. But the bay we had come in was full of ice, too firm to push through, and too rotten to drag her over. However, by a skilful manoeuvre of dragging her across a point we succeeded in landing here; we then get on an eminence and mark our course bit by bit for about ½ a mile to open water, allowing for the drift, and not making a mistake in working it out to get safe back to the ship, getting an eider or two on the way. These eiders, especially the drakes, are magnificent birds, wings and back and head black and white, necks green, breast

Sketch 9.6 Kongsfjorden

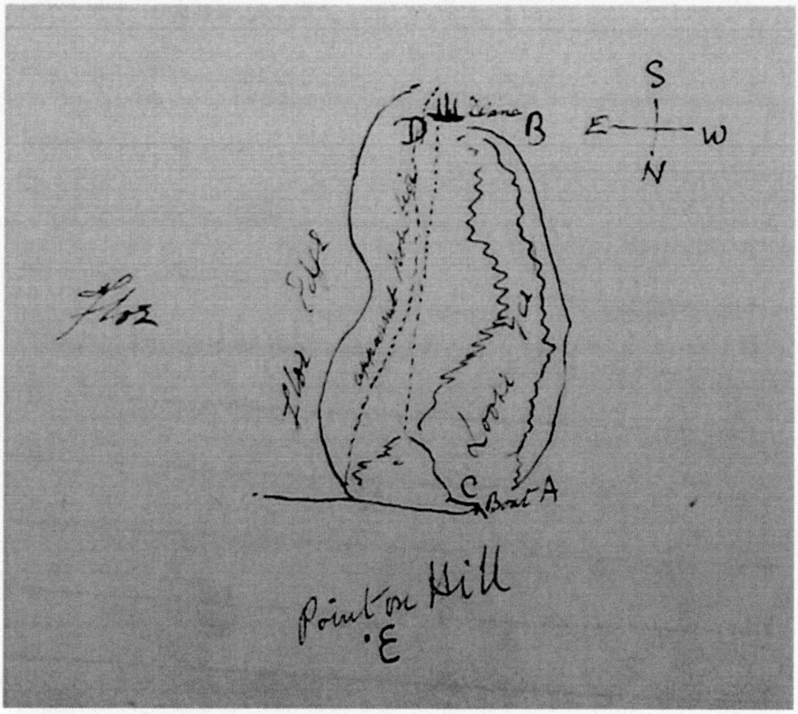

white halfway down and then with a satin black, legs and bill orange (pale).

June 9, Monday.

Steamed out of Kings Bay at 10.30 a.m. and kept along the land northwards past the Seven Ice Mountains. It was a smooth day and we had a charming time of it. This coast, which is laid down in a straight line, appears to be really more projecting. It is not very deep water here, for seeing a berg grounded we took soundings and got bottom in 15 and 17 fathoms. We turned into the South Gat [Sørgattet], passing the entrance of Magdalena Bay, and kept on until we came to or rather opposite Moss Island [Moseøya].

Here the dinghy was lowered. As it is a great place for eider ducks' eggs, although there were thousands about in the water, they have not begun to lay, so our anticipation of laying in boatsful was not realised. We then steamed on and after some time, just as we had discovered that the passage through the gat was closed by fast ice, were electrified by the intelligence that there were seahorses on the ice. These we had seen some time before but had not made out to be such: in fact, I was told they were stones. I forgot to say that we had previously seen two very fine "bulls" in the water. A boat is lowered at once. Luckily, I am ready with a rifle and off we go. Unfortunately, we pull for the nearest of two lots of three, the harpooneer not seeing the others. Had we pulled for them, another boat might have got to the nearest lot almost simultaneously. As we near them and row quite noiselessly, I get a good view of the immense brutes. They are lying all huddled together, and as the outside one puts up his head for a moment, I see the dark mask and long ivories of an old bull. Unluckily, the harpooneer doesn't and takes for the other two, hoping to "string" two of them, as he expresses it. It is now intensely exciting, and I can hear my own heavy breathing as we are within the reach of the foreganger, as the part of the line that is next the harpoon and which is drawn out as it is fired, is called. James has his harpoon gun nearly horizontal (it moves on a pivot). Now he has a sight and we are almost on them. Bang—off they all go. But we are "fast" and as soon as enough line has been taken out, a turn is taken round a bollard in the bows and the walrus allowed to

Figs. 9.2 and 9.2.1 Chermside's 1873 view of the so-called "Hanging Glacier" in Magdalenefjorden. The modern view that reveals the complete loss of the hanging glacier was taken by Tyrone Martinsson in 2012

amuse himself by towing us. Generally, when one is fast, the others often stick to him, attacking the boats. They are kept off with lances and rifles and, if possible, hand harpoons stuck in so as to get another fast. For this reason, I did not fire when they left the ice, as if by any chance the bullet had killed one, he would have sunk. Unluckily, however, these go straight away, only coming up a long way off. As our friend seems disinclined to come up, we help him by hauling and suddenly up he comes with a rush but dives at once. I fired as I was told at his neck and, though it was rather a snap, placed the bullet about right, and an "express" hollow-headed bullet driven by 4½-drs of powder has a wonderfully quieting effect. As soon as he is finished, we haul him to the ice and try to get him out. He is, however, too heavy and we send off to the ship for a tackle. Meanwhile, the other boat is fast approaching the others. Nearer and nearer they get but the beasts who are restless make off. One who lingers gives the harpooneer a long running shot. He is hit, but the harpoon glances. Two rifle bullets succeed in wounding him severely, but they can't get a chance with a hand harpoon. At last he comes up in a hole in the ice about 200 yds from me. I make all haste to reach it, but the floe is rotten, and I am over my boots at nearly every step and fail to get there in time. Meanwhile, another boat brings orders to tow ours whole to the ship and after a short but heavy pull, during the latter part of which one of the unsuccessful boats helped us, we get him alongside.

As I have seen two more about 1½ miles off, which have not been seen from the ship, I am despatched to guide a boat to them. The men seem very doubtful about them. However, as I had seen one more, I felt positive. Sure enough, and before we get up, one takes his departure, as we near the others we see to our disgust that he is surrounded by bay ice, and when we get to almost within shot of him, we are stopped by it. He now puts up his great head and looks at us, and thinking better of it, relapses into his former position. Shall I shoot? No, sir, give way lads through the ice, way enough, but the boat is rocking so from the scrambling sort of pull through the ice that the harpooneer can't get a sight and with an activity wondrous in such an unwieldy looking beast, the seahorse, hurls himself backwards with a crash and is gone. "Back out." "He will come up clear." So he does, sure enough, within five yards of the boat, rearing his great bewhiskered head out of the water and looking suspiciously at us with fierce little red eyes. Too quick for a shot with the harpoon gun. Alas! Too quick for me to drop the bow oar and have a rifle shot, which I fancy would have done the business at that distance. We wait in vain. He comes up again and again but never near enough. At last he comes up at a hole, right out of our reach, in the floe, which is too rotten to bear one's weight walking, at 100–120 yds off. "Well, sir, you can have a shot if you like, it may drive him out." I stand upon the thwart and try to steady myself as she rocks a good deal. He is looking straight at us, his head alone above water, not quite to the tips of his tusks. I am at the water line, *crack*, down he goes. It is too far to see for certain, but if the shot didn't hit him, where did it go? Because no one saw it strike elsewhere. The question is soon solved. Up he comes, "you hit him hard sir," says the Swede who pulls stroke oar, and up again, now I see, shot under the chin, through the windpipe, and he blows great gulps of blood every breath. I didn't fire again, because if I kill him, he will sink and we are in the hopes of his coming up near so as to get a harpoon in. He is up and going to take the ice, now we shall get him. D_____, the ice is so rotten he can't get out; it breaks away and away and he gives it up as hopeless. I am now on the look out to give him a body shot, to rouse him up, but he can't or won't because of the ice. He turns round and round and seems very bad, and at last disappeared for good and all. We then proceed to hunt some white whales that come opportunely on the scene. However, they are very shy and as the "Jack" is up on the ship, we pull back to her. She, it seems, had anchored with an ice anchor to the fast ice, but finding it rotten changed her position. We were, it seems, recalled to hunt another seahorse—the one wounded by the other boat, which seemed unhappy and kept taking the ice. He was, however, gone before our return. We anchored in the South Gat for the night. The walrus proved to be a young bull. They are very big brutes, with a tremendously thick hide but nothing like such thick blubber as a seal. I believe their flesh is very fair eating, at any rate "Rover" the bull pup seemed to think so. After a feed and smoke we turned in about 2 a. m., a rather cold misty and foggy night.

June 10, Tuesday.

A cold, foggy morning, and no seahorses to be seen. The passage through Smeerenburg [Smeerenburgfjorden] being blocked with ice, we have nothing for it but to return the way we

Figs. 9.3 and 9.3.1 Chermside's 1873 view of Buchanbreen in Magdalenefjorden. The modern view was taken by Tyrone Martinsson in 2012

came. The harbour we are in is the southern end of Smeerenburg, which runs nearly N and S here. The bay I speak of taking a turn to the eastward seems a fair harbour and well sheltered, and the hills around it have a very majestic defiant appearance and enclose the fiord completely. They are still very snow-covered, but I think that suits the hard crust look of the land here better than any other would. The name is Dutch and means simply "Blubber-town" ['smeer' = blubber, or speck], and from 1614 to 1660 the Dutch had a factory, warehouses &c here, boilers, coppers &c, and all the means for converting blubber into oil. This was of course in the days when the fishery was carried on close to Spitzbergen itself. The ships moored in the bays, which abounded in whales for the summer, and sent out boats to fish. The whales, at one time so

Fig. 9.4 Another view of Gullybreen in Magdalenefjorden, taken by Chermside in 1873

abundant, have long since left the bays, fiords and harbours and are now only to be found in the open sea near the ice edge or among the fields and floes of the western ice. I find from reading Scoresby's book that frequent and serious quarrels having arisen among the ships crews, belonging to the various nations, who frequented the fishery, it was resolved in 1618 to divide the extent of the western coast into several portions. The bays of these portions to be fished solely by the nation to which they were allotted, but the open sea to be common to all [see Scoresby, *An Account of the Arctic Regions and of the Whale-Fishery* (Edinburgh: Constable 1820: 32–38)]. The English had first choice, and so on; hence the names on different parts of the coast are English Bay [Engelskbukta], Kings Bay, Prince Charles Foreland, and further N Smeerenburg, Amsterdam Island [Amsterdamøya], Hamburger Bay [Hamburgbukta], Danes Hand [Danskeneset], and Danes Gat [Danskegattet], Norway Islands [Norskøyane], Biscayers Hook [Biskayerhuken] (for the Biscayans who were apparently the first whale fishers sent casual vessels here), and so on.

About 10 a.m., steam was up and we came out through the South Gat. This time we took the passage between Moss Island (the place so many eider ducks breed in, a small island in the middle of the Gat, just South of Dane's Island [Danskøya]) and a rocky flat island lying in the middle of the channel, as this bay was specially surveyed by lieutenants Franklin and Beechey in 1818, and the soundings were marked as over five fathoms on the shoalest place.

The captain took this passage with confidence. We were, however, sounding out and the water shoaled very suddenly to three fathoms. Then, after a cast or two of "by the mark two" "mark above water twain" and though the engines are reserved, we run quietly onto a rock at 10.45 a.m. The rock is chiefly visible amidships where one can plainly see it in 1½ fathoms. It unluckily has a sort of bight or loop so that after making way astern we again got fast—the result of too much hurry before finding our position. That, and the extent of the rock by sounding from boats, as though the water is very clear there is too much ripple on the surface to see far, so boats are lowered and grapnels got out and all are soon very busy hauling. I soon go off in the dinghy to fix the position of the rock and to measure and check that of the low islet I have mentioned. This I find to differ materially from the chart as most rocks, islands, and coastlines do here. Whilst rowing off I see a fine pair of king ducks [king eider]. The ship soon gets off and retired up the gat again to anchor. A boat comes off to me to sound for me and assess my sketch of the

Figs. 9.5 and 9.5.1 Chermside's view of Waggonwaybreen in Magdalenefjorden, alongside a modern view of the same area taken by Tyrone Martinsson in 2016

sunken rock. After taking a few soundings we pull for the ship, which is again steaming out of the gat, having found "no bottom" in the spot marked as good anchorage in this "puckle" [mischievous] new Admiralty Chart. We now sailed N, passing Kobbe's Bay, the appointed rendezvous for the *Sampson* on the 1st July, Danes Gat (the sound between Danes and Amsterdam Island), the outer islets and about 5 p.m. sounded Hakluyts Headland [Hakluythovden], the N.W. point of Spitzbergen.

We then got up steam and came into Fair Haven [Fairhaven], as the sheltered water between Vogel Sang [Fuglesangen], the Norway

Fig. 9.6 A second view by Chermside in 1873 of the so-called "Hanging Glacier" in Magdalenefjorden

Islands, and the mainland is called. To our right lay the high dark hills which surround Foul Bay [Fuglefjorden, the name in Norwegian actually means "Bird Bay"], a bay which well merits its name, as the water on that side of the haven, as well as in the bay itself abounds in sunken rocks. The mouth of the bay is split by a vast ridge-backed island, or rather rock. To our left rose the dark overhanging cliffs of Vogel Sang, an island with a very marked profile of its own, with a charmingly wild effect, rightly named too, for it is not a rugged uncouth looking rock slumbering peacefully when the wind is low in these Arctic solitudes, but a fitting summer home for myriads of birds. The animation of the scene is of course due to these, above all to the rotches, who come and go with their quick graceful flight, in such a hurry, and busy screaming and chattering, in numbers too perfectly bewildering. The water is, of course, too covered and the well-known Irish gossoon [serving boy, or lad] who answers your query for snipe or cock, that at such and such a locale they are so thick you can't see the ground for them, might almost make the statement with truth of the water immediately below the cliffs. We steam quietly and are now rounding the southern point of Cloven Cliff [Klovningen] island and turning next E, into the sound between the Outer and Inner Norway Islands. Down goes the anchor. We all land (that is ourselves), Potter and I leaving in the dinghy rather after the others. Eaton, equipped with an Alpens Pack, a straw knapsack like a carpenters tool bag, numerous vials, corks, instruments &c, an *enormous* pair of boots, a glass bottle with a hole at the bottom of it to carry his shot and a rusty single barrel, and followed by a little "Jack," who is apparently an aspiring entomologist, takes the hillside and gets to work as usual, like a sandpiper turning up every likely looking stone for worms and insects, and I expect like them congratulating himself in tones as expressive, ere he consigns each treasure trove to a fitting receptacle. BLS and the Captain, followed by the boat's crew, make for the top of the mountain to get a good lookout over the ice to the N. They had just discovered a cairn of stones which contained nought, and Potter and I, who were making a shortcut right up the hillside, were within a short distance when we were all startled by a tremendous row on board the ship, a yelling, then shouting, then the bell going like mad, a large flag at the mizzen and then the shrieking of steam whistle. Our party of two at once seat themselves on the snow, an expeditious

Figs. 9.7 and 9.7.1 North side of the entrance of Magdalenefjorden with Knattodden as the outermost point to the North Atlantic and mountain Ystekollen (628 m) above it. Mountain Skarpegga (707) looming over the lower ridge more to the right. Recent image from 2017–08-27 by Andreas Umbreit

and pleasant way of descending the common Alp of Spitzbergen, when there are no rocks ahead. Steering, of course, is done and the pace moderated by the heel of one's boots. I can't say I got a good start, but very soon got a great pace on and seeing rocks dead ahead was obliged to "take Potter's water," crossing his bows in a cloud of snow. We were under the impression that it was a bear or whale or deer or something, though the noise seemed a rash method of recall, but saw from BLS's and the skipper's faces that something was up. The "Union" was at the mizzen, flying *downwards*, and they had heard a shot. Eaton, who as amateur sawbones is of course necessary, had not joined the assembly on the beach, but calmly and uninterruptedly continued his labours. He was soon, however, brought up, and the boats rowed off, P and I remaining as the dinghy had been fetched, and we could not be of use. After a long and rather unpleasant interval passed in smoking and contemplation, during which of course (as we didn't like to shoot about) the eiders and such behaved as if the millennium had arrived *ça va sans dire* [it goes without saying], a boat arrived to relieve our suspense. Nothing serious, the third mate had pointed a rifle at the dog and accidentally shot the steward through the arm and the dog through the jaw. The owner doesn't need us on board, so we think it better to let them get settled first and continue our wanderings. P goes off in the boat and crosses to the other island, and I get to work climbing the face of the hill to the cliffs. These are swarming with rotches, dovekies, and loons, and as I should like their eggs, I have a try. Presently I spy a "vulp" [Arctic fox] and having no compunction try to get him. Vain hope, he goes like a cat and gets me in rather a hopeless mess, as I get pounded for a time, and he gives me the slip, while I had put down my gun to try and rout him out of the rocks he had retired into. Excelsior, toiling across the snow slopes from rock to rock, driving in one's feet sideways to get holding, and digging foothold in the frozen snow with the butt end of my gun which was of course empty, arrived at the "rookery."

My presence excites a good deal of "Billingsgate" [a fish market in London infamous for the foul language of its stalls] and in revanche I turn over on my back on the snow, and keeping my feet in the holes I have dug, wipe over a rocketing loon (with a wire cartridge) who bangs down into the rocks far below. I wait 'til the birds return and begin my egg hunt, but the eggs if any are yet laid, and which I doubt are carefully stowed away in deep holes, and none that I could fathom contained any. Nor am I the only egg hunter at work on the cliff: from time to time I am aware of a young cloud of scared loon-kind (to express all three) making off, and a stately Burgo-master arrives, and after promenading the ledge on which his majesty has alighted, in a most pompous and indifferent way, as if he were never hungry, and as innocent of ovisuction as a snow bunting, but just come to inquire after his dear protégé's the loon-kind and what prospects they have of families, he looks carefully into all the nests he can get at, just to make sure that no luckless bird has been rash enough to lay an egg since he was there last (probably half an hour at most) and then sails contentedly away on his rounds, happy in the certain anticipation of a future meal. It was an unlucky business, that is firing that shot, because it made me look down and I now find I am up a "pretty considerable rock" or say (fossil) tree, as there are such in Spitzbergen. I am at the end of my tether upwards as the cliffs above if anything overhangs. Besides, looking round makes one lose nerve. Of course, I can by picking my own footsteps get down with infinite labour, but I want to mount this Alp. I must try sideways. This in a short time makes matters worse. I get onto a snow slope, which becomes too hard for me to make any impression with my feet. I am side-ways and can't well turn, hampered with a gun, cartridge bag, field glasses &c, and below a mass of bare and projecting rocks. Like an infant undergoing instruction in governing its naughty temper, I turn my face to the wall and count 20 or a mental equivalent, and then get to work hacking with my gun, step by step 'til I get to a rock and so on, sometimes having to retrace my way, and at others to dig holes for my hands as well, and at times I am almost tempted to abandon my gun. Then the jump one's heart gives as one's foot slips a little is not reassuring. However, all things have an end and mine was a satisfactory one, as I got off, not without some dents in my gun barrels, as I had a slip one time and had to leave go of it, to help stop myself. Ascending the Alp by an easier route I spy BLS and the skipper descending. I soon join them and take them off to see some boats that Potter and I had previously discovered. We had also found many footprints in the snow of all of us, and the total recent population appeared to be two men, a boy, and a dog, probably the survivors of the crew of a ship that had been caught last winter. They had often ascended the hill, the prints of feet being of very different dates. The boats were three, fine, large seahorse boats, in good repair, but some showing the hatches of iron plate, where the brutes had driven their tusks through the planks. They had apparently burnt their oars, and some of their lances, the iron of these, and broken shafts and one complete lance, which they had failed to cut up, remained. One boat served as a tent, and we found remains of boiled coffee beans, barley &c. They had then dropped a boat right across the floe, as one could see by the marks, probably in hopes of falling in with the sea horsemen here as soon as the open water appeared. There were two snow graves, at least what we took to be such, but as the corpses would be quite fresh and they may return to bury them, we did not disturb

Figs. 9.8, 9.8.1, and 9.8.2 These two images show the area below Grønfjordfjellet and the entrance of Grøndalen (Green Valley) on the western shore of Grønfjord (Green Harbour), the westernmost side branch of Isfjord. Interestingly, this mountain is called Heerfjellet in the expedition notes, while today, Heerfjellet is a mountain further south. As visible in the recent image (taken from a bit further North in Kokerihamna, compared to Chermside's location), the originally Dutch, now Russian, settlement of Barentsburg covers the shore under Grønfjordfjellet nowadays, while Chermside's two images show the area still in its pristine state, with fresh snow in the late summer season. His two images can be stitched together, though this leads to a doubling of the *Diana*, indicating that Chermside did not take the two photos straight in a row from the same position. Recent image with sailing vessel *Antigua* from 2017-06-03, Andreas Umbreit

them. There were a great many old graves also on the island, chiefly consisting of heaps of stones. At King's Bay, there were 70 together, many of them having open coffins containing skulls &c exposed. There were also at the latter place four, fine, large boats in which one might travel in safety to Norway, round-sterned and heavy to enable the white whale fishers to pay out their nets. It is the custom of these fishermen who come to Spitzbergen in "smackies" to leave their boats about in this promiscuous way. We rowed back to the ship getting some shots by the way, and the bag tonight was in all one eider, ten or 12 "Greenland Parrots," as they call the Arctic puffin, two sandpipers, a snow bunting, some rotches, dovekies. Wallace [the black retriever] had to be executed; the steward is doing well. Turned in at 4.30 a.m.

June 11, Wednesday.

I was by no means in a hurry to turn out, and the others seemed to share my laziness. I was a bit knocked off my stroke by my climb on the hill. You don't catch me egg hunting by myself again in a hurry. A boat was going to Kobbe Bay to leave word for the *Sampson* and BLS went with it, and the skipper. They had a fair wind and stepped the mast, leaving as soon as the men had had their dinners (12) and taking provisions for about 24 h. I set to work at photography to try and find out what was the matter with the chemicals, and by heating them &c, at last got them to work, though of course the photos which were experimental were useless as it was now blowing very hard, and the ship swung a great deal. The ice between the Norway Islands was breaking and came down on us steadily after 2 p. m. It was a glorious evening, though windy and we took a photo of Cloven Cliff, which came to grief in drying, through our slipping our hands and spoiling the film. It certainly is photography under difficulties on board ship, as of course the ship is never or hardly ever perfectly still. Then again there is this difficulty of chemicals, and that of getting pure water. Also, a small jug and basin are not the most efficient substitute for a tap and proper sink. Added to this difficulty is a place one can barely stand up in, very small and very dark, so much so, in fact, as to render the pouring on of solutions harder than ever. We mean, if possible, to remedy this by a candle, and I hope if we stick to it, we shall get some good results. This Cloven Cliff is a peculiar shaped rock rising abruptly from the sea to about 300′ [it is in fact 295 m in height]. There is a dip or collar between it and the main ridge some way lower. The cliff itself is a regular landmark as it has a great cleft in the middle, which can be distinguished a long way off. The sound between it and the outer Norway [island] was free from ice, but it is not an over safe channel. The ice which closes the gat between the two Norway isles is still blocking the way through, but is breaking fast tonight. After tea, the second mate and I started in the Shetland "casey" for the bay between the inner island and the mainland in which the Swedish expedition to the N were quartered for some time last August and September [here Chermside references the overwintering expedition of A. E. Nordenskiöld, which he and the Leigh Smith expedition were soon to meet with]. Soon after starting, the gale increased and we had some difficulty in making way even under the lee of the land. When we rounded the extreme point of the inner island, we had to take across the sound between it and the mainland, in spite of it being blocked right to windward of us with fast ice. A heavy sea was running, on which we jumped tremendously. This made me afraid that the boat in Kobbe's Bay would have to bivouac there, as part of their way at least homeward would be on the open sea; that is as far as Dane's Gat, that would be no joke even under the lee of the coast. Having got across the sound, the thing was now to find this bay, no such easy job when lowland and ice are undistinguishable apart, rocky islet, and headland. I, however, had drawn a sketch of Fair Haven and keeping up along the sound from Foul Bay at last made it out. As we neared the fast ice, we saw swarms of birds, an immense number being eiders. They were very shy, but dropping my paddles, I made a right and left at an eider and a dovekie coming downwind. But as from the pace they had on, they fell far to leeward of the point we had just weathered, with such difficulty, inch by inch and sometimes appearing to almost to go astern, we gave them up. It now came on thick and very cold and the gale carried about endless columns of whirling snow, ground to a cold hard powder, which crept in at every crevice of one's clothing, and so continually fitted my gun barrels that I gave up trying to shoot as a bad job. The Swedes had apparently, in fact, told us so today, lived there for some weeks last summer. They had constructed a stone house and stone walls round the bottoms of their tents. These still remained but a rigorous and somewhat tedious search produced no memoranda. The *Polhem*, the expedition vessel with Professor Nordenskiöld and the scientific people on board, was here last year when the *Sampson* left on the 1st of September, as well as the government brig that was bringing her stores and provisions &c, as the former was to winter. A third vessel, a steamer chartered from Gothenburg, was also expected bringing more stores. All three left northwards on the 4th of September and have been frozen in and obliged to winter. The bay looked cheerless enough now, closed as it was by an immense glacier and shut in by frowning hills still almost completely snow-covered. We hoped to find deer here, as the Swedes had brought 40 with them, which with one exception escaped here. This must have been rather riling, as they brought four Lapps with them for the special purpose of looking after these deer. They had besides this all the sledges and harnesses and an immense stock of food for them in sacks. The deer were supposed, by some strange fiction, to come when the Lapps called them. The latter, although enjoined to watch them, all turned in and away went the deer. All attempts at recapture failed and I suppose they have long ere travelled to Wüde Bay [Wijdefjorden] or to whatever other part of the land, frequented by deer, their instinct may lead them to. Back across the sound again, the wind of course though not so strong, having backed sufficiently, head to us slightly. After a tough pull we land on inner Norway Isle and seek for cairns &c, finding, however, only eight or 10 empty water casks. I shoot two "Tam o'Shanters," or whatever they are, for the mate who wants specimens, also one of my dear "dovekies," who came by in "such a desperate hurry." When near the ship, we see that the dinghy is away in pursuit of a fox, which they are trying to cut off from Norway Island. He has some breed in him and no mistake, "Gone away—" as he takes over the ice any pace you like, jumping the cracks &c beautifully and does them (i.e., Potter & Co.) to some tune. Looking up this sound, we saw some icebergs far away to the sea, moving a great deal in the strong sea that must be running there. One of them, after jumping about rather strangely, turned over three times in succession before our eyes. I had a splendid view through the glasses and the clouds of foam &c which were a grand sight. The boat is not back of course. The ship is a good deal surrounded by drift ice, but the strong tide soon takes it by her. It is, however, blowing so we have out another anchor.

I turned in about 3 a.m. and was much relieved by the arrival of the boat about 5 a.m. They had left between 12 and 1 p.m., followed by the conventional three cheers, running before what was then a nice breeze. Going down Smeerenburg, they viewed two seahorses on a piece of ice to the westward of them and sailed down on them but did not get near enough for a shot. The brutes, as happened with me, suddenly hurled themselves back with a crash in the water. Sailing on, they found Smeerenburg blocked across with ice, and had to drag their boat across the low S.E. promontory of Amsterdam Island into Danes Gat. Down this they reached and got on a good pace as they carried a tide under them, a well-deserved respite of toil after pulling a whale boat, with only six hands (all told) over a mile of land. The first part was uphill and very hard work but getting up to a cairn of stones, they hoisted their sail and lashing oars across so that a man at each end can keep her on her keel, got her along famously, rounding the northern point of the island. Rowing had to be recommenced as the land which they coasted runs far more easterly than shows in the chart, tending E into Kobbe's Bay. Arrived here, they landed on the N shore near the Russian Gat [possibly Linnévatnet], the remains of which stand on a point in this bay. Getting under the lee of the overhanging shore ice they rigged a tent by stretching their sail between two hummocks, and BLS proceeded to indite [write] his dispatches to the master of the *Sampson*, this being his preconcerted Post Office. Three dispatches were prepared and put into bottles, one bottle left in a conspicuous place directing them where to look, i.e., under a cairn of stones 12 yds to the N, itself fairly conspicuous in case the other bottle be removed. This tent was also the scene of their lunch. What a glorious picture such picnics would make. They then rowed to the island and landing, deposited the other bottle. It was now blowing very hard but, taking out the spirit of sail and running up the head, they kept the beach in their hands aft, making a jib-headed fore and aft sail, the surface of which they could easily regulate. At the N point of Dane's Island, the wind came off the cape that they had to give it up, as it blew them some way out to sea, and had a most awful pull up the Gat, hugging the land and keeping from islet to islet. They took two Burgo-master's eggs and one or two eider's. Arrived at Smeerenburg, they tried pulling the boat across the ice instead of the land but found it a terrible drag. At last they got across and while working up at Smeerenburg came in sight of the ship about two hours before they reached her. However, by dint of pluck and four glasses of rum per man, the return journey was safely accomplished and the boat was at the davits by 5 a.m.

June 12, Thursday.

After winding chronometer and reading barometer, &c, I again lay down in the cabin, the sofa of which forms my bunk for a few days. Now the steward is "out of the hunt" or *hors de combat* [out of action owing to an injury], as they say in Frank land, and none of us got up 'til late having our first meal at 11 a.m. We get the anchor up before noon, and steam up, and go right at the ice which closes the channel between the Norway islands. Slowly and surely, we push through the large broken float masses, breaking them ourselves where the late gale has not done it for us, and after half a mile to a mile of this work, sounding as we go, we get into open water. There is a "smackie" ahead. From where he could have come is uncertain, but if he is bound for Moffen Island [Moffen] for seahorse, he is mistaken if he thinks he can get there first. So, we steam up to him, hoist our ensign, and invite him on board. Charlie the Norseman does interpreter. He proves to be from eastward along the Northern coast of Spitzbergen and was yesterday at Mossel Bay [Mosselbukta] and there lie the Swedes beset, the expedition and the two accompanying ships. This news indeed is a great triumph for the judgment of BLS as regards ice, currents, winds, &c, as in his letter to the *Times* in the winter, and ever since, in spite of many contrary statements, he has predicted that they would be lying there. This he takes in his usual quiet way. But, nevertheless, a steam is got up and we bowl along full speed. The Norwegian captain has been rowed off to his smackie in the dinghy, in company with a bottle of rum. We, of course, drank with him and while we were drinking sherry, I carefully filled him one of those tumblers of wineglasses holding at least ¼ pint with that rum. He stuck to it bravely, though it made his eyes water. The Swedes, he informed us, were, as we expected very hard up for provisions and he had given them some reindeer he had killed at Red Beach. So, we steam on knowing we shall be welcomed, as BLS brings them a large present of provisions. About 6 p.m., they are sighted from the masthead, and sometime later, up go all our flags, giving her ladyship a very gay appearance. We are now off Red Beach: some way off, as the water is shoal on the northern coast. We sight Welcome Point [Velkomstpynten] and Grey Hook [Gråhuken] also. Red Beach is a long strip of flat foreshore running E from Red Bay to a point and takes its name from the colour of the stones of which it is composed. It is a glorious evening and the golden sunlight on the snow slopes softens the hard, black cruel water of which these north-western mountains are chiefly composed. They form a magnificent background to Red Beach, the low strip between Biscayers Hook and Welcome Point as we come abreast of Welcome Point, the western extremity of Liefden Bay. We are about midway between it and Moffen Island, a low island about six feet above the water, a great place for seahorses, ducks, and other sea-fowl's eggs. We steam on until we come off Grey Hook, a high promontory with mountains rising abruptly from the shore. We get a fair view down Liefden Bay, of which it forms the north-eastern extremity. It is, in fact, the apex of or point of the wedge of mountainous land, separating that bay from Wüde Bay. Looking across this bay, you see at once how the character of the land is changing to the western side. There, tall, sharp, mountains still line the coast, and prevent one's seeing anything of the interior of the country, except the peaks of taller and more distant hills. But to the east you have "weald" and "feld" of snowy ground, some of it indeed highland, but softer and smoother, and with the hills of rounder and less angular profile. This Wüde Bay, never much over ten miles in width, runs nearly due N and S and is over 40 miles long, separating near its southern extremity into two fiords. It is a great place for reindeer. As we approach the fast floe that encloses almost the whole of Mossel Bay—that is, that part of it which can be called a bay for it is far wider and more exposed, than is shown on the chart—we make out the unwilling inhabitants on their way to the floe edge. Their ships are much further into the floe, close in to the land, and all fly the Swedish ensign. Another half an hour and we are now at the ice edge, penetrating with ease through the broken ice that separates us from what is fast, as although the pieces are of large surface, and only separated by cracks for the floe is only beginning to break. It is new ice and of no great depth and the *Diana* pushes them contemptuously aside with her bow. Over with the ladder forward [but] no need of that as these hardy Swedish sailors swing themselves on board. While our own people are securing us to the floe by warps of ice anchors, we entertain six of our guests with schnapps and tinned meat in the forecastle. We smoke, drink, and chat on each other's health and had a quite spirited party. One is struck at the little they eat, owing I suppose to their not being accustomed to it. About 1.30 a.m., they take their departure homewards across the ice. There is still a glorious sun, but as we expect to have lots to do tomorrow, we turn in about 2 a.m.

June 13, Friday.

While at breakfast the first lieutenant of the brig arrives with a boat crew dragging a small sledge, this is to go with them in one of the *Diana's* boats that they are borrowing to Grey Hook, where they have left some provisions. Mast and sail are put into the boat, and after breakfasting with us the party start. We are busy all day photographing and have a chapter of accidents. I hope, however, we shall get some results in time. The Swedish doctor [Axel Enwall] brought out his apparatus, tent &c, or a sledge drawn by two Lapps, and took several views of the ship, apparently much on par with ours. He lunched with us and soon after, BLS and Eaton, who had gone to visit the Swedes and inspect their ménage, returned with some of them, to lunch. It seems that they have been making a very complete set of observations of all sorts, and Eaton was much pleased with their collections of seaweeds &c. Professor Nordenskiöld, of whom more hereafter, is away N, with the expedition. This is one of these boat sledge businesses, and in spite of all their hardships (½ rations all the winter &c), a party of 16, if not more of them, made their way to Parry's Island [Parryøya]. Professor N took with them 40 days of provisions. They left for the N on May 3rd (or 23rd) and the remainder returned. They are expected back to take away the provisions BLS is giving them. It was a very animated scene, watching the loading of the sledge these men dressed in semi-Arctic costumes, with curious canvas boots with cork soles, our crew lowering away the stores from the decks, and the Lapps in their curious hats and long blue frocks (apparently adorned with pagan devices and symbols) and once bright leggings, smoking their pipes and looking on most unconcernedly.

After dinner, Potter and I started off to visit the settlement, carrying with us my small camera and some "dry" plates. It, however, came on thick, and then *rained* hard. It was heavy walking over the rotten surface of the floe; after three to three-and-a-half miles, we reached our destination. The country at the head of the bay was still almost completely snow-clad, but the side of the house in which the "expedition" proper lived was on some low rocky ground, covered, where there were no rocks with a layer of large loose stone, to a depth of some feet. A more desolate spot it is difficult to imagine: to the left, the same low rocky barren land stretching away to Verlegen Hook [Verlegenhuken], behind a snow- and ice-covered arm of the bay, and beyond that rising ground sloping up to the bleak fells above the bay; this is of course as you approach from the sea. To the right rear of the house lay the ships, at about quarter of a mile distance, firmly imbedded in the ice, and almost in the arm or inlet I have already mentioned. A more desperate position could not have apparently been chosen, though how far they originally occupied it through necessity we have been unable to make out. The bay, which narrows opposite the side of the house, is here wider again so that the ice in it—even should it break up—will have some difficulty in getting out to sea, unlike the ice of the greater part of the floe which, I think, is four feet thick at most, and in many places about two. That by which they are beset is seven feet thick, and between them and the thinner ice lie ground ice and several large hummocks. The ships lie on a curve, concave to the sea; the chartered vessel, the *Onkel Adam,* being at the centre; the *Polhem* on the left; and the brig *Gladan* on the right. The brig is a smart-looking craft, but not the sort apparently for this work, being amongst other things copper-bottomed. The house was made, I believe, in Gothenburg and was brought with them. It is a large fine solid wooden hut, though due to stress of circumstances put together rather too hastily. It is, I think, about 50 or 60′ long, by 25′ broad and about 12′ up to the eaves. It has a strong wooden truss roof which has stood the gales wonderfully. It has a loft above formed by boarding up the spaces between the tie beams. The sides are formed with vertical halved and tongued planks, not weatherboarding. Then there is a little wooden observatory, painted red, a small octagonal erection among the rocks 150 yds seaward of the house. In this are their astronomical instruments &c. They have a small, roofed house for their thermometers, with double walls, the walls and floor being louvred, and another wooden observatory for their magnetic instruments. A tide pole, fixed in a hole in the ice in the bay, observable through a telescope from the house, completes their outdoor arrangements. Stop, I have left out immense piles of sacks, built up pretty regularly and stretches like sandbags. These contain the immense stores of moss brought for the truant reindeer, and covered with sails and tarpaulins make very good general stores for provisions &c. There are also two or three wooden sties, for the "sveen," or domestic grunter, who are now no more. This represents the ménage of the expedition proper, the whilom [erstwhile] passengers of the S. S. *Polhem*, who lies deserted in the ice. The others live in their ships in the bay but are often up at the house. Though very desolate now, it doesn't seem so bad to live in. But one shudders at the thought of the long dark winter, when it was so dark you couldn't see the sight on the gun. The fells behind and to the left rear of the house seem to have been very unproductive, and they are cut off from those on the right by a long high chain of hills rising like a wall abruptly above the gully which separates them from the fells, and running eastward from the sea edge, completely shutting in that side of the valley, and though fair enough now with their white gilt summit no doubt, unless they knew passes an awkward enough obstacle in winter. We sat for some time inside, chatting, but the atmosphere was so hot for our habits (the *Diana's* cabin ranges from 35 to 40°) that P and I had great *sommeil* [sleep]. The Captain, however, who had joined us, stood it well enough, talking to them in "gude braid Scotch" and capped or corroborated their statements, of the ups and downs, hardships and pleasures of Arctic life, by yarns about "Davie's Strayait" especially in "October Month" and I was convulsed by hearing them talking of, May Month, March Month &c. After looking at their photos &c and smoking for a long time, during which we learnt that they had only got two bears during the winter, and one or two deer, but had killed a "snowy owl," a bird seen but not killed before in Spitzbergen, we went on board the brig. Between 11 and 12 p.m., the party from Grey Hook returned. They had been successful in finding the provisions and had brought back with them as far as the *Diana*, one of the fine boats left by the seahorsemen there. They brought on their sledge from the *Diana,* besides the provisions &c, the skin of a fine old bear they had shot when last at Grey Hook. Between 12 p.m. and 1 a.m., we started home. It was blowing a gale of wind. We, however, made very good way, following the sledge tracks until we caught sight of the *Diana*. Photography had been out of the question. Turned in about 2.30 a.m., and had a remarkably testy night, as the continual shiver of the vessel as the ice had struck her stopped my sleep effectually. None of the others much minded it. There was a good row, too, besides the howling of the wind.

Account of the Swedish Expedition 1872–1873, in search of the N. Pole and for observations, on the climate, tides, magnetic currents &c of Spitzbergen. (This does not profess to be accurate but is an attempt at connecting the information obtained during conversation and statements most conflicting often occurring into a continuous story. The Swedes have of late years been extremely zealous in the exploration of Spitzbergen and the Polar Seas around it, and there have been recently by expeditions, i.e., in 1861 (this expedition extending over the summers of 1861 to 1864), 1868, and 1872–1873. It is principally from the first of these expeditions that the recent and best chart of Spitzbergen has been prepared, and they have done more than any others towards a complete arrangement and catalogue of the fauna and flora of that country, as well as determining the geological and physical features of the land and sea, having done a great deal of dredging in many of the western and northern ways, besides tidal and other observations. The expedition of 1868 succeeded in sailing or rather steaming as far N as 81° 42′. That of 1861–1864 was under the direction of Dunér and Nordenskiöld, or at any rate they were members of it, and the chart is due to their observations. Professor Nordenskiöld is at the head of the present expedition, which has also received the support and assistance of the government. The original intention of the expedition was to leave Sweden, in the summer of 1872, and proceed gradually N, dredging and collecting, in their way N in some of the bays of the western and northern coasts, but not I fancy doing any surveying. They were then to make their way northwards and establish themselves for the winter at the "Seven Islands" [Sjuøyane] one of which was to be their base for further operations, I believe Parry's Island, but this programme was of course liable to be altered by circumstances and they might have to winter in one of the harbours of the N.E. land. Their vessel was the *Polhem*, a steamer of about 200 tons, and as they were to winter, the government sailing brig *Gladan* was detailed to accompany them, being loaded amongst other things with their house &c. A third vessel, a steamer out of Gothenburg, was also chartered, but by the month, [and] only to help transport their necessary stores provisions &c and was to join them in Fair Haven at the end of August 1872. All three vessels were then to sail N in company and on arrival at their most northern destination, i.e., Parry's Island, their stores, provisions &c were to be disembarked, their house erected, and ere their winter establishment was completed; the other two vessels would be homeward bound, the *Polhem* alone remaining, with the Professor and his lads. But, as often arrives, "*L'homme propose, Dieu dispose*" [Man proposes, God disposes]. They executed the Ice Fiord part of the programme greatly to their own satisfaction, they then took up to Fair Haven, and building a stone house, and erecting tents with the walls of stone round their bases, as I have already described, in the cove eastward of Foul Bay, patiently waited until the *Onkel Adam*, the Gothenburger, should arrive. The ice which this season was unusually unfavourable, being packed tight down on the N coast, should allow for their pushing onto their destination. They explored Foul Bay pretty thoroughly and took a series of photos of it, but I don't think did any sounding, little use there certainly, but highly desirable, in other parts of Fair Haven. They busied themselves dredging &c, and it was here that they performed the brilliant manoeuvre of losing all their reindeer. It would appear from their chart that the unsatisfactory state of the ice caused them to give up several attempts to proceed that they made, as well as the non-arrival of the *Onkel Adam*. On these occasions, they returned to their old quarters in the cove under the glacier I have before described, and this Fair Haven being as a rule a rendezvous for the ships bound N among the ice, who when stopped congregate here, and like us ascend the hill on the outer Norway Island daily for look out.

In 1872, when the *Sampson* arrived here, on her way N, there were, I am assured by the owner, as many as eight or ten smacks, and

Norwegian seahorsemen waiting here, and several both on her outward and homeward bound voyage in 1872. On the latter occasion, they had been waiting to penetrate to Wüde Bay for reindeer but had apparently given it up and were homeward bound. It was here that the *Sampson* fell in with the Swedes last year and promised to return and look them up, a promise opportunely kept by the owner on board the *Diana*. The *Sampson* left on 1st of September, and shortly after the Gothenburger put in an appearance, all three vessels left. On 6th of September they arrived in Mossel Bay, and on the 7th the ice had come down and they were beset, as they remain now. It was, I have remarked, an exposed and unfavourable position, and certainly it is a pity that they could not get to Treurenberg Bay [Sorgfjorden] where their expedition of 1868 remained so long and from a cover in which (Hecla Cove) [Heclahamna] Parry's memorable boat expedition started in 1827. Here, however, they prepared to winter and got up their quarters, observatories with all haste, as darkness was soon upon them. The house brought out for the expedition was made in Gothenburg and the wooden observatories in Stockholm, and appeared to be at once simple and efficient, with the exception of the roofs of the smaller buildings. Their total strength was 60. There were three ships, but the *Polhem* wasn't occupied as quarters. I believe these lay a little eastward of the house, which stood on a promontory, the *Onkel Adam* in the centre, the crew of that vessel, and that of the brig remaining on board their ships. The western or red observatory contained a small passage or transit instruments, sextants, horizons &c, they also had apparatus for observing atmospheric electricity, a Thomson's electrometer &c. To the E was another wooden observatory, containing three magnetic instruments—magnetometers, I suppose—to measure electricity, the horizontal and vertical forest of intensity of the earth's magnetism. I have never before seen these instruments, but the principle seemed simple enough: the readings were reflected from a mirror and read by the aid of a lamp and a telescope off a graduated card. Hourly observations were taken to correspond with those taken at Stockholm and on the first and fifteenth of each month, observations were registered every 5 min of the 24 h. A tide pole was fixed above a hole in the ice in the bay and the observations taken at the same hours and with the same frequency on the first and fifteenth of each month. During the winter a good many astronomical observations were taken and very good results got with a spectroscope from the Aurora Borealis. These lights were invariably to the S of the observers. A long series of observations on refraction were taken, with what result is of course not yet known. In a small erection a few feet beyond stood the thermometers, protected by a roof, and in sort of cages like a meat safe, consisting of an outer and inner case, each completely louvred, floor, walls &c. The average rise and fall was three feet, with two-foot neap tides and four-foot spring. But the tides were extremely varying being, to a great degree, dependent on the ice and winds. The readings were taken from the house by means of a telescope and are expected to be of great use in determining the lunar influences. Darkness came on very suddenly and by November they were in total darkness, except the light due to Aurora Borealis, and the moon. But they often had long spells of cloudy weather. They experienced a tremendous succession of heavy gales and had, except in the bay, open water during the whole of November and on many days in December. By 21st January, all ice had cleared out of the bay and they were only deterred by the darkness from setting out homeward. The "Gothenburg" vessel wished to start home, but for reasons that did not transpire was not remitted to do so. I understand that, but for the darkness and the heavy sea running on these occasions, they had several opportunities of getting out. It was not 'til towards the end of January that the ice began to form, and the cold to become excessive. Up to this time, the weather had been singularly open and the cold anything but intense. In fact, their sufferings cannot have been so far very great. In February and March, which were the coldest months, the ice formed rapidly and is still (June) seven thick around the beset vessels. The greatest cold experienced was −35 °F (−37 °C). The gales were not so heavy now as at the end of the year, when on one occasion the anemometer at the roof of the house registered 4500 revolutions. On 1st of February, the light began to show at midday. On 28th of February, the sun should have been visible, but was not so, I believe, owing to the state of the weather until 3rd of March, though very cold things began to mend from then and their chief suffering was want of provisions, as during the whole winter they had been on half rations. They had had a good many cases of scurvy, but had only lost one man through that, and one had got lost on the weald. They kept the men in plenty of work all through the winter, maintaining discipline and giving them as much outdoor exercise as possible, though they seem to have been very poor sportsman, as 'til within a few days before our arrival they had got no birds, which for some time past, if not along Mossel Beach itself, for it is hardly a bay, have been swarming in all the neighbouring bays and inlets (with the exception by the way of a snowy owl, until now but a doubtful member of the Spitzbergen fauna). Of beasts they had, I believe, not killed a single reindeer. The skins of those that they had, I suspect, being obtained from the Norwegian skipper (he who informed us of their whereabouts). But they had killed two bears, one close to the house, and the other on an expedition to Grey Hook. I certainly can't imagine Englishmen wintering there, and although half starved for want of food and suffering from scurvy for want of fresh meat, not trying after deer, when within ten or 15 miles of one of the best localities for them in the whole of Spitzbergen. They saw seals and I believe seahorse but, I fancy, didn't try to catch them, and simply scoffed at our suggesting that if they had caught them, they would at any rate have had fresh meat. From one or two hints let drop, there seem to have been some hitches at least during the winter, and we expect that the Gothenburg skipper [Captain Clase], who seems to have a grievance, will have something to say on his return. Four Norwegian seahorse vessels were beset at Grey Hook and two at Welcome Point. They seem all to have assembled at Mossel Bay, but not liking their fare (bread with flour apparently supplemented with reindeer moss, without, stupidly, the bitter ingredient taken out) took themselves off. The crews then separated: 18 men and a boy taking down Wüde Bay in a boat, to strike across the mountains for Ice Fiord. These have not since been heard of. Of the rest, the vessels at Welcome Point got free and took off all but two: Mattilas, a well-known old seahorsing skipper, and a man who stopped with him to guard the four vessels. These men died of scurvy and are still lying at Grey Hook unburied, a horrible sight. Their journal, which I saw, ends on 5th of February and they probably died a few days after that, and their boats remained on the beach. The vessels were subsequently utterly smashed up. The Swedes, wishing to correspond with home, attempted to send to Norway Islands either to leave a letter for the government or to get some smack to take down intelligence of their need of relief. The boat, however, was too heavy laden and 20 days provisions had to be deposited at Grey Hook. The boat then went on, but owing to her not being strong enough for the ice—or on account of the appearance of open water (we could not well make out which), which would necessitate the immediate presence of the officer in charge of the crew at the vessel to give it up—the provisions remained there safely until brought off by the boat lent them by the *Diana*. This boat, as I have said, brought off also one of the sea horseman's boats, as of late they have been short of seaworthy boats owing to those that have been taken away by the expedition. This expedition was, we have seen, to start N after the winter, but had now instead of starting from Parry's Island, first to make the Seven Islands, and having made a depot there continued towards northwards. The expedition, like Parry's, was to be a sledge boat affair, the boats being very light, one as little as (250 lbs?) and the others rather heavier. They were wooden frames with a canvas covering and whatever their weight might be easily lifted by two people (at least the lightest). They were named the *Sofia*, *Louise*, and *Anna* (I have only seen photos of them). During all April they were busy making preparations and started at the end of that month of or beginning of May. They took four boats to Hinlopen Straits [Hinlopenstretet] leaving two there the party of 16 or 18 made their way to Parry's Island, crossing from N Cape [Norkapp] on the ice. The Professor then started N with 40 days provisions, sending back a return party with letters and news to Mossel Bay. He is now away on the Ice and not due back at Parry Island 'til 23 June and probably will not be back 'til later. The boats still remain at Hinlopen Straits, so they were very glad of the one they got from Grey Hook. It seemed a first-rate boat. The owner [Leigh Smith] has now accomplished one of his great objects in hurrying up here, namely relieving the Swedes. They ought now to do very well until the end of July, before which time should they still be fast. They ought to be relieved as their new king would have been crowned. His next object is to relieve Nordenskiöld, who in all probably will stand still more in need of relief, and he hopes the *Sampson* will arrive at the Seven Islands in time to take them on board if necessary, and to carry dispatches for them down to Hammerfest. In fact, so far as we have seen, in spite of difficulties and dangers, the expedition so far has been brave, and as far as it lay in their power, carried out the objects for which they were sent out. The astronomer and botanist seem to have every cause for congratulation on the successful result of their labours. Every expedition that has been sent into the Arctic regions has, I believe, made grave, serious, and often fatal mistakes; it speaks very well for the pluck and discipline of those men with such a large strength—the greater number of whom, too, were unprepared to winter there—they should have only lost one man through illness during the whole winter, and that too when suffering from such privation in the way of provisions. They are an unusually fine hardy looking set of men, especially the officers, who are very heavy and large limbed, and almost all stand over six feet. The finest men of the crew we hope yet to see, as they are away northward on the ice with the professor.

June 14, Saturday.

The last sledge carrying the "Rum," having made its trip last night, had now nothing to detain us. There was far too much swell on among the ice for photography to be of any use. It was, however, a fine sunshiny morning, tho' still blowing hard. BLS and I started over the ice to say goodbye to the Swedes and to get "Greenwich Time" with the pocket chronometer took with us the small camera, in hopes of getting a successful view of their winter quarters. We got back on board about 2.30 p.m., the last part of the road being from piece to piece broken ice, as the wind had smashed up the floe a great deal. We got up steam and got through the ice. The wind unluckily soon dropped and we had to get steam up and lay a course for the Seven Islands. Passing Verlegen Hook, we could see into Treurenberg Bay, a cove of which is so famous as Hecla Cove, and then again looked right down Hinlopen Straits above which was a very "watery" sky. We now keep along the coast of Nord Ost land [Nordaustlandet], only however making out Low Island [Lågøya] from the deck by the grounded ice, all around it, as it lies extremely low. We are now, as BLS remarks, on classic ground, and the names of all the striking features of the land are those of former Arctic explorers. The water as it has been all day is very open, but latish on the edge of the fast pack is much out, about the Seven Islands. About 9 p.m., we come in sight of Walden Island [Waldenøya]. We can also plainly see the North Cape. Gradually we make out Parry's and Phipp's [Phippsøya] islands and before I turn in, all except Martin's [Martinsøya] and Nelson's [Nelsonøya]—that is five out of the seven, the other two being hidden. We can also see the pack edge running away W from Parry's Island. During the day, we spoke a Norse "schooner" (seahorses): he hadn't much to tell, had got 23 seahorses at Moffen Island. I stayed up late watching the islands with a strange interest: no wonder, on finding oneself within sight of Ross's Island [Rossøya], a "wee bit rockie" certainly, but the most northern known land. Turned in about 3 a.m.

June 15, Sunday.

At 8 a.m., we are pretty close on Walden island. We leave it on our right and coming in through the drift ice and a stream or two that be to the westward come up to the edge of the fast ice between it and Parry's Island about 10 a.m. and make fast to the floe with ice anchors. Our original intention had been to get, if possible, into one or other of the harbours in Parry's or Phipp's Island, which give on the channel between them, as that is a sheltered enough position. All the islands, however, are still girt by fast ice. So, we moor to the floe as I have said about three miles from Walden Island, and soon after started in a boat to climb one of his peaks and get a view over the ice to the N. We sail down on him gaily enough with a fair wind. The owner, Skipper, P, and I landing on the ice gain the shore and commence our ascent. We saw no seals on the way though the ice looked very tempting. In fact, the only one we saw was one that came up near the ship about eight this morning. He was what they call an "ogdok" (an Esquimaux word I think) and was an enormous brute for a seal, with a light grey, in fact almost white hide. Unfortunately, the boat was lowered after, instead of before, I shot and so we didn't get him. My rifle seems to kill the brutes too dead to get them, knocking all the wind out of them, as ordinarily they do float for a few seconds. This poor beast, who wasn't more than 35 or 40 yds, had none of the front part of his face left at all. The floe at the base of the island proves very bad for walking as the snow is soft and deep and there are numerous crevasses in which you sink as far as the top of the thigh. The island consists of bare rock, though of course there is a good deal of snow deep too in the rifts. The rock is granite and I think syenite in places. We find a great many marks of bear, none very recent. The captain saw one at a very long distance off on the floe, early this morning, and though bones were put on the fire to attract him, the wind was in the wrong quarter. We also find the marks of human footprints, a good many together, too, ascending the hill, possibly the Swedes for a look out on their arrival, and following them, make our way to the summit of the third highest peak. From this, we take a round of angles and determine the height &c. We are all annoyed of a good deal by dovekies &c who scream and fly round us and do their best to prevent our observing: not an easy matter in such a high wind with the thermometer at 27 °F. On our way, I had attempted to shoot loons with indifferent success, getting only one or two out of several lots I fired at trying to get as many heads in line as possible, and of course, unless you kill them dead it is as easy to shoot fresh ones, for they carry away the shot. I can't make out, as I several times killed the feathers off one and as the gun was as low as possible, it cannot have gone over the others, nor do I think they minded the shot. I hate shooting birds on the water, so Potter and I never do it if we can help it. On this occasion, however, they were required for the larder. Two dovekies were especially annoying and I vowed if they came 'round again I would polish them off. So, as they wouldn't take warning the next time, they came. Snatching up my gun from the foot of the instrument, it being the only shotgun we had brought, I sent them on a voyage of discovery, far away [onto] the rocks below us, where they were subsequently picked up by two of the hands, who moreover scaled the next peak in search of birds eggs, as there were [?] loon-kind about the cliffs. The highest cliff was apparently almost inaccessible, and birds' eggs were not forthcoming. Meanwhile, our observations are over and we see that for the present at any rate our bolt is shot. The islands are still connected with the N Cape by fast ice. Along the whole coast eastward and to the northward, as well as to the N of the Seven Islands, stretch long fields of ice. There is a great point to the N.W, and away from this runs a line of ice which is the pack edge running away W and S to where we have come from. The captain has, meanwhile, made out what he states to be a bear, on the ice near the N Cape; it is, however, too far for us to be certain. While we are still lying about the top, for somehow after all it doesn't feel very cold, we are aware of a commotion at the boat below. They had been fetching in under sail, having been standing out after the loons I shot, and which we did not pick up at the time. Now, however, the sail is down. Soon a flag goes up in the stern and we hear two shots and a great shouting. Besides that, before hoisting the flag they had been making frantic signs to us. They can have no more ammunition as I had left my rifle and only two cartridges. Down the hill we go as best we can. It is a "Brunie," says the captain, and the boat seems somehow unable to get away from the ice, which makes matters look worse. We are now in the snow at the foot of the rocks and keep tumbling into holes &c, and the captain goes to the front with a rush. We can now see the boat and they shout frantically "in front the guns," "come on the guns." I fumble in my cartridge bag as I run, as I have some wire cartridges [a type of cartridge developed in the nineteenth century that enabled quicker loading and increased velocity], but of course No 8 Snipe Shot of which there are only about three, keep coming to hand. "There he is shoot," from the boat, bang goes the Captain's rifle and as he goes down again, we have time to see that they are fast to a fine seahorse. A few bullets more and it is *whoo-ooop*. A boat puts off from the ship, and we soon have him on the ice and flensing commences, and getting up the gun harpoon which had drawn, and was foul of a tongue of ice, and was not extricated without a great deal of trouble, it was uncommonly lucky that they had him fixed as well by a "toggle" hand harpoon. It seems that this brute had come up near the ship and gone down not reappearing until close to the boat. Tom, a naval reserve man, acted as harpooneer, got his harpoon in, capped his gun &c and Mitchell sculling from the boat's stern, and represented the crew. Thus prepared, they awaited their opportunity and it is certainly not every pair of hands that would under these circumstances have dared attack a full-grown male seahorse. Up he comes near the boat and they fasten him. Pulling on the line they get fast with the toggle harpoon, and give him two shots, one behind the flipper, not much use, and the other which was in the head unluckily breaking the imbedded part of one of his tusks. I proceed to shoot loon-kind, which fly by in considerable numbers, thinking we could pick them up afterwards in the boat. However, after shooting about half a dozen, I gave it up, as there is a strong tide or current here, and they soon become a prey to Burgomasters and malleys. This, however, afforded an interesting episode in natural history. On one occasion, I fired at three loons but only killed one, the other two being only winged. These were at once attacked by "malleys," who followed them wherever they dived and attacked them open-mouthed and shrieking on their reappearances at the surface, taking care however to keep out of reach of their long powerful bill. One malley who got hold of a dead one, proceeding backwards getting his wings outspread and a good purchase with his beak, dragged him a good quarter of a mile and disappeared round a promontory. As soon as the birds were a little distance off, the affair complicated itself, as the Burgo's now showed a strong inclination to assist in the discussion of the "nice pieces." The malleys didn't see no-how, and managed to hold their own though as soon as the common danger was averted; they resumed their perpetual family feud. I wanted a snowbird but was unlucky enough to lose him. After lying apparently dead for a long time on the water, as he floated nearly in reach of the boat, he got up and flew off and "towering," fell among the rocks, a long way up the hill, whither I followed as I thought he looked a good specimen with unruffled plumage. But the Burgos &c at once surrounded his corpse getting up some time before my arrival, so I was unable to find him. We lunched in the boats and then sailed and pulled back; the walrus had very fair ivories, about a foot that is the exposed part, the upper part of one, being unluckily smashed by a bullet. We didn't get back much before dinner, and it then came on thick and snowy. We determined to leave the "kreng" unmolested 'til morning in hopes of a bear. It was unluckily too thick for photography. Towards evening it began to blow and by midnight was blowing a gale from the SE. Turned in about 2.30 a.m.

June 16, Monday.

Soon after 4.30 a.m., I was roused by the noise of the propeller, which was going round at a tremendous pace, by fits and starts though, as the signal bell in the engine room which is connected with the masthead by a wire, was treating us to solos as well as running accompanied, with one, two or three tinkles at a time as the case may be. I hazard sleepy suppositions: the gale has broken the floe and we are penetrating into the haven where we would be &c, but curiosity getting the better of my disposition to slumber. I was soon on deck and find that we are much where we were, in fact making no way at all. The wind has shifted, and though not blowing on to the ice, pieces have come down with a current, which seems to come 'round the island and get in towards the ice, and we are in a fair way of being beset. She is held fast by a piece on her lee bow and another amidships, and another piece on her weather quarter. Steam was not forthcoming in desirable quantities and was knocked off for the present. A hole being dug in the ice an anchor was fixed and a warped thence round the capstan, and all hands had a heavy spell at the capstan. This giving no result worthy of mention, a blast was tried, primitive but effectual, a bottle, a boathook and a piece of fuze [sic] being all the apparatus, and we now felt the want of the guncotton that we wished so to bring. The present blast succeeded in completely breaking the piece under which it was fixed, but this was not of much use as there were several other pieces requiring to be similarly treated. However, steam is again up, and the propeller takes us a little astern, round goes the capstan again, and up go the headsails, and as these fill her head gradually comes 'round. She is at last slewing and, once turned, all is plain sailing and in 15 min more we are free. Two alternatives now present themselves: either to go down Hinlopen Straits and look for reindeer, seahorses, &c and see if they are open as far as Cape Mohn [Kapp Mohn] and try exploration in that direction, keeping in the neighbourhood of the Seven Islands, or to strike along the ice right away for the whaling ground. The latter is decided on as BLS wants to give the crew the chance of getting a whale. But it comes on thick and snowy and some drift ice intervening between us and the pack edge we give it up for a bit. However, dodging inside being interesting we go at it again about lunch time and by the help of steam get through easily and away we go, with a fair wind, westward ho! Turned in about 1 a.m., very early for Spitzbergen habits.

June 17, Tuesday.

Like yesterday, we go sailing along the edge of the pack with the north coast far away on our left hand, but very clearly discernible. The islands in sight last night are no longer to be seen. In the morning, I start after a seal in the dinghy. After a very long and tedious boat-stalk, I approach him from *leeward* and coming 'round a hummock get my rifle up, at his head, which is looking the other way. But alas, he is off, owing, I fear, to his seeing, hearing, or smelling, another boat which had put off—judging from my manoeuvres that I could not see him and which in spite of P's endeavours to keep them concealed steered much too straight at him, as our sailors appear to me, invariably to do. He is one of these enormous ground seals and at one time we thought he must be a seahorse. It is disappointing and of small consolation making a long shot at him (I had two shots) in the water, as it is too far, and we just miss him even with the harpoon, so down he sinks and the chase is over. We took some soundings this evening, but the water was shoal. We, however, got a piece of coral in the swab we sent down. Turned in at 1.30 a.m.

June 18, Wednesday.

We no longer see Hakluyt's Headland on our port bow, as we have rounded it, tho' far away at the ice edge, and all day take along, passing abreast of the Seven Ice Mountains, which we see at a distance of 50 to 60 miles. The breeze is fresh and we bowl along famously, exploring bight after bight with no success. We lose sight of the land in the evening, and it becomes cloudy and blows hard. We, however, heave to and take a sounding, but the rising wind makes us drift so, that we only let out 400 odd fathoms, which is equivalent, I suppose, to about 300 direct. We all turned in soon after 10, as the cold snowy deck is not inviting. We, of course, drank "in memoriam" of the glorious 18th [of June 1815: the defeat of Napoleon Bonaparte at the Battle of Waterloo by the Duke of Wellington].

June 19, Thursday.

A whale was sighted between 1 and 2 a.m. and, after some time, two boats were lowered away, as it excited the men tremendously seeing the fish playing at the ice edge. "There she is," "There she blows," and they said she was unusually playful, as she was lashing about her tail out of water. There was a heavy sea on at the time and the owner felt very relieved when they returned safe, having had a shot, as had they got fast. In all probability the fish would have taken the floe and they would very likely have lost all their lines and had all their boats swamped as well, as it is a lee pack, and the wind soon increases to a gale we have to run. For the greater part of the day, we are hove to, and the steadiness of the "little *Diana*" under these circumstances forms a strange contrast with our usual experiences of these frequent gales. We are only hove to as long as it is cloudy and the ice obscure, and then we go exploring the edge and the successive bights. We are now again in horrid ultramarine water; no whales can be expected 'til we find green water again. We have had every sort of colour since the Seven Islands and the colouration of these belts is very curious. First you have ordinary greenish grey sea water, and ordinary looking blue water, then different degrees of this ultramarine kind, that is, of various intensities of blue, then the last is again both turbid looking and clear blue; colours apparently independent of the state of the sky. Then there is brown, and brownish green, olive green, and pea green; the latter, like the ultramarine, in different places looks turbid or clear, though always in reality clear, as can be seen from the submerged ice-tongues. Towards night the gale abates. There is a moderate breeze and heavy swell, and up 'til now (12.30 a.m.) a bright sun obscured at times by heavy snow showers. We take a sounding (no bottom) of 500 odd fathoms and have a loon *chasse* [chase]. They are, however, very shy and cunning. We get five brace, and an "ivory gull or snowbird." As I go up to smoke a last pipe on deck, an excitement arrives, though it ends very disappointingly in the mist. Towards the ice edge a blast is discerned, and is visible at intervals, until we come to about a mile distance. A boat is now lowered and towed astern and we make up through some very hard heavy "washed" ice towards the pack edge. We at last get a good view of the fish and away goes the boat, "call all hands" and another boat is manned and I ruthlessly rout up all the others. From the bridge, we have a very good view of the fish occasionally, head, back and once the tail, and at one time he was nearer to us than the pursuing boat. Unluckily, though clear at the deck, it was misty aloft, and it was not 'til the second boat was lowered that the captain discovered it was a huge hump of a Finner, though some of the hands on deck had suspected it from the high and powerful blast, which is a characteristic of these brutes. The boat, which was very near the fish, is recalled, and after a lunch of bread and cheese and stout we turn in at 3 a.m.

June 20, Friday.

Went on deck at 7 a.m. and shortly turned in again (8.20 a.m.), breakfast at ten, very dissipated habits. We cruise along with a fair breeze but see nothing but a few seals in the water. At midday, we come across a very heavy neck or isthmus of ice, the continent of which trends out of sight in an easterly direction. We get up steam and even then have tremendous difficulty in getting through with that. We get into this water beyond, be it "hole" or "bight," I do know not which one. As the winds keep changing, we beat as soon as steam has run down, tacking across backwards and forward, in fact 'til we get a midnight observation as we hope to in an hours' time. We don't know much where we are. This wind, however, ought to be good for scattering the ice. We get an observation in a bight and turn in about 1.30 a.m.

June 21, Saturday.

It was rather thick about 8 a.m., but soon cleared, and turned into a lovely day. It was soft and mild and the thermometer showed up to 40° and the "western ice," as called by all, looked at its best. We felt to its fullest, the charm of cruising in the polar seas in fine weather. The wind got westerly and we tacked for some time, in fact, until getting near up to the edge of the pack, for we are sailing more in interior lakes among scattered ice, than in a proper bight. We found that, to penetrate into the "floes," which is of course our desideratum, as it is there that we should seek, bears, narwhals and whales, the lanes and passages into which this westerly wind is opening, we must go southward as well as westward. For this the owner is unwilling to do, as it is endangering the success of the expedition to get among the floes, where we are liable at any time to be beset, should it blow from the E, and whence it might require a large consumption of our coal to extricate ourselves. So, we again take as near N as the ice will allow us. It is a glorious afternoon and the water still and clear, and we glide noiselessly and evenly along, slipping through the water faster than one would imagine. The scene seems like one from fairyland. There are a good many seals in the water and I lowered away after a wounded one in the dinghy and had a most exciting hunt. I don't want to fire at him as he is not over good at diving, so row after him. One can see him diving and turning below in the clear blue water, but he can't keep down long. After a long chase, I got him into the boat alive and he began snapping like a dog at the hack pick. Whilst becalmed we had great fun getting the mallies to flight for bits of blubber, fat &c. Towards evening the wind came on easterly, so it is perhaps as well we are not in W! About 2 a.m., we weathered a point of ice and without warning came into a heavy lumpy sea.

June 22, Sunday.

A long heavy rolling cross sea from the S.E. has effectually stopped our making way to any extent today, especially as we are beating. This wind is unlucky, as it will close up the ice, at any rate down here, and is very likely not prevailing N, where it might be of use, as these gales seem very local, and it would appear from the sea that the late gale has very likely continued here ever since we left it. There is also, we fancy, an exceptionally strong southerly current, but we have no ready means of testing it. A deep-sea sounding of 800 fathoms gave no bottom, and we got plenty of exercise hauling in the line. Turned in about 1 a.m.

June 23, Monday.

A light breeze but still we have to beat. It is again, however, a glorious day, and we are on deck almost the whole day. I have a good lot of work in the way of observations to arrange &c below, but the swell, which is still heavy, being against it, neglect it utterly. There are a great many loons about, and we sail up as near to them as possible and then give them a volley on the ice or as they fly off, generally getting two or three. The wounded birds cannot, of course, be got, as it is waste of time and ammunition. The captain was shooting with BLS's gun and found a great difficulty in getting it off owing to his inability to manage the safety spring. On one occasion, there was a tremendous patch of loon, apparently not so shy as usual; the Captain and I were ready to deliver our fire from the bows. As we neared them, heard a click and found that he had succeeded in getting his gun to half cock. He again got ready and again the same manoeuvre was repeated. At last we were getting too near for the shot to scatter, and with a "now captain" I fired and was nearly convulsed to see his face of mortification as he again failed to let off his piece. Sometime after, he went away in the dinghy and approaching about 20 loons together on a piece of ice came to the "present." We were all in expectation of a tremendous bag from such a pot shot, but the ambitious skipper, in attempting to fire both barrels at once, failed to get off either, and away flew the loons midst jeers from the spectators. He, however, subsequently got over all his difficulties and had good enough success and as he is, I believe, a good rifle shot, I will have no doubt get on. As the wind fell completely, though the swell continued, we got up steam about 2.30 p.m. and ran for five hours. The owner and second mate went away in the dinghy after a seal on the ice and polished him off with great success. This was lucky, as he was on such a small piece that the roll of the sea sent him off it, soon after he was shot. It was a fine bladder (female). It came on thick about 9 p.m. and the slight breeze which had got up from the N.W. is dying away (11.30 p.m.). We took a sounding (no bottom at 600 fathoms) an hour ago. The ideas on the subject of shooting entertained by the men are most amusing: that is, as regards fowling pieces, being mostly in the whaling and sealing trade, they are accustomed to rifles, at any rate to seeing them fired and some of them are very fair shots. Of shotguns they are totally ignorant, but apparently believe that their range is little if anything inferior to that of a rifle, and are of opinion that the scatter of the "drops" as they call the shot, to a man, embraces a field of proportionate extent, and having pointed you out two loons on the water at, say, the modest distance of 100 yds and not more than five to ten yards apart, look on your statement that it is "too far" with incredulous dissatisfaction.

These calm days, when the crew have little to do, are productive of most marvellous and amusing yarns. They almost relate all to the "Straits" or Davis Straits as the whole of the water of the fishery on the West Coast of Greenland is called even up to the Smith's Sound. Of course, some of them have been shipmates before, and as our whaling fleet is not very large, they are known to each other before and, of course, know all the ships. So it is always: "when you and I was in the *True Love*, Tom" or "the night the *Husky Maw* (as the *Esquimaux* is always called) got nipped in the flaws" and "her hatchway crushed into a rhombus (in which state it remained until the whole ship was restrained into its proper positions) (a fact)," "we was alongside her in the *Canipadoon*" &c. The most marvellous yarns are thus, by a code of honour, certain of corroboration. So and so: when a boy having, while the men were at dinner, got fast to and tackled a Uni (of incredible size of course) in the dinghy and after the fish was secured being walloped for his pains, or: So and So: while fishing in the Greenland waters as the Spitzbergen Seas are called, having struck and lost a fish, which was picked up next day by the *Wildfire* or by some such ship, of which his own brother was mate in Davis Strait, the wounded fish meanwhile have either made a journey of 150 miles round Cape Farewell or, as suggested by the narrator, would have entered Scoresby Sound, penetrated right through the heart of Greenland 600 miles by an unknown (or subterranean?) channel. These are among the least startling of the stories, and we have had many minds to compile a yarn book. Then there are the endless tales about the natives, Yaks, as they call them, and of bear hunts &c, many of which are doubtless quite true. One of the men, David Cobb, the boatswain, the cook, a great character, were on board the old *Diana* who was caught and wintered one year and was lost the next on her homeward journey. As might be expected, now they are here, Davis Strait, with all its hard labours &c, and its long spells of watching by the ice edge in boats for ten or 12 h, its fogs in which for a time, sometimes altogether the luckless boats watching for fish, lose their ships, and the never to be sufficiently commemorated dangers and terrors of October Month is a sort of Paradise, a happy hunting ground to this. The sky is clearer, the air warmer, the fish more plentiful, and the birds finer, ptarmigan ("why Lor bless you, you can get to brace a day in the straits, Snowy owls of course four foot high"), then the bears ("why Sir, I seen 'em on the ice, like a flock of sheep on a common"). Of course, the inevitable answer to a query as to why a battle did not at once take place is, "Oh, we were after a fish at the time" &c. Then, I should think, the fair lassies of Dundee would hardly relish the praises lavished on the flat-nosed native beauties, nor the high commendation of their skill in dancing, but "thus we pass the day" while we journey on our way. "Oh, gaily goes the ship when the wind blows fair."

June 24, Tuesday.

A calm hazy day. But, just as steam is going to be got up in despair, a slight breeze comes on and we get slowly along, though greatly stopped by the still prevailing roll. There are thousands of rotches about today, flying screaming round in flights like snipe, and we shoot four or five, as they are the greatest delicacy to be had up here. We also get out and inspect the net (for salmon) and the tent &c, and a tremendous painting of the dinghies goes on, as they have not been hitherto of the uniform white tint of the rest of the boats. We don't stick so much to the ice, as although we are among drift ice all day. We are again bound for the Seven Islands as soon as possible. A sounding gives bottom at 500 fathoms (clay). Soon after midnight, the wind gets more easterly but freshened. Turned in at 1 a.m.

June 25, Wednesday.

About 2 a.m., we came out of a bight of the ice in Lat 80° Lon 6° 30′ E and began our dead beat eastward. Land has been visible since 8 p.m. last night, clearly enough at times, but often hidden by the haze. It is a fairly strong breeze and we have for a few hours reefed topsails, mizzen and foresails and take in the jib. It, however, soon moderates and we don't make over much way, as we have the benefit of a heavy sea, and her ladyship jumps as much as ever. Still, the whole thing is charming and I am afraid will spoil me tremendously for channel yachting. I feel the cold, too, so very little. I hardly ever wear mitts or gloves while in the deck; moreover, fall asleep there after lunch with great comfort, with the thermometer at 34 to 32°. The steward, who is doing well, is beginning to help again. On his reappearance in the pantry, the mystery of why "little Jack," who has been his deputy, and the office is no sinecure for such a small boy, in spite of all the hard work, has so enjoyed himself, has transpired: Jack has apparently lived solely on marmalade, jam, and sweet biscuits. We also heard him assuring the steward that the smashes, several of which he had perpetrated, were committed by us. He is, however, a very good handy boy and very keen about sport. He cannot, as Potter says, well avoid being right about seals, loons &c now and then, as everything he sees whether shadow or black ice &c he announces as seal or loon, and I think hardly realises that the brutes, of which we shot over 20 brace two days ago, are only shot when required for the larder, and when we can stop to pick them up. Sounding gave bottom at 300 fathoms. Turned in at midnight.

June 26, Thursday.

Not quite such a head wind, but not enough of what there is, as the sea is very short and choppy, and right ahead, so we make way slowly on our course at two and a half to three knots per hour. "Little Jack," who once omitted to call me, was therefore admonished, and was so impressed with the mystery of winding the chronometer, at which he was present, and the awful grief the ship was to come to, were it permitted to run down, that he has made my morning hours a burden to me ever since, often turning me out before 7, which is of course useless as it should be always wound as nearly at the same time as possible. Finding of late that I refused to budge 'til more certain of the hour, as his notions of time are of the haziest, he had recourse to a ruse, and rushing down the companion in a tremendous hurry informed me it was nearly 10 o'clock (at about 7.45 a.m.) and got me out of bed with great success. It has been a charmingly soft warm day, with a beautiful light this evening on the distant hills. The ice limit along the northern coast is far further N than when we came down. Got bottom at 360 fathoms.

June 27, Friday.

As the sea was not running so heavy, we got up steam about 6.30 a.m. But it soon came on to blow again and it was knocked off. So, we commenced tacking S and NE. At noon we found we had made very little way since the day before. It now came on to blow strong and we had a bout of making secure below as the cabin was soon strewed with books &c. There were renewed smashes and the inevitable milk and soup tins as drinking vessels loom nearer. It was a charming day on deck and we sat in a row for some time, under the lee of the weather bulwarks (tightly scotched up in various ways as she lay over a good bit), talking and singing. Our "Jackdaw Parliament" was, however, brought to an abrupt termination by a sea, which succeeded in wetting the deck and every one of us besides more or less. Lunch and dinner were managed with *e'clat* [brilliance], the soup alone taking charge, but we are very wary by this time, and stick to our plates and glasses in grand form. The evening was cold and raw, and Donald (a boy from the training ship) at the wheel was quite a success as he succeeded in landing three seas down the companion in about 20 min. He was suppressed after that, and certainly it is wonderful in a stiff breeze, a difference there being that you have a watchful and clever hand at the wheel. We turned in soon after 10 p.m.

June 28, Saturday.

Potter and I had a look out about 4 a.m., and certainly it was not cheerful: cold and damp, blowing harder than ever and that strange "eerie" whistling sound in the rigging. So, we turned in again expecting another day of it. By 6 a.m., we were, however, out of it. Through it, I fancy, for we got into smooth water very quickly. The gale apparently continuing behind us and, I think, the surmise that it came down Hinlopen Straits was probably correct. Soon after 8, steam was up and we made to the Seven Islands, much in the track we left them by. We pass through a good deal of drift ice and see two "sea horsemen" plying about the ice edge to the W of the North Cape. On passing Walden Island, we find, nor is it an agreeable surprise, that all the fast ice around the islands is much as we left it. The owner is very anxious to get to Parry's Island to see if there is news or trace of the Swedes. A party consisting of himself, the captain, second mate, and Tom, a great man on this ice, starts across the floe, and the start proves an awful business. They flounder along, sinking through the snow and getting into the numberless cracks often up to their middles. The going, however, improves, and soon they only appear as four black dots, moving in a "bee line" for Nelson's Island. I was very anxious to go with them, but as it was fine had decided to try and get some photos of these islands, the most northern *known* land. Altho', of course, at the distance the nearest points are off, much beyond an outline could not be expected. But *che sara sara* [what will be, will be] and no sooner had the first been taken than "the blinding mist came down and hid the land," or rather in this case clouds, so the camera which had been erected in a good position on the snow for a circular series had to be brought back, and it was all over for the day. The birds were as usual, when one is observing or focussing, very annoying and inquisitive, and several dovekies and loons were polished off. One of the former, who was left dead on the ice, proved a great attraction to Burgos and Snowbirds and several of these, as well as a Boatswain or two, were knocked over. The difficulty is to get a good specimen: if you use small shot, you don't bring down the birds or, even if you do, they soon recover; if you use large the blood spoils the feathers. This is especially the case with that most beautiful Arctic birds there (here at least), the snowbird, a gull with a plumage of the most spotless white. The habits of these gulls are most amusing. The Burgo is, of course, the "boss," but "Malley" and "snowbird" are both bold and voracious, and insist on having their share of any considerable spoils the former in the water the latter on land, although never absolutely asserting their rights *vi et armis* [by force and arms], but hovering round and apparently succeeding by their importunity. Then there are the poor "kitties," the same bird as the smart beautiful kittiwake we see in England. They are in subjection to everyone and seldom or never taste *kreng* in any quantities, though they will eat it fast enough. They are, however, very industrious fishers, but have to pay a heavy toll on their captures to the "boatswain," their principal and unrelenting foe. You see a lot of kitties hovering over a pet fishing spot, harried by a couple of boatswains, who rob bird after bird of their spoil, often not without a long chase, and capital courses you see with many a turn and wrench worthy of Master McGrath [a greyhound famous for hare coursing that died in the spring of this same year. A sculpture in County Armagh in Northern Ireland commemorates the dog]. In the evening, P, Eaton and I sailed down to Walden Island in a whaleboat. We ran down (about five miles) in a very short time, as we had a clipping breeze. About halfway down, we saw a seahorse, but were running too fast to get down sail, and a harpoon ready in time. He had splendid tusks, but a couple of bullets, one in about the right place in the neck, were not good enough to get him out of hand, tho' from a sort of shudder he gave, it was very likely his death wound. We had soon difficulty in getting across the broken ice round the island foot, as it was in small uneven lumps very angular and carefully concealed by a level coat of snow. Eaton took to the rooks for lichens &c, and P and I along the foot of them for the low ground at the eastern end. Arrived there, we found not the expected eggs. In fact, nothing but an old coffin full of stones and a bone or two, and two boatswains who, I believe, had a nest. They were stalking without success, but somehow forgot and came past a large stone behind which we were standing and were disposed of. One will make a good specimen. We now had a splendid view across the fast ice to the other islands and to Cape Platen [Kapp Platen], the North Cape and North East land, and this *land* ice which is, however, only young ice, is certainly level and to be sledged over. We now took round the southern side, and soon came into a different kind of going. The snow had melted a great deal, in many places almost completely, and on slopes between the rocks which are granitic, I believe, almost solely primary, though in parts there is a laminated black rock in layers amongst the other which is, I should expect, metamorphous. There were large patches of mosses and lichens of most beautiful colours, green and orange mosses and lichens through every shade from white to black. It was often boggy and the pools of water knee-deep. Further on, we took along the cliffs at some distance up, as the bottom was a precipice, and the ice below too rough for travelling. In the pools among this, which were of the dirty green colour of a duck pond tho' transparent, there were a few eiders &c, but they took themselves off without giving us the trouble of trying to get at them. Thro' a small hole in the ice that formed the bottom of a pool far, far below us, a seal thought proper to put up his head. But, as he showed no signs of getting out, in fact from the size of the hole, I doubt its feasibility, he was saluted with a bullet, which was nicely placed in the centre of the hole, so I suppose settled him, but it was much too far off to see. It was a safe climb, well enough on the moss, but anything safe on the shelving slopes of bare rock that had often to be crossed, hampered as one was with firearms. The slopes up here were really, for this land, covered with luxuriant vegetation, endless cushions of a little smooth leaved green plant like a shamrock, scurvy grass with its heads of countless white flowers, yellow flowers of the size of the common buttercup in flower, and many others, all of which with their proper Latin names (far too long to be inscribed here) written down, with all care and precision, on the lids of the nice little wooden coffins prepared for any rarity vain enough to bloom, and pining for sepulchre in a museum, by the indefatigable Eaton. A little way on and five great birds rise. An easy shot, but my rifle comes down, as I remember it is not a scattergun, just in time. "Down," I shout, and we all crouch on the moss, and as they come round a bit, Potter lets fly with "number one" shot, which he luckily had in his gun, and with one of the tallest shots I ever remember to have seen drops two fine rein geese, far into the rocks below, one at least falling right away on the ice. Another good shot soon after fails to get one, as such very long bowls with a bird that size cannot be expected always to fetch them down. We now descend to the floe and are soon joined by Eaton, and in another half hour of very heavy walking and climbing succeed in completing the grand tour of the island and regain the boat. Luckily the wind allows us to fetch back to within a short distance of the ship, and tho' rather wet, songs and yarns make a lively enough journey. We reach the ship shortly before the other party who are pluckily toiling thro' the last heavy bit to the ship. They had gone, as we could, see straight for Nelson's Island, the travelling improving a bit after the start. Arrived there and seeing nothing, they pushed on, almost as far beyond, to the south-eastern extremity of Parry's Island. Here their search for the Swedish depot—or marks, traces, or tidings (in bottles)—commenced, as it was at this point that we were informed their depot was, and the party had letters from their own people to leave. Not a trace, however, old or new could be seen and if, as it is to be hoped, they have succeeded in returning to Mossel Bay, they can have hardly touched here (where they were due on the 23rd), without leaving a sign. This was, as may be imagined, intensely disappointing, after ten miles or so of travel more laborious than one could imagine possible without experience, the monotony of the deceptive distance, in which one never seems far off, but never to get nearer, totally unrelieved. They are, however, determined to go through with it, and continue their search along the farther side of the island, until coming to a low neck of land between its two ridges, at a point where it is almost divided into two islands, they took across it, and took the floe homewards. Like Walden Island, the place seemed a perfect bear walk but they never saw one. One had, however, come down from Nelson Island, and taking up the scent, "run heel" in their footmarks towards the ship, and they were in hopes of either falling in with him on their way back or that he had reached the ship and been secured. The brute, however, tho' seen from the masthead had turned off halfway. We feed, heavily too, and turn in at 3 a.m.

June 29, Sunday.

It is, alas, another grey day. As the clouds are heavy on the islands, photography is at a standstill. So, an expedition to Table Island [Tavleøya] and Ross's, if possible, is organised. By about 1 p.m., Potter and I, the mate, and a boat's crew start. Unluckily, just as the breeze drops, so away we pull, and after a mile or two come across a large ground seal. Though not fired at until to wait would have lost a shot, he floats a second too little, and as it takes some time to start a heavy whaleboat with her lines in, we can just touch him with the harpoon and see the enormous brute slowly sinking far below us in the clear water. Never cry over spilt milk, tho' these ground seals do "fetch" me. On we go and soon far ahead two seahorses are visible as black spots on the ice. As they are dead to leeward of us and there is no way of getting 'round them, great care must be used. So, at about a mile and a half off, we get in the oars and sail down on them with a light breeze. As we approach, even at a great distance, the one who is on the lookout becomes very nervous again and again, so he raises his head but on seeing nothing suspicious drops it again. Now he is reared up and with his nose as high as he can get it sniffing, but seeing nothing, concludes it is all right. Now we are only 100 yards from him, and turning his head as far as he can, he looks bang at us, nothing but a white boat and a white sail to be seen moving no faster than he has seen ice travel and down goes the sleepy head. Only 20 yards off and he looks round again and now the affair begins to interest him and as the harpooneer stands up and swings his gun on its pivot, he is watching us intently. Seven yards off "a sight" along the gun, and before he has done well more than turn 'round, his companion who is asleep is woke by a report and cold steel in his innards. The height of the ice on which they are and the position of the furthest render it impossible to fix the other with the hand harpoon and we back astern just to keep them from jumping right into the boat, and give the outsider a bullet in the head, just to teach him to keep a better lookout. We then brail up the sails and await the reappearance of our attaché. The brisk conversation that ensued, *en attendant* [while waiting], was a strange contrast to our long-enforced silence, broken only towards its close, by a loud and continuous whisper from "James" to the harpooneer: "Tak time," "Tak time, man," &c, and our attempts to silence him by *ssshes*. Now, however, with an aplomb seemingly strange, but natural enough, we light our pipes, and have some time to enjoy them before our friend reappears. When he does, he is very lively and fire is opened at him, with singularly little effect. After a shot or two, P and I desist as we do not wish to waste ammunition. He is much too quick for fatal shots, and besides four of us huddled up forward have little room to get a steady aim. James, however, loads and fires, loads and fires, like a man shooting against time until his ammunition is out, and the harpooneer keeps it up pretty hot. Meanwhile, the other has appeared and seems unable to get off a tongue of ice onto which he has got. Now is our time to fix him with a hand harpoon and giving way we try to get the boat near enough. The "fast" lad, however, strongly objects and as he has been taking a good breather tows us in the other direction. So, we must be contented for the present with what we have got, and splendid sport it proves. My vociferations for a hack pick are answered by my being presented with a whale lance and hauling in on the line we get close up to him. Thrusting at a hide a good inch thick, and driving a slender lance through it, are however two different things, and failing to accomplish this satisfactorily we bully him a bit, as he won't turn and give a tight back but keeps his head towards us, blowing and snorting, rearing up, and hurling himself over backwards giving a complete view of his underside right to his hind flippers. After being pricked a bit about the head and snout with the lance, at which he snapped in a way that convulsed one with laughter, he came straight towards the boat, and succeeded in coming up underneath it, making it vibrate a good deal. He was on his back with his head close to the side, and one felt almost inclined to catch hold of his tusks, not being able to manage a long lance at such close quarters (we had no proper stiff seahorse ones). I took up my rifle and, holding the muzzle against him, fired and down he went, coming up soon by no means done for, as I had no aim and had not selected a vital place. In a few minutes more a couple of bullets from Potter and myself settled him, and he was soon on the ice, and flensing with pocket knives as the proper ones had been left commenced. During this operation, I shot a snowbird with No. 8 shot so as to ensure his being a good specimen, and tied him carefully up, one handkerchief round his wings and body, the other round his head and neck, and knotted as tightly as possible round the bill. When about to start again, after the skin had been got securely into the boat, I went to pick him up off the snow where, in true Spitzbergen fashion, he had been reposing in company of guns and rifles &c, with an adroitness worthy of the Davenport brothers [nineteenth century American magicians and supernaturalists] our defunct snowie shook himself free from his bonds, and leaving them almost exactly as they had been before, sailed triumphantly away. Another barrel from P settled him, but the spotless plumage was smutched. Another curious incident occurred: a boatswain in full chase of a kittie came straight towards us. Shooting when they were in line, I killed the boatswain, but not the kittie of which I was glad enough. After a long row, we arrive at Table Island, distant from the ship about _____ miles, and landing on the rock off its northern extremity proceed to lunch. As there are no eggs, the crew go off to the main island to search, and we remain to take a round of angles from the summit of the rock (100 ft). This done, we take a long look at the scene before and around us as it is probable we may never set foot on land again so far N and there is a fascination in looking away over the ice towards the mystic pole that defies approach. To the northward lie Little Table [Vesle Tavleøya] and Ross's Island, the former at present hiding the latter. Away to the W trends the ice edge, hard, heavy large ice, and from our feet stretches away ice right away N, at first indeed drift ice, but beyond and to the E and N of Little Table island, pack and, I think, fast ice, at any rate so closely packed as to be undistinguishable. Such ice, too, far different from the endless level fields I have heard described as running hence northward and looking down at the rock foot when to a height of 30 feet from the water is a half-welded looking mass of flat slabs of which are piled irregularly one above the other in endless confusion. One shudders at the giant force that must have been rampant here to reduce to such morsels an unbroken floe, or as James remarks, "Bless us, there hev been the awful ness's here." The frequency and closeness of the hummocks as far as one can see (miles) is astounding, and the horizon is a broken jagged line. These hummocks are so hard, angular, and blue, and in places so devoid of snow that I suppose not until they have melted and become rounder and completely snow-clad will they present the appearance of the ordinary regulation "seal pattern" hummock. Far away, one a little westward and the other a little N of Little Table Island show two patches of dark water sky, tho' they may be false sky. Away E and S lie the other islands, the high rock of Table Island (750′) rising abruptly above us.

Their southern sides show little snow, here and there, veins of white against the dark grey and red purple of the rocks, while the northern slopes of which we see little more snow. But they altogether have hard and sharp appearance on the western coast. Just clear of the group stands out in bold relief the flat-topped outline of the huge rock called Nelson's Island. Away beyond stretches the level floe, flat enough this side to the capes, bays, and promontories of the N. E. land, and in the gap showing between the North Cape and Walden Island of which we have a very fine view lies a small black speck, which we know for the little *Diana*. It is almost with a feeling of awe that one turns away from the outmost (or inmost) threshold of the dread inapproachable tract of frozen ocean. But as we light our pipes and commence a descent to search for a Burgo, who had gone tumbling inwards over the edge of the rocks, we regain our tongues, and as we await the boat, sitting on the rocks at the water edge, we marvel to find how much we feel our ordinary feel. Meanwhile, a wreath of cloud comes slowly winding round the eastern precipice of Table Island, the ascent of which is made for a better look out, and by the time the summit is gained, it is so thick as to prevent anything far off being seen, and as soon as all the party, need I say eggless, whereat James waxed very wroth, had assembled at the boat, it was so cloudy on the islands. Since it was clouds not fog that little Table Island was no longer visible. All day, [over] Phipps and Parry's Island white wreaths of cloud were unrolling themselves, while slowly on the venerable head of Nelson's rock there descended a three-cornered cloud, which required no vivid effort of imagination, to liken to the ghost of the cocked hat in Trafalgar Square. It, however, succeeded in "connecting him" and, after a time, his dark cranium projected above its covering. James, who appeared really upset at the failure of the egg hunt, declared it dangerous to attempt to reach Ross's or Little Table Island, which was our original objective, as the ice from a tide or current seemed to have a decided set that way and, with reluctance, recognising the force of his statements, we started homewards. The fair breeze that had enabled us to sail about for some time, while the detached units of the party were assembling and conducting a vigorous loon hunt, having subsided, there was nothing for it but to "put our backs into it," and send her along, seahorse and all, and a calm was accepted as better than a head wind. It certainly was lovely gliding on the smooth water among the floating masses of ice, with the casual excitement of a "ground seal" or a "dorothy," none of which gave a shot. But they look so sweet, as they raise their sleek shining heads and with large wondrous dark eyes looked at the intruders of their peaceful lagoons. One soon gets accustomed to pulling and keep on mechanically half asleep, roused only at times by the perpetual chaff kept up between James and the engineer, the engineer chaffing and James growling at the other's steering, erratic certainly and so admirable that it can only be likened to the hue of beauty (repeated *ad lib*). Then at times the shock of running against a bit of ice roused you again, as every opportunity of so doing—purposely, I fancy—was seized. From the bow: "A piece of ice dead ahead," "You are steering straight at it," "Now you ae on it," and so like moving dreamers we moon along, until the steersman, who by agreement is taking us to the whereabouts of the seahorse *kreng* to get some Burgos &c, remarks: "there is a "Brunie." No one, I fancy, believed him as for some time we didn't look round, but Potter soon made him out from the bows, and handing over my oar to James I get my rifle out of its case. What struck me the most now was the beauty of the picture to the left: the fast floe and beyond the dark islands, all around us drift ice, and on the right a large broken floe piece composed of these blue hummocks with their piled up slabs, in front ice and a bend of the floe, and to our right front and separate from the floe piece by a channel 50 or 60 yards wide, a piece about 1/3 acre in extent with a central ridge of blue ice on which were numerous Burgo's gorged, I suppose, and sitting in state on the highest pinnacles. In the foreground lay the huge bloody *kreng*, with a circle of busy and hungry Burgo's and Snowbirds sharing the feast with the "Brunie." This beast, tho' small (about the size of those in the Zoo), looked very handsome as he shook the spray from his yellow coat, and picked and ate like a dog, apparently not over hungry too, or about "through" for from time to time he played with pieces tossing them about. The whole thing made an Arctic picture, as we remarked at the time, not to be surpassed in beauty (of its kind, i.e., a group and foreground). Brunie soon sees us and, leaving the *kreng*, walks along the edge of the ice to the point nearest to the large floe piece. We, however, had at once told the steersman to make for the edge of this piece, not straight at the beast, as being large and over a mile in length and very broken, once there, it would have been easy to lose him for good. But he has his doubts about leaving the *kreng* and returns to it again, then he looks at us and sniffs and sniffs, "hang it, discretion you know" &c and biting at his paw in which, I suppose, some snow had balled, he makes again to the extreme point. But he has been wasting golden moments, and he goes on doing so, as instead of taking a plucky header, he sways his head and body about over the water, and finally ignominiously turns round, and lets himself down hind legs first, and turning heads for the large piece—too late, *mon ami*—for although we have succeeded in fouling the only two pieces of ice in our way, we are pulling along the edge of the piece and have cut you off. So round he turns as he recognises this, and gets back to the other piece, and after a great deal of unnecessary fuss and clumsiness lands with great deliberation, and apparently being at the end of his resources, waits for an idea. We are now taking across straight at him and the only one that occurs seems to stand with all four feet together like a goat and his head towards us, right on the ice edge, with what object never transpired as the rest was very tame and rowing to within 20 yards a couple of bullets (head and shoulder) and all is over. He is soon in the boat, and meanwhile I land and get a snowbird, and find that Brunie landed on the other side and tracked my footsteps across from where I killed a snowbird this morning. They have a nasty habit of doing this. We pull home merrily enough and find the ship rather surrounded by ice and have some difficulty in getting the boat up to her. Two seahorses with a young one had passed today, the old ones when pursued pushing the small one along with their noses. A long shot at them missed. They had, however, succeeded in getting a large ground seal. Eaton had been to Walden Island in a whaleboat collecting, and had been away some time, his proposed half-hour on shore becoming three or four, with satisfactory results however. We were decidedly hungry and feeding &c kept us up 'til between 2 and 3 a.m. Total bag today was one seahorse, one bear, one ground seal, one whitey coat, 13 Snowbirds, one Burgo, one Boatswain, 41 Loons, and some Dovekies and Rotches.

June 30, Monday.

Nothing to be done again in the photography line as the clouds are still on the hills. The ice, however, [is] coming down on us and about noon we get up steam and get off and proceed to dodge. We spend the day in arranging our respective arrears of observations &c, and in general idleness. We are among the ice and there are a good many large bits of scruff &c about, some disintegrated parts of floes that bear those unmistakeable marks before mentioned, of the awful grinding and crushing that has gone on. We keep going about at annoyingly short intervals, which decidedly spoils one's nights rest, and although I don't turn in 'til between 1 and 2 a.m., I don't get much sleep.

July 1, Tuesday.

It is decided to make for Treurenberg Bay, getting into Hecla Cove, if possible, to water and a walk over the hills thence to Mossel Bay. [This] will give us a chance of hearing news of the Swedes. We are still not far from the Seven Islands, having been in the night as far N as 80° 50′. About 4 o'clock, a large ground seal is seen on the ice, and two seahorses further on. The owner goes in the whaleboat for the seahorse and Potter and I on the dinghy after the seal, which gives us a long crawling stalk over the snow with infinite trouble to keep ourselves out of sight, but unfortunately does not give us a shot. Hearing one further on we pull on. Then follows a volley, or rather a continuous fire, like what James the other day called the "bombardment of Bomarsund" [in 1854, during the Crimean War, an Anglo-French force attacked a Russian fort at Bomarsund on the Baltic Sea], and we after a sharp pull arrive at the scene of another great seahorse hunt. The boat is one side of a piece of ice and the seahorse the other, the line running underneath. After failing to get fast with our seal harpoon, we drive the brute onto a tongue of ice and one of the whaleboat's crew landing on it gets in a "toggle" harpoon. As the other seems on the point of drawing, after some splendid sport, it is at length all over, and the usual obsequies performed. We got back by 7 p.m. to the ship, which had meanwhile been flying the ensign at the masthead, a vessel having been discerned far away along the ice edge, which is pronounced to be the *Sampson*. This is capital news as today, the first, is our appointed day of rendezvous, only at Kobbe's Bay, not here. We got up steam, and steamed down on the schooner, and as she proved to be the *Sampson* took her in tow and kept on our course for Treurenberg Bay. Captain Walker came on board and from him we learnt the doings of the *Sampson*. After leaving Lerwick, she made her way N against a heavy head wind and made the ice on 22nd of May, about eight or ten miles E of Jan Mayen. She soon fell in with the saddlebacks [harp seals] at Lat 73° N and followed them steadily point to point along the ice up to nearly 79°, finding them very shy, but getting in all over 150. During this time, she experienced weather, almost always thick and foggy, and often rough, and had to frequently dodge for long spells at a time all of the pack edge, meeting a few vessels who as with us spoilt each other's sealing as far as lay in their power. One of them was commanded by a Norwegian skipper, Mack, who is very well known in these seas. He was the one who had caused us such annoyance and by way of adding insult to injury felt himself very aggrieved as his mate and crew complained that our bullets came very near them, *tant mieux* [so much the better], though I don't believe it was intentional, but owing to their running right into the part of the ice where we were busy seal shooting. On 19th and 20th of June, the *Sampson* must have been close to us, but a little S, the gale we experienced on that date caught her very heavily and as some fears were entertained of her foundering, she bore up and running over to Spitzbergen was obliged to lie to, and dodge about for several days off the coast of the N of the foreland. It at last, however, cleared, and getting into Magdalena Bay, which was clear of ice, she found very snug anchorage. Here she remained five days, sending away a boat to Kobbe Bay, where besides receiving the dispatches from the *Diana* she found a letter from the 18 Norwegians before mentioned, who had instead of taking across the hills from Wüde Bay to Ice Fiord as the Swedes supposed, performed the journey in boats. Magdalena Bay was full of birds, and as the Westley Richards lent Captain Walker by BLS was apparently unprovisioned with cartridges, members were killed with an old "Brown Bess" [British Army muzzle-loading musket]. There were numerous seahorses here, and they got five in all and one ground seal. After leaving Magdalena Bay, they stood along the coast, getting by the way a lot of eggs, off Moffen Island, 100 dozen in all. They passed pretty close to Moffen Flat, but did not stop there, and pushed on N to the Seven Islands. They were hunting all along the edge of the ice and had yesterday been someway further N than we saw them. Tonight, it has been coming on to blow, but it is still luckily tolerably clear, and far away on the other side of the straits the hills lie in golden sunlight. There is a very curious refraction visible, the said line of cliffs being much distorted, some of the bases of cliffs being almost level with the heads of others and numerous gaps showing in the line, forming apparently islands where none exist.

July 2, Wednesday.

Early this morning we let steam run down, and with the *Sampson*, still in tow, both set sail. We were now crossing the straits up where a heavy wind was blowing, setting a strong cross sea against us. We were enveloped in a grey haze, caused by the flying mass of grey clouds. The wind was discharging up the straits, on both sides of which it was bright sunny weather. Overhead, too, above this thick driving mass of scud, it was clear enough. As soon as we begin to get into stiller water, the course which had been hitherto on a point but little [westerly of] Verlegen Hook became more southerly and we made towards Hecla Cove, sounding as we came in. This, however, being still shut by fast floe we came to anchor in ten fathoms, about noon, about ¾ of a mile from the beach, outside the rocks that lead from the bounding promontory of Hecla Cove, the schooner lying a little beyond us. Soon after lunch all the cabin passengers, owner included, went on shore to reconnoitre, and look for a watering place. Just inside the foreshore lay a lagoon, shallow and filled with brackish water; the streams, however, running into the sea all along this beach are good fresh water. The lagoon had many eiders on it and for separating we took a round it, but did not get a shot except at one that fell away to sea, and had to be left. We now made for Hecla Cove, taking across the level ground that comes between the foot of the high mountain that ends this range on this side of the bay. This hill is now very bare, almost all the snow being gone but of fine bold outline, the ground at its foot is very even with a gentle slope seawards and extends some miles right away in fact to the north-eastern promontory of Treurenberg Bay. It is very stony, but later on when the mosses and little pink flowered plants [purple saxifrage], which swarm in the spaces between the stones, have had time to come to maturity, it will doubtless present a less sterile appearance. It has a tremendously flat, pressed down appearance, being apparently far better rolled than the average parade, owing to the pressure of its snow and ice covering in winter. In many places, the stones have the appearance of an extremely well laid and regular pebble paving, "laid in screened gravel," as the schedules say. It is intersected by numerous streams running across the plain, carriers of the melted snow from the hills, and in one of these we got well wet some way above our boots. We soon arrived at Hecla Cove, the standing point of Parry's expedition and examined it with great interest, failing however to recognise much likeness between the original and the well-known engraving. A photograph is, however, out of the question, as the clouds have come down on the hills and a thin mist envelops everything. So, after looking at the general features of the place and the various curios littered about as bits of old rope &c, we determined to explore a bit more having hopes of finding some deer and commenced a descent of the rock above the cove. This rock is composed and has curious veins running through it: white, yellow, red, and purple. It was 'til last year surrounded by the flagstaff, left by Parry. Some Norwegians left here without a vessel, nipped it as a mast to a piece of drift ice and proceeded on their way. Our route now took us along the rooks that rise abruptly from the bay edge. Beneath us lay the fast floe and in the pools at that rock foot were some eiders and numerous seals asleep near their holes, on the floe. As we could not get within gunshot of the eiders, we shot two with our rifles and had one or two long shots at seals, almost killing one which was however so close to his hole as to succeed in floundering in. We were now almost under the high hill and as nothing was to be seen on the bare land at its foot, we took across the floe to the other side. It was very soft and covered with pools of water which made two or three miles of it no mean grind. Landing the other side on some rocks, surrounded by a cairn, we soon came on a river a little beyond. We could, however, see no salmon in it and, fording it, took across the low bottoms between the mountains beyond and the shore. On these there were a good many rein geese feeding, and eiders in the pools. We shot one goose and a brace of duck, and desisted shooting as carrying the brutes was no joke. The bottoms were not so stony, and composed of black alluvial soil, well covered with lichens and dark-coloured mosses. They had the flat, pressed look of all the low ground one sees here. Finding no recent signs of deer and seeing none, we again took the floe some way down and, after a long and tiring drag, made the point beyond Hecla Cove, and in about 20 min more our boat, which we had dragged onto the beach, and right glad was I to be quit of the weight of three great eiders. We took in a cargo of driftwood and rowed off, getting a duck or two before starting. In the evening, we rowed along the ice edge in the dinghy getting some eiders and loons, some of the latter, that is the single ones, were killed by hurling a seal club as a tomahawk at them and great fun it was. Rover fighting them bravely as soon as they were secured. Failed to stalk seals. Turned in between 2 and 3 a.m.

July 3, Thursday.

Hailed up the *Sampson* alongside in order to get off her the coals, stores &c, she has brought, and also to bag some of her water that she had filled up at Magdalena Bay, it proved to be rather slow and laborious work getting it there. As it is again dull and the hills all cloud-capped, I start for Mossel Bay after lunch to see what has become of the Swedes. The owner was very desirous of coming, but his presence was required on board to superintend the spoiling of the *Sampson*. As I am not allowed to go alone, which from the length of the journey and the probability of my walking with my companion to a standstill, I should prefer the "hardy Norseman." He is told off for my orderly and his powers of endurance proved better than I either expected or would have been likely to find among the rest of the crew. We were to be rowed across the bay and landed and then make the best of our way, according to our lights and our circumstances, the distance being 10–11 miles on the map, presumably 12 or 13 up and down, hill and dale. Soon after lunch, we started and found a good lot of drift ice on the further side of the bay, and on one of which pieces there was a ground seal, which I missed at about 70 yds, I suppose owing to the swell as the brutes are big enough in all conscience. A large school of white whales then came across us, rolling and snorting, appearing and disappearing with great rapidity, but all making the same way, some black and some spotted, these latter ones being the young. I did not get a chance with the seal harpoon, but shot one, which we could not secure, and so went on. After a rough scramble across the floe edge found ourselves on terra firma and took up a gully straight in front of us. Beyond this, which was good walking for Spitzbergen, came snow and stones, and having crossed the first collar, we came upon another valley, which wound up from the sea very obligingly. This contained a glacier which had to be circumvented and there commenced our work up the snow slopes. These were very steep and the snow were a foot to three feet deep on an average; often, of course, more in places. It was so rotten that you sank right through it at every step, entangling and hurting your feet in the sharp stones below, and about once every five minutes got in up to the middle. It was worse, too, than soft snow, as the top crust bore until your full weight came on it and then let you through with a jerk. As each successive ridge was climbed, a few yards walk was joined amongst the stones bared at the top by the sun's rays. The walking amongst these was very distressing. Then commenced a descent: jerk, jerk, jerk, each leg sinking in just as you thought you had found a place that would bear you, as far as the other would permit it. Then in the bottoms were pools of water some of which had to be waded, an unpleasant operation as their bottoms were of treacherous snow, then to climb the next slope. The land being of that rolling sort like the downs though the hills were a good deal higher (about 2,000′ by barometer), but solely composed of broken stones like the heaps outside a quarry, though as I have said almost completely snow-clad. To add to our difficulties, a thick fog had come on very soon after starting, and it was very difficult to steer by pocket compass, as one could never see a point at any distance ahead, or behind, nor would the obstacles that had to be circumvented allow one to keep a very straight line. We kept on, however, getting about 100 yds each spurt, as to go further up the slopes at a spell was almost impossible. Such a hold did the rotten snow get on one's legs, at times pulling off your boots for you, the interminable length of a journey of this sort where one can see nothing, for even in the few glimpses of clear, our horizon was as in the downs at no distance from one, can only be realised by experience, and at last we began to think we must be travelling inland parallel to the shore. We had been for some time descending an unusually stony slope, with very large holes between the concealed rocks, and a roar of water on our left, from a subnivean [under the snowpack] glacier stream, which gave me hopes that it ran in a gully leading down along the range that forms the southwestern side of Mossel Bay, when the clouds lifted and we found ourselves right above and at no great distance from our destination. Away rolled the curtain showing us first the bay, I was surprised but glad to see empty, though a vision of continued toil, instead of luxurious rest, and a half feeling of disappointment may have for a moment influenced me. We luckily had our own provisions and at about 7 p.m. made our first meal, finding water with some difficulty, although there was and had been a ceaseless trickle at one's feet, as the melting snow runs off between the interstices of the rocks, but often at a considerable depth. On reaching the bottom, we sunk so deep into soft red mud that we gave up the idea of crossing the half mile of it that lay between us and the house, as the windows of the latter appeared completely boarded up, the door barred, and no smoke coming from the chimney, though from a sound as of distant shots we imagined it must still have inhabitants. My companion now declared his unreadiness to return by the same route, wishing to take along the bottoms. I assured him that the distance round the extremity of the range would be double if not more than by the route we had come. He, however, preferred it, and as I felt very fit for walking and hated the toil of dragging one's limbs through the snow, I consented, the more readily as we had seen no living things coming except a Burgo, a few snow buntings, and two rein geese, and several fox tracks, and I was in hopes of finding deer at these bottoms. Leaving these a bit, we crossed over the saddle of a projecting hill and took down a snow slope on the farther side, very steep, with a torrent running beneath it in parts, which, pent up at one place, burst through the slope in a grand fountain. We soon came to the bottoms which were, however, so full of pools of water, and the black moss of which proved so soft that giving up walking, compared to which walking across a wet fallow would be bliss, we had to keep along the broken stones at the base of the hills, going in places for a mile or two, in the style of progression depicted by Doyle in the immortal "Jones, Brown, and Robinson" [Victorian artist Richard Doyle illustrated *The Foreign Tour of Messrs Brown Jones and Robinson* in 1855. He was also the uncle of Arthur Conan Doyle, who as a young medical student would meet Leigh Smith in the Arctic in 1880]. Endless streams and rivers which came roaring from the hills had to be forded, and at one or two regular mountain torrents, we had great difficulty in finding a ford where we would not be swept away. One river had a bed over a quarter of a mile in breadth; this must, however, have been owing to an unusual spate and was not very deep. The rocks and mountains about us would be very interesting to a geologist: Granite, Syenite, Trap, the latter very dark-coloured, then beds of mica slate, dipping at a tremendous angle, and above again dark brownish yellow peaks of hyperite [sic] or some of such rocks. We got on quickly enough now considering, though to my horror I found my comrade showing unmistakable signs of getting done, just as the change from the snow had made me feel as fit as at starting. He was, however, very game, but it certainly is annoying to be obliged to keep your eyes firmly rooted on the ground as you pick every step. Every now and then, I took a look thro' the field-glasses for deer, and tho' I saw none of them, made out the *Polhem* steaming along near Verlegen Hook, and a Norwegian smackie off the mouth of Liefden Bay. A halt for 20 min at 11 p.m. and an eider duck egg with rum besides a feed picked up the Norseman wonderfully (we found a goose's egg but smashed it). After a walk where we continually thought we had come to the end of the range but always found a hill further on, we at length after a third halt crossed a low collar and found ourselves above Hinlopen Straits and soon made out the bay and ships. We had seen no deer, tho' fresh traces. We now had a tremendous pull of eight or ten miles to Eol's Cross [Eoluskrossen, the site where a cross was erected by the ice-bound crew of the sealing schooner *Æolus* in 1855], where we were to be taken off in a boat. These bottoms were still covered with snow melted, however, underneath, and the journey over this deep slush or rather through it almost exceeded the toil of our first one. I did not like to stop, as Charlie, on the occasion of our last and third halt had at once dropped asleep; we were, however, obliged to stop once and feed again. We were now at a very low level and the snow melted into streams often up to our middles. I now pushed on to Eol's Cross and proceeded to fire the preconcerted signal. It had, however, no effect, as everyone was asleep, tired after the watering, getting in coals &c. This is a wooden cross standing on a pile of stones at the summit of a rocky promontory here jutting into the bay. The rocks, tho' not columnar, are dark and have a basaltic appearance, and the cross was, I believe, erected in memory of the captain of an exploring vessel named the *Æolus* [see note above]. A little below the cross lie 30 Dutch graves. In less than an hour, Charlie came up. It had now come on to blow from the ship and firing, which had had no effect when it was calm, became useless. The cross of *Æolus* is certainly a fitting name for it can blow on that rook and no mistake. Hanging out signals from the cross, we next collected chips of wood and with these and some paper succeeded in making a fire in a small oven of stones that we constructed under the lee of the cairn. This we kept up with moss, lichens &c and other things likely to make a smoke, but tho' we made plenty of this, it blew chiefly into our eyes and round the base of the cairn and naturally in such a high wind refused to use rise in a column. We could not of course leave this bleak spot as there would then be no chance of their seeing us, but seating ourselves in shelter as much as possible, we lit our pipes for about the thirtieth time, and proceeded to divest ourselves of some of our wet things as a result of fording ice streams up to one's middle. [It] was a very cold and clammy feeling now we were stationary. We warmed our feet by heating stones, and the Norseman soon after dropped asleep. I was rather nervous about letting him sleep, as I have heard it is dangerous, but being a sunny morning tho' cloudy and he out of the wind and near the fire, I let him be, and showing myself every now and again for a quarter of an hour and firing about two shots per hour, passed the remainder of my time in tending the fire and smoking. At about 9 o'clock, I had another look through the glasses. Men were busy at work on the ships as they had been for two to three hours. I piled on the remainder of my moss &c and proceeded to fix my last two shots in quick succession and how the relief of seeing a boat put off in a few minutes and getting up sail bear down on us. I forgot to state that one of the first sights that greeted my arrival at the cross was the Swedish S.S. *Polhem*, lying quietly at anchor near the other two ships. The wind now fell and, rousing Charlie with some difficulty, we collected our traps and made for the beach. The wind falling, we had a calm row back, and "putting our backs into it," got on board in time for breakfast: broiled ham and eiders eggs, fried potatoes &c, a great luxury after a long journey like ours, 30 miles of difficult going, let alone ups and downs and unavoidable detours.

July 4, Friday.

It soon comes on thick, and blows hard with snow, so we are unable to do any photography, dredging &c, and pass the day in paying visits to and receiving them from the Swedes. The *Sampson* is still alongside, and we continue to spoil her. The gale is still raging in Hinlopen Straits and tho' it is calm enough here we can see the dense fog and mass of grey-driven clouds come pouring out of the mouth of the straits. While on the hills yesterday, I could see right above those clouds, which seemed like dark pestilential fumes running across the sea, was bright and clear, and the top of the hills beyond appeared in calm bright sunshine. No one tried the shore, as the others had landed at Eol's Cross yesterday, and taken over the bottoms towards Verlegen Hook, seen nothing, and found the going as I have described, it being in the latter half of my journey, simply execrable. I had seen their footprints and spotted them all, Eaton included. He had done better than the others, getting some satisfactory specimens. They had got some sandpipers and sea swallows eggs, and the crew some eiders eggs. It seems that Professor Nordenskiöld and Lt Palander and the sledge party have returned safely. They had seen from Marten's Island the hopeless state of the ice to the N, as we saw it from Table Island, and left a depot at Parry's Island and their boat at Marten's Island, taken across the ice to the North East land. Along this they travelled on the fast ice and, at times taking over low necks of land, settling several important points by observation. They travelled east nearly as far as Cape Leigh Smith [Kapp Leigh Smith] and saw patches of open water off this land. They experienced very bad weather being absent in all 60 days and experiencing 50 snowstorms. Bears were numerous along the N coast, and they killed three. They also met with some reindeer. From their most eastern point, they took inland across the country, the interior of which is so completely filled up with glaciers that it was like being at sea, and one of the leading sailors cried out "Land ahead" on seeing a projecting peak of rock. Snowbirds, snow buntings, and a Burgo were the sole living things they saw. Travelling was very laborious, and hidden crevasses numerous beside constant fallings through the crust of snow, but by travelling tied together by ropes, accidents were avoided. They slept in a tent, and pemmican formed the staple of their food, except the products of *la chasse*: bear meat they found excellent. They made for Cape Mohn but were prevented from getting within some distance of it by an impassable glacier. They crossed numerous wide and otherwise impracticable crevasses. By the existing snow bridges, they found they made Hinlopen Straits about Wahlenberg Bay [Wahlenbergfjorden] and took up the N side to Shoal Point [Langgrunnodden] where they had two boats, in which they crossed to the Verlegen Hook, where it is only ten or 12 miles to Mossel Bay. The ice there broke up on 30th June and the merchantman and *Gladan*, who had, I believe, been doing a little sawing soon afterwards, started for home, the *Polhem* leaving the astronomer and three or four others at the house, then took N to collect the boats &c I suppose and, if possible, execute somewhat more of the proposed objects of the expedition. She got as far N as 80° 42′ and then took South and, proceeding to Grey Hook, interred the bodies of Mattilas and his companion. They have, I understand, received some provisions from Norway, news &c, and amongst other things heard the fate of the party of fishermen who, as I have said, coasted down to Ice Fiord in boats and a letter from whom, was found by the *Sampson* in Kobbe Bay. They made for Ice Fiord where, it was believed, there was a settlement, as a coprolite stone, giving good phosphate for manure, was being worked by the employee of a company, but tho' a house had been erected and every preparation made the men who were to winter, had backed out of it, at the last, and returned in the ships which brought their provisions &c. The fishermen, however, found the house and an ample store of provisions, neglecting, however, the preserved meats, vegetables, and wholesome things. They lived on salt meat, the effects of which, probably combined with want of cleanliness and exercise, produced scurvy and they are all dead. The *Polhem* seems a comfortable enough ship, as she is not intended to winter in, as she is of iron, which would be awfully cold, but they had their house &c. She is strong, but iron will never do for ice work. Her upper deck, which is completely covered in, is the part chiefly lived in, and the top of this structure forms a sort of hurricane deck, bridge, and quarterdeck combined. We did a little dredging tonight and turned in at midnight.

July 5, Saturday.

Hazy, cloudy, and cold and a good deal of wind at times, after trying the boats with their gear, sails &c, in fact at one time there were so many boats about from the three ships, it had quite the appearance of a Regatta. Potter and I select the "casey" and, with lots of ballast, off we go. We take round the bay, shooting, and the *Polhem* soon after steams out. She is going to have another try N before she is homeward bound, doing some dredges and sounding in Hinlopen Straits, if possible, on her way down, as the wind has moderated and shifted blowing down the straits. We go in for getting as many birds as possible at a shot and, after some good sport, join the whaleboat in which the owner is dragging with the net at the river mouths for salmon. We run back with a fair wind in time for dinner and soon start again afterwards. Dragging our boat on to the ice, we soon make the land and explore right up to the head of Treurenberg Bay on the western side. We fail to get the desired geese, as they were very wild. There were a tremendous lot of ground seals on the ice, equally wild, and I tried some long shots at 500 and 600 yards. But the "sighting shot," for elevation, usually frightened them, tho' I hit one (with a ricochet). No deer to be seen, fresh marks though. We got some eiders and sea swallows eggs and a few ducks. Saw a brace of pintail but couldn't get near them. When we got to our boat, we found the floe a good deal broken, and our ballast, which had been deposited on the ice, previous to dragging her up, had divided itself by two or three. Blowing hard, so we reefed our sail, and though it soon moderated so as to let us shake out the reef, we carried a good breeze home and got on board, supped and turned in by 2 a.m.

July 6, Sunday.

Worse weather, cold, and clammy. The *Sampson*, who had been paid off a bit yesterday, lying inshore of us, and the owner and I soon board of her and chat with Captain Walker and look at his charts &c. A few of the men go on shore today and collect whalebones &c, ancient and very bleached, but numerous enough. We hoped to have a *chasse* of white whales today, as by observation we know their way in and out of the bay, and hope, by stretching a net in their course at the right time of tide, to get some. They, however, don't put in an appearance, and as it is a bad evening, I put off my intended dredging with a trawl, as we are to be under weigh early tomorrow if possible, and it seems hard on the hands to take them out now. The two Captains have been to dinner today. It snows hard tonight and tho' I wait up 'til 2 a.m., as it is calm enough for a photo if the clouds would lift, snow and haze continue.

July 7, Monday.

4 a.m., fog and snow. 6 a.m., ditto and ditto. 8.30 a.m. seems clearing and at 9.30 round goes the capstan to the tune of one of those endless songs with a chorus melodious enough at a short distance. The *Sampson's* crew, by the way, sing that glorious Yankee song "Georgia." She is to follow at her leisure. After steaming a bit into the offing, we knock off steam and, with a strong breeze in our favour, go cruising along the land. The haze lifting in an hour or two, we have a splendid view of the straits, parts in the strong bright patches of sunlight, others with their commanding hilltop cloud capped or haze hidden, for it is a cloudy sky with bursts of sun. We keep pretty close to the western shore and pass along not far from some of the glaciers where projecting ends rise abruptly with vertical walls from the sea. The fractures show crevasses and endless fissures: however, less green and more snowy than the Swiss ones. These glaciers draw on our devoted heads or rather our admiration of the said glaciers, tremendous yarns of Davis Straits: there they rise abruptly thousands of feet from the sea &c &c, the seven ice mountains, to use their own expression, are but a speck to them. Then, when a large mass breaks away with a crash, falling into deeper water than is usually the case in the Spitzbergen seas, and floats away as a fair-sized berg, their contempt is supreme: a bit "bergie piece" like that "why in the straits you see them one, two" and it soon mounts to seven miles long. We run gaily on and, getting the log overboard, try our hand at surveying as we go down.

The straits are here about 15 miles wide, varying to five in the narrowest part, as we approach our destination Loome Bay [Lomfjorden] on one side, and on the other we open up Wahlenberg Bay. The land 'round this consists of low rolling land with huge glaciers and there is a low island lying in it, and another called Hyperite Island [here Chermside is apparently referring to the two small islands at the mouth of Wahlenbergfjorden that were later named Gyldénøyane after a Swedish naval officer and member of the 1901 Swedish-Russian Arc-of-Meridian Expedition, Hans Olof Fredrik Gyldén] consists of the curious dark stone, which I have spoken of just outside it. Cape Fanshawe [Kapp Fanshawe], the southern promontory of Loome

Figs. 9.9 and 9.9.1 Cape Fanshawe on the east side of the entrance of Lomfjorden (Lomme Bay). This area was visited twice by the expedition, for about 4 days following 11 July, and again on 30 July, and in both cases, they met Norwegian hunting vessels, as visible on the images. The different snow situation on the images in the slopes of Cape Fanshawe allows us to allocate this image to the first visit, with Chermside's position most likely on Hingstsletta a bit south of the ice front of Valhallfonna on the western shore of the entrance of Lomfjorden. The matching recent image with sailing vessel *Antigua* in about the same position as the *Diana* 125 years earlier was taken on 2018–07-04 by Andreas Umbreit

Bay, is a cliff of bold profile, its rocks being of grey mountain limestone, some of the upper strata which look as if fossil producing being very plainly marked. Along this cliff beyond the bay is the Loonery, and on the ledges of the cliffs, sit simply myriads of birds as James remarks, like a "regiment of sojers." Four or five miles down on the same side, the rocks are of a different formation, and rise in vertical and basaltic looking columns. We saw a great many white whales some coming very near the ship. In or rather on the edge of the fast ice of Wahlenberg Bay lay six white whalers (Norwegian smacks). The straits were closed just below Fosters Islands [Fosterøyane] by land floe stretching right across them and, running through the drift ice at the edge, we hooked on. We were abreast of and were about a mile from the central one of this group of islands. Surveyed in 1827 by Parry's expedition, they are of dark rock and the largest is two or three miles long. They are very bare and bleak-looking and still covered with snow. We were anxious of landing on them tonight, but it comes on to blow and we have to give it up, just as all was ready for starting. We have a very good view of Cape Selander [Selanderneset], the southern cape of Wahlenberg Bay. It is a very rounded promontory, that is on plan, the elevation being abrupt enough with curious seams, and fluted groovings running down its face in vertical and slanting lines, its face rises abruptly enough but the cliffs are of uniform height, and the land above table land, sloping southward at a gentle angle, until some miles down, the limestone disappears beneath the most dark layer of rocks. The surface of the flat land or, more correctly, limestone slope is still very snow covered, the rock showing only here and there, and the dip and strike of the strata seen from the ship are very well defined. Away along the same coast, we see Cape Torrell [Torellneset], and in the centre of the channel at about 12 miles distance Wahlenberg's Island, and many others of the same group; on the western side a succession of cliffs and glaciers. Stretching towards Thumb Point [Tumlingodden], and up the straits Duym Point [Tommelpynten] and Cape Fanshawe, a succession of low dark rocks and islets was lying off the coast. We see a good many seahorses and ground seals in the water.

July 8, Tuesday.

Blowing very hard, and we are in a by no means bad berth, all things considered. We give our minds to making at the same time cartridges and a tremendous mess in the cabin. During a lull in the wind this afternoon we got up steam and put out to seek a better harbour, but not such a heavy head sea that we were glad to put back to about our old position. We saw the *Sampson* today, but she hauled her wind when she caught sight of the land ice and beat up the straits for a harbour again. Nothing much goes on as we can't get to Foster's Island for a look out. In the evening, an old and a young bear appear on the floe, but refuse to be beguiled towards the ship, not even by burnt offerings of loons and eider ducks.

July 9, Wednesday.

No photography to be done, no nowt, it blows harder than ever and is very cloudy, and there is a good deal of motion on even among the ice. I go for a walk on the drift ice in company with a boat hook to inspect a "bit bergie" that is an object of intense and unpleasant interest to the Captain, but as I suddenly disappear thro' the ice when nearly at my journeys end, I soon return very wet and with my tail between my legs. There is another bear excitement tonight and *faute de meilleur* [for lack of a better term], we all make it our amusement to watch him through glasses. He is a very deliberate brunie, but for all that his soft flexible limbs and broad feet get him along a great pace. By the bye he gets near the drift ice and then his proceedings are ridiculous. He never goes ten, seldom five paces, without falling in, and having got out with a great deal of exertion at once repeats the process. Then he has better luck and catches a seal. Then, when his din is done, a little further on he gets wind of the nice din we have prepared for him on the engine room fire, and so in an hour between successive immersions he sits down on his hams and sniffs. Then he has a roll on the snow, and then a scratch, then he points like a dog and one gets a good look at his black muzzle. Then as he draws in a pace or two, he tumbles in again, and when he is out and has had a shake thinks better of it, and goes clean off, getting a good way on as soon as he gets a hard floe.

July 10, Thursday.

Blowing very hard, and we cannot get on to Forster's Island, as we are closed in for more than a mile by drift ice closely packed, that assembles no one can tell where from in this country, like a street crowd at home. Among it are a few small boys, and there are one or two pretty near us. The Captain is perpetually in a state of excitement, about one or other of them coming and taking away our boats or something. There was a "bit bergie" that afflicted him tremendously all yesterday, and about midnight he rushed on deck in his shirtsleeves, with the thrilling intelligence that it had moved two handspike lengths. We worked at photography today, taking instantaneous ones as it was blowing a gale, with, I hope, fair success. We also did a good deal of dredging in 40-odd fathoms. Tonight, it is fine and the wind dropping and as soon as the tide slacks off the drift ice, we are to get under weigh. Another distant "brunie," who catches a seal, and so is proof against enticing smells tho' well to leeward. The floe is so rotten that unless they are obliging enough to come to us, we can't get them.

Figs. 9.10 and 9.10.1 Recognisable scenery details on this image are sparse, but the situation with the vessel off a low ice cliff of what looks like a long and smooth sea front of an ice cap that fits very well to the situation on the south side of the Valhallfonna ice cap on the western side of the entrance of Lomfjorden, where Chermside was on shore with his camera on the occasion of the first 1873 visit of the *Diana* for some days following 11 July. The accumulation of snow and fallen ice pieces all along the visible foot of the ice cliff indicates very shallow water, which matches with the situation at Valhallfonna well, and was in contrary to most of the ice cliffs of the ice caps terminating in Augustabukta and Vibebukta, where the *Diana* visited as well in 1873. The low perspective of the image indicates that Chermside must have been either on a beach or on a safe ice resting on a shallow sea floor near the shore with his clumsy photography gear. Chermside mentions the hunting vessel *Sampson* anchoring very close to the glacier front in Lomfjord and this fits very well with the photography. As Valhallfonna has lost thickness and a bit of extension since 1873, the depicted glacier front is of no help for trying to determine the position of the photographer more precisely. Recent image with front of Valhallfonna and sailing vessel *Antigua* was taken by Andreas Umbreit, 2017–08-22

July 11, Friday.

Steam up early and we are nearing Cape Fanshawe. By the time we get on deck, we find the entrance to Loome Bay much clearer from ice, and as we come on deck after breakfast are steaming into it. The *Sampson* is at anchor right under the glacier, which runs in from Ice Cape, and two smacks nearer in. We get to anchor in 11 fathoms about ¼ mile from the beach ½ a mile from the floe which with a strip of ice, only a mile or two wide with open water inside it, closes

Figs. 9.11 and 9.11.1 Hecla Cove (Heclahamna) in Sorgfjord (earlier: Treurenberg Bay) was visited by the *Diana* from 2 to 6 July and on 2 and 3 August 1873. For a British expedition to Svalbard, this was a famous site, as its name goes back to the expedition vessel *Hecla* of the 1828 expedition by Parry, which anchored here 45 years before the visit of Leigh Smith. Chermside's image shows the mountain Heclahuk (Hecla Hook, 434 m), ascended by a Swedish expedition (Torell, 1861) just 11 years before the arrival of the *Diana*. Heclahuk was also the name-giving mountain for the rocks of the older geology of Svalbard. Chermside used the opportunity of this visit also for taking the images 013, 014, and 015 from the area of this bay. The recent image with an almost identical view by Andreas Umbreit, 2011-08-31

the bay. This bay is a long and very beautiful fiord: huge mountains of grey, brown, and dark and purple rocks, with their snow-striped sides and crowned head rising above the white floe, dotted with its blue pools. Above the still clear depths of the open water, sea, and ice, hill and mountain glacier and fell, and long fertile looking mossy bottoms luxuriant in the short-lived, quick-growing vegetation of a Spitzbergen summer, all at once enhance the beauty of the vista of a long fiord rather narrow compared to its length, which in about eight or nine miles branches into two, the arms of the bay winding out of sight. The *Diana* was, I am glad to say, brought up for once at a convenient distance from the shore. The captain of one of the smacks comes on board for a sawbones, one of his men having shot himself in the arm, and Eaton goes off to look after him. He has got five deer here, and the *Sampson* has a boat away after them. After lunch, the owner, the Skipper, P and I, and two hands start for a quest in the Shetland casey. After reaching the floe, we have to drag her about two miles across it, in which we are much

Figs. 9.12 and 9.12.1 View southwards from Hecla Cove/Heclahamna to the lower end of Sorgfjord with Dunérbreen at its end—taken by Chermside probably on 2 August 1873. The recent image taken by Andreas Umbreit (2011–08-31) from Crozierpynten on the north side of Heclahamna. Compared to 1873, where its clear ice-core moraines reached partly all the way to the shore, Dunérbreen in the background and its moraine have melted considerably

assisted by her sail and one or two hands from the ship who then return. We sail down towards a landlocked cove, above which rise some very tall dark-jagged hills almost completely free from snow. We pass the mouth of a river that lies pretty deep and runs in a narrow sunken channel. The water at the mouth is very turbid and looks a good place for salmon; the Norsemen have caught a few here at various times. This bay later on is a great resort for white whales, and the low sandy beach is a very convenient point for hauling through on; their abundance is testified by the endless skeletons all along the shore for some miles. Landing here and taking the usual precaution of getting our boats wet at the outset, we take up a gradual rise of bare rocky ground, and in about half an hour come upon a large valley buried in the hills, the outlets being only small rivers running in deepish chasms. The bottom is for some distance a swamp or shallow lake, and it rises towards the glaciers that surround all its farther sides. This gives way to flat ground intersected with endless shallow streams and dotted with ponds. We take along the side that bounds it to the eastward. On the slope of the hills, the sides of which tho' stony and rough at the top are of an even tho' deep slope near the foot, and more fertile than I had imagined this land ever to be—mosses of every shade of green, thick soft and swampy, yellow, white, and pink flowers altogether very desirable ground for does, horns and hoof marks abound—but the deer of the country does not show (no wonder as the Norsemen and Sampsonites have already had the same beat). But, beguiled ever onward by the freshness of traces, we come at last to the end of the valley, the outlet being a narrow passage between the high vertical moraine of an immense serpentine glacier, from which are pouring numerous white gills and cascades. The steep hillside on which we are climbing, on the reverse slope of this our forces which had been separated reunite, we get one ptarmigan, but a heavy snowstorm coming on prevents a search for now and choosing a convenient rivulet we have a feed. As the snow is very thick, we have to give up the idea of trying some long likely bottoms sloping up from the shore of a very deeply indented bay, above which we have come out. I determined to take home along the shore of this, working round, while the others take some likely ground back that we have not yet tried over. On a stony ridge above the shore, I come upon a vertical pillar of flat stones, and some inscriptions in Norse. Almost immediately below are some geese in a pool and two of these it appears cannot fly, doubtless a safe and easy prize: are they? No ostrich ever ran like those evasive geese (to miss the old joke) and away they went across the floe towards the bay head, right the wrong way. Away went I, getting a good ducking on to the floe, besides soon finding myself done to a turn, and the geese "forrard on" harder than ever. Being on the floe, I gave my mind to trying to get some pin-tailed ducks, rather rare up here, which were disporting themselves therein, but after being enticed by degrees some way further I gave them up and was glad to regain the shore in about half an hour. Along this I now took, following perforce all its bents and sinuosities. The walking was, however, good and in about an hour and a half the blinding snowstorm gradually crept off and a fine bright sun lit up the fiords, gradually driving off many of the clouds that dispersed on the hills. I was walking close under an immense dark range, and the ground at the bottom was fertile and excellent walking for Spitzbergen, tho' the torrents that carried down the melted snow had such a little distance to travel and such a steep incline to do it on that there was some tallish fording to be done. I shot a very pretty snow bunting for a specimen and encountered endless ducks and geese. Of the former I had no need and so left them undisturbed and the latter had no idea of being shot, the majority of each flock flying off, and the remainder who were unable or unwilling to fly ran at a desperate pace to the water's edge and then put out to sea. The bay was alternately composed of strips of open water and ice from the number of pools that dotted its surface in an advanced state of decomposition. There was always lots of open water at the mouths of the streams, but I could see no salmon. By degrees, the hills declined more sharply to the water's edge, ending in low cliffs some distance above it, and I had some roughish climbing. The water below was intensely clear and the varied colours of the bottom plainly discernible for a long distance out. The hills on my left were very abrupt and apparently inaccessible, and here and there in the deep shadows ran back with small glaciers or roaring white gills, and the scene brought back vividly a winter one among our own English Lakes. On reaching the point that formed the promontory of the bay in which our boat was left, seeing no sign of the others, I took back a bit, and climbing the hills in a practicable place I had seen before up the course of a stream reached a table land hidden among the hills. This was about a mile across, and my stream proved the outlet of two lakes, one of which was still nearly completely ice-covered. These were replenished by streamlets from the face of a glacier closing the gap between the two highest hills on the left. On the right, the ground sloped, gently up to the next of the cliffs which overlook the bay where we landed and in front lies the dip between two hills through which I descend. I am suddenly startled by a yell. At first, I thought it must be a shout from the others. I soon, however, make it out to be a fox, a very fine big beast, perfectly black with a grey brush, who barks his very life out at me. I bark back but fail to take him in, also to stalk him, as directly I disappear to get my next point. He at once changes his position so as to keep me in sight, a more distant yap informing me that he can still see me. Out of vexation, I fire a long shot, and then as he goes away, give him a long view *holloa* [a very loud sound] and gone away, and as the echoes bring it back, I see him away over the rocks at a tremendous pace. I can now see the party below making for the boat, and the surprise my foolish vent to my feelings has caused, I gain the beach not without difficulty as for a duffer at it, the climb down the rocks is dangerous and giddy. Nor do the presence of two firearms assist me. The last part of the descent was quick enough being performed on the summit of an amateur young avalanche, which obligingly carried me down with great steadiness. Potter and the owner are on, and bravely ford, the river. We row the boat as near the ship as possible and dragging her on the beach clear of the ice get back by 11 p.m. The others had tried home blank.

July 12, Saturday.

It is blowing hard all day and is rather cloudy. We take a photo or two from the beach, and a good many voyages are made to and fro to the *Sampson*. Both of us do some shooting for the pot: guillemots and some Arctic terns, sandpipers, boatswains, &c. Eaton, when on shore yesterday, got some boatswains eggs and some plants. The Norwegian patient is visible today. The skipper is a fine good-looking young man and the crew seem very hardy efficient sailors. Late in the evening, the owner and I go for a walk on the glacier, a very good imitation of a treadmill, an endless gently rising slope, with snow knee deep at every footstep, a good two miles up this slope, giving no view, or probability of reaching the top, and a snowstorm coming on we return home, not omitting to wet our feet just as we get to the beach. Turned in about 1 a.m.

July 13, Sunday.

A lovely day but clouds on the hills. There is an expedition on to Foot Island [Foottøya], about six miles off, and another to Loom Head [Lomfjordhalvøya] and the islands beyond but, deciding to stop for the chance of photos, I prepare my apparatus and take the camera ashore. The redoubtable Captain Mack has arrived in his smackie, a good-looking English-built yawl, and having a sick man on board Eaton goes to visit him. After lunch, the owner, P, the engineer, and the captain start for the island hoping to be able to penetrate beyond it to a good piece of reindeer ground at the head of the bay. On reaching the boat, they set sail with a fair breeze and soon get across the open water to the next strip of floe. They have lovely weather and a fine view right up the fiord and the hills on both sides and are much struck with the beauty of the bay and hills round whose feet my road lay two days ago. Dragging their boat on to the floe, they take across it towards the island. The ice proves very rotten and they have to make continual circuits to avoid the numerous holes. By the time they get within 200 yds of the island, it has become so rotten that they can't proceed; the owner and the engineer both having been in up to their necks. The land on the eastern side too appears impracticable, as there is a glacier with a high moraine to be crossed, and not without difficulty do they get back to their boat, and so home to the ship by 6.30 p.m. I have been taking photos but with little success, as I came across a batch of "dry plates" that had gone bad, so my carefully selected views came to nought. Some heavy drift ice has come in from the sea and is all around the fleet consisting of the *Diana*, *Sampson*, and three smackies. These latter first drag and then have to weigh and one is nearly driven onshore; this is sea not bay ice that come in from the northern pack. This morning, we were attacked by a "bit bergie" which has previously overturned itself and performed sundry other annoying evolutions so close to a smackie that the latter had to weigh

Figs. 9.13 and 9.13.1 View from above the eastern shore of Heclahamna to Crozierpynten on its northern side, photography by Chermside probably on 2 August 1873. Twenty-five years later, this location became the main Swedish base of the Swedish-Russian Arc-of-Meridian Expedition (1898–1902), which erected a number of buildings here, along with scientific measuring instruments, to determine the exact shape of Earth so far north—with the first precise triangulation maps of much of the archipelago as a side result. The recent image (Andreas Umbreit, 2011–08-31) shows almost the same perspective, with some remains of the Swedish base visible as ruins, while Chermside's image shows the same location still in its pristine condition

anchor and shift his ground. We managed to keep him off or rather to swing ourselves clear off him, and away he went, dead in the winds eye, the wind having from the small proportion that the exposed bears to the immersed volume, but little effect compared with that of a current. The first mate with Eaton and a boat's crew have meanwhile sailed away past Cape Fanshawe—on which the Mallies, themselves so common, and their eggs so rare, appear to breed—to where, from beyond the glacier that fills the space between it and the grey limestone cliffs of Cape Fanshawe, there was a craggy and shelved face of dark basaltic-looking columns of hyperite. Along the foot of the cliff a young fox was trotting like a terrier, who not having learnt the hereditary cunning of his race fell a ready prey to the *chasseurs*. The cliffs simply swarmed with loons but there was a rather heavy sea on at the foot. The loons assembled on ledges and a great hunt ensued; a gun, stones, seal clubs, &c being all in requisition. The methods were simple: sweep a ledge with a boathook &c, and those below grabbing them as they flew off. Eaton meanwhile got a good many satisfactory specimens of plants. Then they fell to taking eggs and

Figs. 9.14 and 9.14.1 Heclahamna, with the Norwegian hunting vessel *Sofia*, as depicted by Chermside, probably during their second anchorage here from 2 to 3 August 1873. Observe the little dot on the hilltop to the right, which is probably the base of the flagstaff set up here by the Parry expedition in 1828 (see also images 012–014) as the only clear human sign in an otherwise pristine scenery. The corresponding new image (Andreas Umbreit, 2011–08-31) shows almost the same perspective, 138 years later—now with parts of the ruins of the 1898–1902 Arc-of-Meridian Expedition Swedish main base, which in the meantime has been both set up and then given up again

collected a large number. They found a good many Burgo-master's nests and stood on to one of the dark rocky little islets. This was very bare and there was still much snow upon it. The ducks were, however, breeding here, and they collected a great number of eggs and proceeded to feed off biscuit, tea, coffee, and eggs, as they had managed to find some driftwood. They did not come in sight 'til this morning and had to pull the last few miles and experienced some difficulty in reaching the ship, which they did by 9 a.m.

July 14, Monday.

A party was dispatched early for the casey and when they came on board at about 11 a.m., all was ready for a start, when Captain Mack, who has a patient under Eaton's charge, comes on board. Like all these Norwegian skippers that I

Fig. 9.15 Cape Fanshawe seen from Hingstsletta on the west side of the entrance of Lomfjord (Lomme Bay) during the first visit of the *Diana* (ship to the right) from 11 to 14 July 1873, with two Norwegian hunting vessels in the picture. For a modern view, see Fig. 9.9.1

Figs. 9.16 and 9.16.1 View from Hingstsletta on west side of northern Lomfjord (Lomme Bay) southwards over the inner part of the fjord. Chermside took this photo during the first stay of *Diana* in Lomfjord from 11 to 14 July. Most of the bay is still covered by dense drift ice or winter ice, with the ice border visible in the water a few hundred metres into the picture on the left side. Much of the lower parts of the gravel-covered plain behind the beach wall is still covered with ice and snow from the winter, while the corresponding image from 22 August 2017 by Andreas Umbreit, about 6 weeks later in the season, is free of ice and snow, both in the fjord and on the lowland

have met, he is a well-informed, superior class of man. He seems to know a great deal about these seas having been far to the East in high latitudes as well; as in 1871 he penetrated as far E as 82° in Lat 75′ and almost as far E in 77° 30′. It is, in fact, these Norwegian smacks that coming here year after year do most of the exploring, rough and ready as their discoveries are. Still, they find new bays, islands, and anchorages, sailing in fact anywhere where the ice will let them, and where they can pick up anything: seahorse, seal, white whale, or bear that will help to make a voyage. As for instance, last year while the *Sampson* was undergoing a tremendous buffeting among the ice in vain endeavours to get through Hinlopen Straits, three Norwegian smackies, who have come round the other way, sailed east in tolerably open water and sailed completely round King Charles Land [Kong Karls Land], an island the east coast of which had up 'til then only been seen from Mount Middendorff [Middendorffberget]. This voyage repaid them well, as reindeer were numerous and large, and bears and seahorses were plenty. Of late years, however, they have taken more to a regular plan: white whaleboats—of which these are two or three very good specimens here, and their powerful but expensive nets, with all the necessary gear of grapnels—are left at their fishing station from year to year, until these are free from ice and their boats were accessible. Also, the proper moment arrived and with it the whales, which are not supposed to live in the bays, appear at certain times—coming and going generally by a pretty well-defined route, at stated periods of the tide; all of these points must be ascertained by observation and experience on the spot to enable any fishery with a far prospect of success to be attempted. The earlier part of the season is spent in picking up what they can. Captain Mack, having been at the sealing and subsequently looking about, and pretty profitably too, stuck to the "saddles," after we descended them, but did not do much. Then he came over and looked in at Ice Fiord. He went up as far as the company house in Ice Fiord and there discovered the dead fishermen from Grey Hook. They had got along well enough until February when, owing to darkness &c, they gave up hunting, the results of which had been two bears and a few reindeer. The mate had kept the journal—the composition and style of which is, we are assured, exceedingly good. It ends in April, the entries for the last few days being illegible. He buried all but one, whom he could not find. One he found dead in the yard, the remainder, on forcing the door, were discovered in the house, mostly in bed, an awful sight from the ravages of the disease fatal to them all, and terribly wasted. He picked up a seahorse or two, and then putting into Grey Hook, discovered, by chance, he said, the blubber of the vessels wrecked there, which had been buried. This, of course, was a treasure trove. Thence he took a trip down Wüde Bay for reindeer, desirable as cheap provision, besides being from its freshness a highly prized luxury at sea. The Norwegians indeed, I believe, salt down a good deal of it and take it home. It is said, however, a very great delicacy in this condition. But then they are real economists and utilise everything: the down from the eider's nests and the skins and feathers of the birds they shoot, their wages, too. I mean, those of the A. B. seamen are about half those of our own sailors, and the provisions coarser and less, and yet, without these, the said sailors are a strapping hardy-looking race, tall and fine looking, and with a seemingly simple but high tone about them, often well dressed and gentlemanly looking. In fact, when I met one the other day on shore, who touched his hat to me, as I floundered my way towards the beach, wet, draggled, dirty, and habited in the garb of Spitzbergen, that will get torn, and bloody, and greasy, and painty, I couldn't help thinking it ought from appearances to have been the other way. These skippers who explore are encouraged to make observations by their home authorities, who seem really "right down go ahead," in pursuit of scientific enquiry, and are, I believe, in cases furnished with chronometers, barometers, and all other meters but a gaso [the gasometer, a structure for holding a large volume of gas, was a commonplace in Victorian England]. Mack himself or Mack *père* has or has had numerous ships, one of which is the *Sampson* found deserted in dire case two years ago. But then she was fully insured, cargo and all, so one must ask no questions. A steamship is now building at Bremen Haven with composite engines of 100 H.P., on the model of the *Vega* (already described) and her twin ship. Of this he is to be master, on behalf of a firm or company. He will then, I fancy, go to the young sealing at Jan Mayen in "March Month" and thence come up here, as the Norwegians no longer compete in the whale fishery, though now that they can build and equip ships as well and as cheap if not cheaper than ourselves, man them with equally good sailors at a lower rate, and provision them at less cost; it would seem inevitable that firms will combine, subsidised doubtless by English Coins, into Companies, destined to send forth fleets to cut out our trade at any rate. But penny a living about underbuying and underselling English labour and manufacturers is not good form, and time is required to mature anything, and outlay on experience, experience gained too often, by failure that cripples the undertaking for which alone it is of use. So, wait 'til it comes, and say it is kismet [fate, from early nineteenth century Arabic]. Meanwhile remember that, as yet, the harpoon gun is to the Norwegian, a weapon newfangled and dangerous, not to be let off on any consideration, and that Harpooneers are "*fiunt non nascuntur*" [made not born]. (Apropos of harpoon guns, the owner of the *Sampson* in his first expedition was obliged to demonstrate their harmlessness to his Norse crew, by filling them with small shot, and shooting loons.) One farewell bumper and the *Sampson* having got her sailing orders, Captains Mack and Walker depart, and at 3 p.m. we start thro' the ice, leaving the remainder of the flotilla struggling with the more important "bergie pieces," that *will* drift broadside on and among their cables. We push thro' the ice with an unusual amount of steam, just, I conjecture, to show them what the little *Diana* can do. We steam right across the straits that, for a wonder, are nearly calm and take in to the N side of Hyperite Island, seeking a channel here depicted on the chart and anchorage. The position of certain rocks and islands as seen in the sea and on paper not being reconcilable, we lose trust in the representation of this part of the coast. It may be all right, we dare say, but we can't make out any but a fancy resemblance to the locality and so content ourselves with splitting the difference, i.e., the chart is fairly right, and we are rather stupid. But still from time to time the question arises which *is* Hyperite Island, and it is not 'til after dinner that this question found its solution. We were then anchored in 15 fathoms or so to the southward and under the lee of some land, with some islets about it, with the slight inaccuracy of the said islets being, if we were anywhere where we expected, S instead of N of it. These were at once invaded by the egg hunters. It certainly is strange that tho' nearly all the eggs we get are hard sat, and we have over 20 dozen on board, the first enquiry on approaching an island from the mate down to Jack is "whether there are any "jukes" about." Their success was I believe on this occasion partial. Well, to this land there repaired in detachments, the owner, master, Eaton, P and myself, and Jack and Rover, and after a good deal of walking about, it was discovered that the long sought Hyperite Island was the land on which we were promenading. Only it wasn't an island, and so it was no great wonder we couldn't get 'round it. What it might have been in the days when fossil trees grew here one can't say, nor how long ago it might have been an island. Certain it is that it is now connected with the mainland by a nice handy communication with a glacier, or "*glaziere*" in the vernacular of the straits. *On dit* [one says] that sheerness was the last place made assuredly Hyperite was a postscript, imaginary or rather *figurez vous* [do you figure] as they always say in Frank land, a low bare peninsula, showing in equal parts patches of broken and sharp black rock, patches of soft slushy brown snow, and dull shallow pools of stagnant looking water, when inland the view confined exclusively to these items, vegetation black like the rock scanty and barely discernible by the naked eye, animal life *nil*.

Across Hinlopen Strasse the view is good: Cape Fanshawe, Loome Bay, and the glaciers as before. This side, however, it is blank desolation, a wide bay with low rolling and snow-covered hills enclosing it, except when it is bounded by glaciers hardly distinguishable from the hills, and stretching inland with a dull drear monotonous slope, ocular evidence confirmative of the Swedes statement that the interior of the North East land was an endless glacier. The distant Cape Selander and the islands are the only relieving feature and they seem round and naked and flat, so after taking a few angles we return. Wahlenberg Bay is still full of ice, the limit of which is not far from our anchorage, and tho' dotted here and there with seals, no seahorses are to be seen. We get one seal today. Nothing worth photographing to be seen.

July 15, Tuesday.

Soon after breakfast steam is up, and we are bound "on a lark" until such time as a return to the Seven Islands is advisable to see how the ice is getting on. As we near the mouth of the straits, a fog comes on but lifts in two or three hours. We come among a lot of heavy but widely scattered ice from the northern pack. As we are off Treurenberg Bay, we find the water absolutely covered with guillemots, chiefly Brunnich's [thick-billed murres], the common sort. What rookery they belong to we can't say, but their numbers and closeness are astounding. There is a very large school of white whales, some hundreds playing and sporting about, but they take good care to keep out of rifle shot of the ship. It is a beautiful day and very hot, and there are a good many of the large Spitzbergen seals on the ice. We lower away three times, but tho' all shot, only got one. We steam into Mossel Bay and come to anchor no great distance from the Swedish House at 8.30 p.m. How different on a calm sunny summer evening like this, it looks from the dreary icebound haven of four or five weeks ago. The ice is gone from the bay, most of the snow from the hills, the streams are full of water, the bottoms and valleys of vegetation. There is a smack at anchor in the bay, newly arrived, too late to set her nets this tide for the numerous white whales that are playing about. Round the bay they go, seemingly with nothing to do but roll and play, as they go past, close, so close you might touch them from the land, to the rocky promontory on which stands the house but in reality, steadily beating their ground for fish. I went on shore to take some dry plates. Unfortunately, however, as they are two years old, these, as often occurs, were a bad batch, and after a lot of work and trouble I got no result, or rather the streakiness &c of the plates spoiled it. There were some Lapps and a Swede or two in the house, but all is packed and ready for a start as the steamer is expected back tomorrow. I write a note. As is invariably the case, Eaton has a patient in the smack.

July 16, Wednesday.

After four hours in bed, I am roused by Eddy, who informs us that there are 30 or 40 reindeer, "a washin of theirselves in the water." We can't quite make them out from the deck, but I ask if he is quite certain they are not rein geese. Still, in spite of the indignant rejoinder, after a long pull in the dinghy to get 'round them, another boat with the owner going the other way, rein geese they prove to be. P and I landing, go after a seal on the ice in a sort of lagoon, who as seals mostly do, "shuts the hatch" before our arrival. The bottom on which we are wading is apparently mud and stones and seaweed grows on it with great success, nevertheless it is false and Potter suddenly follows the seal's example and shuts the hatch with startling celerity, which so convulsed me with laughing that it is some time before I can offer him any assistance. After an instantaneous photo, away we steam, and keeping the E coast come to anchor about midday off Aldert Dirkses Lakes [Dirksbukta]. These are the celebrated salmon lakes of Spitzbergen and we land to reconnoitre. After climbing the rocks that fringe the shore below the lagoon, we had a stony walk to the first two lakes of about a couple of miles. These were still so ice covered that it seemed useless to bring the net so far, nor from the number of large stones—at any rate near the beach—were they adapted for dragging. After a goose hunt, in which the geese tho' runners were one too many for us, getting on to the large area of ice that formed the centre of the lakes but at no place came near the bank we separated a bit, to beat for deer. It was a glorious day, and the scenery charming, grand, but not so barren or cruel, but little snow, and large patches of green

Figs. 9.17 and 9.17.1 View across sea ice pushed into shallow DeGeerbukta (DeGeer Bay) on the northern west side of Lomfjord onto Faksefjellet mountains and further mountains in innermost Lomfjord. This image was probably taken by Chermside during the second visit of the *Diana* in this area at the end of July, where she anchored in almost the same position off Hingstsletta as during her first stay here from 11 to 14 July 1873. Some of the mountains in the background are the same as on Fig. 9.16, but with less snow. The recent image shows a similar perspective, though without packed sea ice in the bay, taken on 22 August 2017 by Andreas Umbreit

plants, moss &c, in the low ground, high frowning but warm-coloured rocks rising abruptly above the lakes to an immense height but tho' wild and rugged giving the valleys a warm sheltered appearance. P and I soon separated as not expecting deer from the look of the ground I was more bent on exploring, thinking from the configuration of the ground that in the laps and folds of the hills, there must lie more lakes. Nor was I disappointed in my search. I do not know how many are known but only two small ones are shown in the chart and by the time I reached my furthest point I had seen eight. One of the most distant of these was of considerable size, four to five miles long and only separated by a narrow neck from another long and narrow one, the head, I imagine, of a different series finding a more northern outlet to the sea. There it lay asleep in the bright sunlight, dead, frozen, silent, not a pool of water, not a living thing, duck or sandpiper, the only sound of life the grunting of the ptarmigan, numerous enough, and the distant twittering shriek of the rotches, who inhabit every high cliff in Spitzbergen, without the least regard to its distance from the sea. It certainly was a fine subject for contemplation so I lit my pipe and contemplated how strange it ought to seem to be away in these speechless solitudes on a rocky craggy hilltop, alone with the works of nature, and with no more immediate connection with the busy turmoil of the world than if I looked at it from another planet; quite true, but it doesn't seem strange only charming. It does seem a little strange to be sitting here not 100 yds from a hard, unmelted snow slope, not so far above a frozen lake with a still hard form, solid covering and yet panting for a breath so hot as to lie gratefully on a mossy rock as if on a lawn in England under the fierce July sun. But my lake, so hidden, so still, buried among hills, with no outlet but through a dry gully and that through high ground and itself 20 or 30′ at least above its present surface a gentle curve all through its length, the line of beauty and high warm-coloured hills all round, it bears a striking resemblance in all but the colour of its surroundings to the *Lago Di Nemi* [a small circular volcanic lake near Rome, from the depths of which two Roman ceremonial barges would be recovered in the twentieth century]. Meanwhile as I lie and dream and smoke a thick white mist brooding over the waters has pushed out a feeler like a huge octopus, down this bay. It has caught our ship, and all the fiord in sight seem to have penetrated right down it, lying at rest on the water like a heavy pall. Above all is warm bright and sunshine, here and away across the fiord, as the mist cuts a waterline (*ainsi dire*) [so to speak] sharp and clear about two-thirds of the way up those, dark and jagged peaks, so far snowier too, but all bathed in the same flood of light. I gain the shore of the lake and walk across it and through the outlet, itself some hundreds of feet above the sea, and gain the shore of the next lower lake, ere I enter the confines of the mist. The walking is rough and strong, and I find walking in the lake at the edge the best, as from the pressure of ice, and the bottom is nice and flat, the said action is by the way a great blessing, much of the walking here especially on the hills has to be done on loose stones of considerable size, and from the effects of their covering of frozen snow they are most of them comfortably fixed in their own place, and don't turn over just as your weight comes on them, and throw you down and sprain your ankle, or else make your heart jump in anticipation of a fall, as other loose stones that I know do. From lake to lake and so to the last, then down the river, well wet by this time, the river runs N for some way and see it turns back, passes close to a beach on which stands a neat little log built hut, once the abode of one of those Russian hunters [Pomor], who were in the habit of wintering here. A great place for that natural British inscription, one's own name: "Ah, Sir," as James who subsequently visited it remarked, "you ought to have cut your name." Then down a lagoon filled with sea water, then a lower lagoon, chief products, ice, drift wood, and eider drakes, and then to the boat. The others were there and we went off to dinner and afterwards, as the fog put a stopper on photography, landed to see the mouth of the river drawn, blank, except seaweed. We took some sea swallows eggs here, and two nests of red-throated divers, and shot one bird of the same species.

July 17, Thursday.

Got a seal this morning close to the ship. The fog is gone and I undertake the photography (wet plates) while Potter goes on shore to expose some dry ones. Mine necessarily not very successful ever, at least not what they might be, as instantaneous photography is not easy and two out of every three fail: thro' the ships motion if time enough is given or else under-exposure. Sometimes, when the light is not hazy and filled with the air vibrating from the heat, we get results that will be, I hope, successful, but our difficulties would fill a volume. We are today to go on and anchor below the third of three glaciers that jut into the sea, on this side of the fiord a few miles along. I have to go in the casey and start as soon as I like and beat between first and second. Potter is to take from the second to the third. Below the third we can do as we like: the ship will be somewhere there, where she can find a good anchorage. A hasty lunch, some biscuits and cheese into the boat and away I go with two hands for my crew. We have a fair breeze and soon get down, the ship being by this time underway. But out in the centre of the fiord and nearly up to us, as we come 'round the furthest point of the glacier,

Figs. 9.18, 9.18.1, and 9.18.2 After an ill-fated attempt of sailing deeper southwards into Hinlopen Strait around mid-July, the *Diana* managed to reach the southern end of Hinlopen Strait with Augustabukta (10 August) and even Vibebukta (22 August, both in Nordaustlandet), navigating through fairly dense drift ice. These photos were taken by Chermside from on board *Diana* probably around 17 August, when they were in the ice up to somewhere east of Thumb Point (Tumlingodden, on eastern Wilhelm Island (Wilhelmøya)). The drift ice helps calm the waves, thus minimising the movements of the ship, which was essential for sharp photographies from on board with the long exposure times required. Stitched together, both images show the complete northeastern scenery of Wilhelmøya from North East, from Kapp Breusing (right) to Tumlingodden (left). However, different weather conditions on the two photos indicate that they were taken not immediately one after the other. The recent image by Andreas Umbreit, taken on 30 June 2012 from Torellneset, is somewhat more northwest than Chermside's corresponding positions in 1873, but shows the same coastline of Wilhelmøya from a slightly different angle—and with no drift ice at all in this part of Hinlopen Strait, contrary to 1873. However, ice conditions in Hinlopen Strait vary also nowadays a lot in summer. Not in every summer, but this area is often free of ice in late June nowadays as here in summer 2012

and haul our wind to go in short, we come into a sort of bay full of birds, chiefly an immense flock of 100 eider drakes with only one duck among them. There are also several pairs of King ducks, ducks and drakes, and several single drakes, and one bird by himself, apparently a sheldrake. I want him or a king duck, but the boat is sailing so fast and we are on them before I can get an aim with my rifle and they all fly off. I proceed now with one of the hands to beat all along the bottoms and slopes from the hills to the sea. These are in places stepped or terraced, in others long gentle slopes and then again steep slopes and flat ground below, the moss is rich, watery, luxuriant, green and red, saxifrages, potentillas and all the rest abound like the common buttercups of Spitzbergen. Stream after stream and gully after gully is crossed still no deer. Then I come to a valley or ravine, with a torrent bed but half-full tumbling into a lake ice-covered to a great extent. At the end is the inevitable glacier, at the sides the equally inevitable high dark cliffs, rent asunder or seemingly so by the glacier. Beneath them, the slopes with here and there a meadow, with thick knee-deep moss, and grass, absolutely green grass, but nary deer, only burgies and boatswains. I cross the lake on the ice and tumble in but keep my rifle clear and so on along. At last, after crossing with difficulty a sort of chasm containing a narrow deep and remarkably cold stream, as I surmount the other side and take the customary look. I drop like a setter and soon signal to the hand to do ditto. About 1000 yds off, I can see two deer grazing on a sort of meadow on a rocky terrace. I signal to him to stop and away I go. The wind is all right and it should be an easy stalk. I never see them again until I creep cautiously up to the rock I have marked. There they are but have moved so I have a little more stalking now then I peep over. All over with them—they have heard the click of the safety spring or something and got up—but are standing at about 60 yds like rocks so there's no chance. Over comes one and out flies the cartridge and before the other has gone 15 paces I have killed my first two reindeer. The boat presently arrives and the hands come on in her as I have gone on alone. I beat up to the glacier but see nothing more and so take the boat and on again. After a feed, as the wind drops, we have to row. We are now caught by a fog like that of yesterday that has been creeping up and we row along by the land. The whaleboat which went for Potter has, to our surprise, never passed us or even appeared. We of course leave their beat undisturbed and in due time arrive at the third glacier. The other, too, had a large front that is perhaps as extensive and as high, but in respect to noise they couldn't touch this one. He was going off quite promiscuous, a field battery on a review day would be nothing to him, and for the mile or two we were coasting along him he kept it up as hot as ever, sometimes large masses, at others little bits not bigger than one's body, but as they struck the water they all added their quota of noise to the din. Below this, I again landed as it looked good ground. I had not the vaguest idea how to find the ship as one couldn't see any distance, and it wouldn't do to leave the land in case they sent along it. I proceeded to beat along some mossy terraces. In a few minutes the hand below shouted: "Run, Sir, run," he had stumbled bang on a reindeer in the fog and seeing him start off did the right thing: shouted. Away I ran on my terrace and the deer on his. I saw nothing until nearly blown, when he suddenly took up the hill and appeared on mine about 50 yds off. I was so blown that I couldn't shoot and seeing me he broke from a trot to a gallop. I caught my unruly breath and following him with the rifle, by a fluke shot, tumbled him, head over heels at about 100 yds to the men's and my own intense delight. Soon after I met the owner, he had seen four deer and got one. We beat on a mile or two to the ship but saw no more and got on board by midnight. Potter had a tasty time of it, bad light and shade for views &c, and when they were finished, he came running down the hill in time to see the *Diana* gaily steam off. He and his companion found an empty whaleboat on the beach and in this they waited a tremendous time, firing shots to bring back the truant crew, who when they arrived after an apparently fruitless search, which had they stuck to orders might have been avoided, informed him they were in search of him and Eaton, the latter having been on board long since. This caused another delay and more shots, and by

Figs. 9.19 and 9.19.1 Unidentified areas of ice fields

the time they got underway, the breeze was gone and the fog on them. Hunting was a fluke and they found fresh traces and nothing more. They then landed below the third glacier and explored a valley running up from a lake. It was here the owner had got his deer and they found his cartridge cases, nothing more. They were, however, in bright sunshine, being above the fog. They did not get on board 'til 4 a.m., bringing five ptarmigan with them.

July 18, Friday.

Early breakfast and away again. I take the casey and the same hands, and try back to the glacier, as three deer had been seen there besides those we shot. It is awfully hot and I have a heavy walk and blank beat, the deer from the tracks I see having gone clean over the hills. I take right up the valley to the glacier and back to the lake and do it all conscientiously, get very hot and lose my field glasses, a nuisance as I got them as a cadet for a prize. No good to draw up between the second and third, I see, as the good ground he's far up near the second. After a long pull towards part of the ground to be left for us to beat, we are hailed from the shore and find the owner, a hand and a deer. He had seen another and I think had a shot. I walk on as this ground has not been beat and as I come to the next hollow see a deer far below in the middle of the torrent and bed, which

Fig. 9.20 *Diana*, with signal flags flying

is stony and sandy, and composed of endless little streams running seaward. He is a big stag and coming our way but is deterred by two ruthless skuas who turn him and drive him up the opposite slope. I stalk the place where he disappears, wet up to my waist by the time I get there, but he is a traveller and on "*forrard*." By the bye, just as I knocked over a ptarmigan with a stone, when with a great deal of persuasion it had got up, hind and fawn come by, from behind, not knowing that they have been frightened by the man behind who didn't shout. I omit to fire a rather easy running shot and they disappear in due time far on. Then just as the boat puts in, I get a shot at three deer or rather two shots. Unfortunately, they are rather far off and I only got one. The owner and I beat on right down to the ship and up a valley opposite it to a lake past which we can't go as the hills that bound it are too steep, unless you climb them lower and keep the high ground. Potter after a good sleep had gone on shore with the second mate. They soon saw a deer on a snow slope and at about 200 yds if not more opened fire on him, P being informed it was impossible to get nearer. Luckily, he didn't understand the shots and after four or five shots was wounded and secured. They killed another afterwards in the same manner, by a great slice of luck as it is not often a beast stops and be made a target of until someone makes a happy bull's eye. They see no more except across a stream running in a deep narrow precipitous gorge or ravine and impassable for some miles in either direction. They get back by 8 p.m. and we all enjoy the doles far outside of a lovely evening on deck.

July 19, Saturday.

Steam up early and we all turn out. Potter and I are to go away in the casey and beat some splendid ground, the owner intending to put off when he has got the ship anchored. Being as pilot necessary on board is, however, as we say, the "Ranger" of Spitzbergen; he can at any time put his hand on a deer. By seven we have dressed, breakfasted, got our guns, field and spyglasses, ammunition, baccey, pipes, knives, &c ready, food on board, and away goes the casey. We land in about half an hour, near the gully that brought up the advanced guard so abruptly yesterday. We got a deer very soon, and having reconnoitred a bit determined to beat in long beats running inland and back again to the shore, and the hand who carries off the deer proposes that the boat shall wait at a gully two miles on—to which I assent as he had been purposely sent with us from his supposed knowledge of the ground. I ought to have known better (but luckily no ill came of it) as for the last two days I have been pointed out all sorts of spots where the owner shot deer last year; on ground he assures me he never set foot. Then a Russian hut still 15 miles or so down the fiord was close to the glacier near which we were caught in the fog, and had we not found it, the ship would have been our sleeping place. It is terribly hot, so we are dressed lightly: under shirt, flannel shirt, knickerbockers, long boots and light jackets, and only rifles; we now make a long beat keeping the gorge on our left, seeing three or four deer but failing to get at all near them and, swinging round, come at length above a long inland valley. The bottom of this consists of a large wide torrent bed or plain fed by numerous streams from the glacier at its head. It is closed seawards by two high rocky sandy hills, thus forming with the funnel mouth of the valley three lesser valleys, through the furthest and southernmost the stream discharges itself in a strong torrent swollen by its numerous tributaries to a considerable size, and running in parts in a deep hollow channel below a range of beetling black rocks. The central valley contains a lake, the outlet from which, tho' now nearly dry, runs straight inland, splitting into numerous winding stony channels on the sandy slopes and running back towards the glacier bringing its waters into the stream already mentioned. The third valley and the one above which we now were contains two lakes close together, the second about a mile from end to end, and these curiously instead of running down into the sandy plain discharge themselves by a channel cut in the hills seaward running in exactly the opposite direction. That is at present, but were their surface two feet higher they would as they have done and I daresay still do at times from the dry bed, discharge both ways at once, and it seems to me a most curious topographical phenomenon to find water collected in a place where the equilibrium is so nicely balanced that it is a question which way water would run. Near these lakes are bits of patches of green and moss sometimes an acre or two in extent and on the sandy slopes are smaller patches. The hills on the left are sharp and stony and rising from the point we are at. The crest is some hundreds of feet higher three miles up above the glacier at the valley. The slopes soon become steep, and then inaccessible, with large heaps of stone at their bases, and but little verdure between them and the sandy riverbed. On the opposite side, they rise at first in long soft green even slopes, in rugged terraces and then in steep stony hillsides with a patch of snow here and there, but underpinned lower down, where the swishing stream wheels round right under them with the dark cliffs I have spoken of. In the far left hand corner, there is a sort of fissure in the crest of the highest hills steep enough here and through this pours a grand cascade of about the same calibre as that of the Dargles (Wicklow) [Ireland], then the ordinary glacier of Spitzbergen closes the valley, and he indulges in a variety of young cascades, and behind him rises a bold outline of dark serrated peaks. Down the hill we go after a great council of war for there are so many deer feeding we have an *embarras de richesses* and the question is to select those that can be polished off with the least chance of alarming the rest. There is not a patch of cover in the plain but the dips of ground, so we travel on our hands and knees and, by stalking and posting a gun and then getting round and crawling in view so as to drive them, but quietly, as they stop to look at the crawling object, ever motionless when they look, and so at last are standing unsuspectingly in range of the other gun, we soon get four. Then we take up the left of the valley, along which we find great numbers of geese who, however, run much faster than we can and only cause us heat and vexation. Someday we mean in revenge to have a regular goose hunt and, until then, I forswear the race. At length we see four fine stags and drive them up into a corner under the cascade, Potter creeping under the stones at the feet of the cliffs and I far out wading in the river to do the driving. At length I turn them and back they go and bring up within 20 or 30 yds of the rock behind which he is so effectually concealed that he sees nothing but the tips of their horns. They wind him and turn and, breaking the line, take the water. A long shot from me at the foremost to turn them, as we often find a turning shot at 300 or 400 yds like this very effective, does no good and he only gets two running shots as they make away across the riverbed here perhaps 1000 yds broad, wading, plunging, and swimming according to the depth of the stream. I now beat down the long green slopes on the other side and get three deer, which rewards me for my ducking and also for being nearly carried away. Then back again, across the stream and the long sandy plains, the hot vibrating air over which soon takes off the chill of the icy glacier stream and makes it seem more like the Algerian desert than Spitzbergen. We now make for the boat, for three objects: food, cartridges, and deer carriers. We meet the Captain and third mate coming down and find they are in search of us and our deer. The ship having failed to find anchorage further down and, turning, brought up opposite the stream running out two or three miles below the third or southernmost valley. Our boat, it seems, is here also, having passed the gully and gone six miles instead of two. We get to her by 4.30 p.m. and finish off four more deer on our way making our total 12 so far. The captain has got a stag and the owner is out on the ground around the two high sandy hills above the lakes and first two valleys. Refreshed by lunch and with my reserve ammunition, I start again alone, and ford the stream across one of the two large sandy lakes it forms in two successive and almost parallel bends after or rather below the rapids running out of the main valley. I call these lakes, tho' they are only broad shallows and taking the dotted track along the unvisited side of the river I soon come on three deer and get all, one standing, the others as they run off. Further up, I again cross and take back much the same beat as the morning. The sport is so exciting that I feel no fatigue and I really think this must be one of the grandest sports the world produces. You roam in perfect freedom where you will with no check upon you, every day a new beat amongst these striking and grand solitudes, new scenes perpetually unfold themselves, hills, cliffs, streams, glaciers and bays, then there is the bright uncertainty of where the best ground lies, and if the game will be there, no monotony, and "this is usually a good beat, where can the deer be?" or "we turn back two miles further on at the gully above the Achilish road" &c. Then there is the sport of the game itself, wild and shy enough to please anyone, from the way the Norwegians have hunted them. You may lay it down as an axiom, that unless you see a deer before he sees you, you don't get him, next to not seeing you he must not hear you, and lastly comes the smelling. If you approach him noiselessly from upwind, you *may* get him; he smells you, but does not always make off, unless he sees or hears you. I do not know if it is the same with Scotch deer stalking. One advantage is that if well concealed, the deer often stand after a short run when you have fired and give you another shot; then the fawns stop with the dead hinds until you show yourself as a rule. The owner came back about 8 p.m., having shot four deer, one a very fine stag. He had seen a herd of 16 together. The largest number I saw was five: one stag, two hinds, two fawns. I only had two cartridges at the time and shot the stag and did not fire the running shot. I was lucky today in twice successively coming across three deer and bagging the lot. In fact, I was delighted with my rifle as, with the exception of the turning shot, I did not miss a shot, standing or running. I got back a little after midnight having got nine more, bringing up the days bag to 26. A Dutchman had arrived and anchored near us and got seven or eight. A fog has come on.

July 20, Sunday.

I go off in the casey with a lot of hands to pick up deer. After a long pull in the fog, missing the gully, climbing the hills to get above the fog, we make out our whereabouts, and sending down the boat a certain distance back, we walk along the hilltops in the glaring sunlight, with the canopy of cold chill mist below us. When near the other deer, I stalk and shoot a live one, as the men are anxious to see the sport. Those that we had seen before, I refused to touch owing to the difficulty of carrying them so far. Soon after the deer are all collected, except three which the Norwegians had taken by mistake, we call a halt. Said Norwegians arrive and apologise; the missing deer are at their ship and shall be sent on board. I am very glad of this as the men have been abusing them, and they come up of their own accord and tell us. We fraternise and smoke the calumet of peace, and as they are inspecting my rifle, a deer comes walking up the gully below us. I ask the third mate to go and stalk him, easy enough from the ground. His method is simple and consists in walking straight at him in a crouching position. He is favoured by the sun and for a wonder the beast allows him to get within about 50 yards. Two shots are then heard and he comes or rather both come from the gully, the deer a good deal faster than he went up. I manage to wipe his eye from my position in the group and the astonishment of the Norwegians at the point-blank range of an express is intense. I own to myself that a running shot at that distance is an uncommonly tall fluke: my best flukes always do come off when there are a lot of people looking, but as I carefully shoot out the empty case [I give] a puff at my only pipe, and look or try to look quite unconcerned, as if that was an Englander's everyday shooting. Then, when the deer arrives, there is more astonishment at the immense shattered hole behind the shoulder (I have no solid bullets) and then we part on our different routes. In the evening, the owner and I take up the valley across the river to show the whereabouts of some more dead deer, the remaining three, and we get four more letting off 20 others as there are no carriers. One of the four, an old stag, gave us an awful stalk. We were obliged to try from upwind, no go, we couldn't get beyond certain points and he was very uneasy. For more than an hour we crawled and grovelled with the express as thro' lying so still one was in capital wind, but I didn't get the office to fire so didn't risk it. At least he disappears over a dip and we tear like lunatics in different directions. A shot from the owner, another, would have fetched him down, but unfortunately I show myself and he is away, hit tho' in the foreleg. He takes the water and we take it one above and one below him and after another long stalking chase, the death shot sings out. The river here ran widish deep and strong and the bottom being composed of rocks. It was at one step a foot and a half deep, the next nearly four feet. I thought the owner would have been swept away in one of these holes where he stood, gun and coat &c partly immersed, struggling with the stream, the rifle slung, and holding on by both hands to a sunken rock below the surface. At this juncture, I became as usual paralysed with laughter. We got back about midnight. I went on board the smackie today and while there one of his boats arrived with a bear, they got in the water close here, while bird shooting near the point. Eaton has been very successful here with the plants &c. We got some ptarmigan and two geese today. We also got a redpole, a bird that we have seen here before besides one at the western ice. This concerns Scoresby's reputation for accuracy, as Maluygen [sic] and Prof. Newton have expressed themselves rather strongly, to say the least of it, as to his having really seen it here. A snow buntings' nest was also found.

July 21, Monday.

We steam down past Cape Petermann and come to anchor in East Fiord about three miles past __________ and one from the land, close to an island that lies over towards the E side of the fiord. It being fine and the views good we go in for photography until dinnertime getting some results, tho' the wet plates from the ship have to be abandoned and both cameras taken on shore. After dinner, the owner, skipper, P, and I land and beat from different points until we meet, blank! Deer, however, are descried the other side of a valley just like those already described. The warm colour of the rocks that frown above it being here owing to red lichen, tho' on the other side of the bay, the colour of the hills with their hard angular, chisel cut ridges and watercourses is that of the rock itself a red purple laminated shale. Meanwhile, the Rubicon is crossed and we are in our normal wet condition. Potter and the Captain are posted and the owner and I commence driving the deer. This is executed very successfully, two deer being driven to the Captain who being, as I have said, a good rifle shot, polishes them off very cleverly to his intense delight. But altho' he can shoot fast enough and well enough, his lights on deer hunting are but dawning, as judge of our astonishment when later on, we were informed with great gusto of the masterly execution of the two deer we had been trying in vain to stalk, we who had walked openly over the plain to drive them, neglecting an easy stalk of four others as the owner was anxious for him to have the shots. It was simply killing, and so we were all in good enough humour tho' for different reasons. I unfortunately after a long round drive a stag close to a stone deserted by a skirmisher who didn't know what I was after. I then come across a head of six, uneasy and on the move from the shots that are going on, in a long fold of the hills above. I follow them like a red Indian, not firing as I hope presently to surround them with the aid of the others. So as the shots continue, they move from patch to patch. It is easy ground, however, and a good wind and I keep near then, looking out for the chance of a signal to the other guns. At length they debouch on a little flat and take thence down a small green mossy watercourse to the opening of a long fertile-looking plain. Here two little meadow patches on either side of a streamlet are grazing another herd, and when the forces are united all told there are 20 or 21, some of them very fine beasts. I get as far forward as I can behind a stone above the stream hoping to turn them when the other guns appear. More shots and they all move on past my ambush but none of those fine enough to tempt me. On my side of the stream, tho' about five pass in shot. I am well to leeward and light my pipe and contemplate as I watch them graze on the open plain below. Now for a real drive, one gun well forward along the hillside to the left to shoot some and turn the rest if they take mountain-wards, one on each side of the stream up whose slopes, I expect they would come, one to the right to take the beach pass and those that break upon the stream, the hands to row clean by in the boat and drive up the plain in extended order. So, I return and at length get the other guns. The hands are not, however, forthcoming, as they are carrying deer to the boat. It is late, the ice packing in the fiord, and we reflect that excitable sailors are not trained in gillies, will not keep line, do not know when to stand, when to show, when to squat, or act in concert at a distance. Beasts, too, so often in spite of everything will go the wrong way, so after one more look at the promised land, we turn our backs on it. By the time we have our eight deer and eight hands in the boat it is very late. The ice which moves here with a swinging tide proves a labyrinth we cannot unravel and so we turn, twist, go back, sweep round, and finally drag her to shore, then along shore then out again at a little distance, and so by 4 a.m. the journey of less than four miles to the ship is completed, the last part being quickly done enough as soon as we got round the point, against which the ice was packed. We found the ship snug enough, protected by the island. We breakfasted before we turned in.

July 22, Tuesday.

We were not over spry in getting up and made a sort of compromise between breakfast and lunch. Eaton did not come on board 'til 8 a.m. this morning. He had been on shore collecting last night and when he came opposite the ship soon after our return was unable to make any signal, so lay down on the beach, as we had thought of doing last night. It was such a glorious sunny night that it was no great hardship. It is fine today. As there are some grand views round the day is spent in photography. There is such a tremendous swinging tide and the ice drifts about that the ship will not sit steady for a good part of *La Diane en chasseresse* as she is simply hung all over with quarters of venison and bow sprit and boom ornamented with pairs of horns. It is a pity but being black she requires a long exposure. She looks very smart, having been painted all over funnel included, masts scraped &c. Every time that the camera has been placed in a good position and I row off with my plate in the dinghy, I find her swing or swinging, and end on, or else in the shade again, so I have to give it up. We get seven white fronted geese today, a species I believe hitherto unknown to Spitzbergen, and three long-tailed or Pomatorhine Skuas, which seem almost as common here as the other skuas. King ducks, which are not supposed to breed in Spitzbergen, do and are breeding here, but where we can't find out. A few snow birds are murdered as usual as the men keep the skins. The captain who uses the owner's No. 10, does not shoot on the wing, except when a bird is hovering, but kills shots on the ice or water at most astounding distances. I turn in about 1 a.m.

July 23, Wednesday.

By 10 a.m., we are ready for a start and away we go in the gig. We land at the valley where our deer shoot came off two days ago, the same guns and four hands. By the time we get to the next valley, a beat of only three or four miles, we have got two deer, one a very large hind that I noticed in the herd the other day with only very elementary horns, but a very large beast, one of the biggest does I have seen here, who with its long ears, drab summer coat, and large hoofs, looks very like a donkey withal. Both sexes of these deer have horns, but they are only growing now, the velvet still on them, and those of the hinds are mostly short so that they look rather ungainly beasts as the shaggy white winter coat is still on their bellies and lower quarters. I saw four more on the hill above the Captain but, as he was talking, he didn't 'til too late. After crossing the stream four deer are seen on a plain, and by a long stalk Potter gets 'round them and drives them very cleverly to the captain, who unluckily doesn't fire 'til they are past him, a great mistake as though he gets an easy broadside shot it doesn't turn them properly. Luckily, they do turn a bit, but he blazes away so recklessly in the direction of the owner and myself that we crouch terrified in our lurking places. He gets two, which is something, tho' we ought to have had the lot. As he stops to perform the necessary ceremonies, about which by the way he is very hazy, P and I proceed to stalk two more. Getting to the edge of the plain they are on we wait as they are grazing towards us, and signal to the captain to stop where he is. Nearer they come by degrees and at length we get our guns ready, hullo what is up, they can't see, hear, or smell us, but they are very uneasy, edge further and further away, and finally betake themselves to the hills. As we shoulder our rifles and walk on rather surprised, we perceive the "*belli teterrima causa*" [shameful cause of war] to be the skipper, who neither above, below, or round them so as to be the least good, has been stealthily advancing on them with no cover. We have plenty of deer, so it is no odds, and we laugh more than anything else at the strange ideas and ways of people have who are anxious for sport but not accustomed to it. Well, "*forrard on*" and I soon drive two deer to Potter who join a third and he gets two. Then, as the skipper is lying down, I take across his beat and up a valley (by order). Here I find or rather see some deer lying in a little hollow and as I approach the completion of a long and tedious stalk, I see the skipper coming along below me by the stream edge, hooray, he will pass and never see them; in fact, he can't from where he is. Quite true, but he can see me, and also what I am after, and so in spite of all my warning gestures takes straight up towards them. I must look sharp as here he comes. I look over a rise, there they are, lying down, just hidden except the horns, I whistle, and knock two stones together, they stand up and gaze, and luckily can't see the captain yet. Down comes the biggest, crack again, and almost simultaneously the captain opens fire, and with the second shot succeeds in hitting a deer who I had shot running and who had stopped as they do a few seconds before he fell. A tremendous long shot, but he unmistakeably hit him, as I saw both bullets subsequently. I fire my third shot and getting my third deer, leave the fourth to the tender mercies of my companion, crack, crack, he has got him alright. He will never go 100 yds, never mind, he fires away, and puts either one or two bullets into the wretched fawn. Very good shooting at a tremendous range, but not "the game." The owner and Potter get two deer in this volley and are by now across the stream, so we follow. They, however, see the boat and recross to get some lunch, a manoeuvre that escapes our notice. Some of these channels of this river ran deep and strong and the Captain was nearly swept away, but getting firm footing across the central deep of the worst and hauling on his gun to pull him over, we both got safe across. We search in vain for the others, and thinking they are along in front take our own boat. I soon see four deer right in sight in a dell, and can only drop flat at once. I can't budge an inch as there is no cover. My ally is just disappearing far below 'round a corner when looking 'round he sees I have disappeared. After some search, he makes me out and proceeds to join in the hunt. This time I am rather glad, as I cannot stalk the beasts at present, and if he only gets 'round them and onto the rocky brae above we must get two, three, or the lot. He disappears apparently to do this, and I am in great hopes, as I have been explaining the principles of the *chasse aux cerfs* [deer hunt] to him and telling him what not to do 15 min before. So, I wait in peace if not comfort for I am stretched perfectly flat on bare ground and daren't move. To my intense disgust, in about 10 min, from a stone not 80 yards to my left and if anything further from the deer than I am, in exactly the same direction too, so that there has been no attempt to turn them, or get to the rocky brae above them, which I have no doubt I could have done by crawling backwards and keeping my eye on the deer, but which I, of course, refrained from for fear of spoiling the other guns sport, there rings out a shot, a miss of course, and away go the deer up the brae, and their enemy after them. This fairly does me up and, shouldering my gun in high dudgeon, I make tracks *forrard* as hard as I can (the deer have gone back). I then hear a succession of shots (six) and subsequently find that the captain got one old and one young one. The big stag he must have missed or driven away. His account of this very successful stalk was charming and, had I not previously long since recovered my equanimity, it would have restored it, as not only was it evident that his blundering was perfectly unintentional, and any apparent selfishness in getting the shots utterly imaginary, and due to the excitement produced by seeing the deer. But he evidently considered these late manoeuvres a masterly stroke of skilful strategy and accepted all the compliments showered on him with the greatest gravity. The whole proceeding, indeed, he considered a signal success, with only the trifling accident of a stone having caught the "doghead," as he will call the hammer of his rifle, but then, headed with pride, the bullet went right into the middle of the deer, as if he expected one to come down like a shot into the brown of a covey. How I hugged myself that it hadn't come my way. Perhaps this was the shot that incensed me so, we couldn't make out. Long 'ere this I had walked along a great broad green stony track, as broad as the *chausse* is of Saint Germain or Versailles, terraced as it were into the hill, and running up and down for a mile or two, and coming out, at length on the sea, or fiord some way above it. Turning to the left here, I took up the valley. It is a very one-sided one, the far side and end being a high glacier the left on which I was being high stony hills becoming cliffs farther on, and with steep cuts running down their sides with the water rushing down in a succession of cascades. After going a little way up the valley, I come across two stags, an old and a young one, and get both from the same cache, the old one a very long shot, that we subsequently paced. Not knowing what prospect there is of the others arriving, I cut his head, a very fair one, off and make back for the beasts. I have some difficulty in getting across one stream, head, rifle and all, and soon afterwards meet the others, and it is there we are entertained with the Captain's exploits. The owner and Potter had come along the beach and got three stags, two fair heads, and Potter's subsequent description of the Captain's two attempts at decapitation, trying to hook clean through the bone, instead of getting in at the joint, a proceeding that has reduced P's knife to the condition of a saw, was highly entertaining. We were now at the end of everything, except the part of the valley I had left unexplored, the whole of the end of East Fiord being closed by a—or rather two—immense glaciers. These have together an irregular curved front four or five miles in extent. They issue from valleys running at right angles to each other and both debouch on the sea, their line of junction being indicated by a long dotted line of heaps of piled up dirt and stones. We had now 19 deer in all and were only awaiting the arrival of the boat. It was now 9 p.m. and the Captain and I were famished as we had had nothing since breakfast but a morsel of bread he had had with him. *En attendant* [Waiting], we beat the valley and saw four deer but didn't go after them, as they took up the hills, and it would have been too far to carry them.

There are a great number of ptarmigans here, and I have no doubt one could get over 20 brace in a day at least with a little experience, as at first they are not easy to find, but they show no sport, except as a cock shy [a mark in a throwing contest]. We now took along the shore and back again thinking we had passed them and so on, 'til just as at 7 a.m. we had made up our minds to start for the ship. We caught sight of them at no great distance from their midday halting place. One man it seems on his way down, having left a deer in order to go and signal the boat along to a convenient spot; this he had never been able to find again and they had searched hours for it. Then they had missed two others that had been carefully pointed out to them, and altogether generally tuckered it. How they missed them I don't know. You say to a man: do you see that stone? Yes sir. Walk there and thence in a line on that hilltop until you come to a stream, a snow slope should be on your side at your feet, take up to its point highest up the stream, cross it and ten yards across opposite the head of the snow slope, in line with the stone &c you will find on a rock, two deer piled up. Now then: "where have you got to go?" he tells you right enough, but he misses them. I have been absolutely back with a man who remembered and made him do it without being told. But to return to our muttons or deer, as there is yet a long job in picking up deer, the owner P and I decided to walk home. By 2 a.m., the boat has arrived and we have fed and had our glass of grog all 'round and leaving the captain with them we start home. In 3½ hours hard walk we are opposite the ship and the shout of "*Diana* ahoy" rings across the bay and is sent back by the echoes, with such a will that shots are unnecessary. We breakfast and turn in.

July 24, Thursday.

The boat does not return 'til 11.30 a.m. and then minus two deer and a head, the Captain giving up the search for them, as both the hands and the grog were done. We rest until towards evening and at 5.30 p.m., the Captain, P, and I start in a whaleboat and go chiefly for the object of determining the position of three unsurveyed islands low down the fiord. We land at the second stream to beat down to the deer, they and the head are soon found, and we assemble on the beach.

Again, after waiting an hour or so the boat is missing, nothing could have been plainer than their orders, row on until you come past the rocks, and land in the sandy bight elsewhere. They had, however, chosen to select a bivouac by some imaginary order, close to where we had left, had made a fire &c, and got their kettles on, and we had to send all the way back for them. At length they arrived and then we found they had left all their pots &c where they were. This was right out of our way, as our intended halting place was on the islands I wished to survey, so we had first to return and fetch the chattels, provision, basket, &c, up the fiord, and then row down again to the island. I ate my supper on the way over and started to survey at once on arriving at the islands. These, though two or three miles from the nearest point of lands, were covered with fresh deer traces and the deer may have been still on them, tho' I think were not. The two biggest were about one mile in length and more fertile and higher than the generality of such islands, of the same geological formation as the east side of the fiord. There were some rypers [ptarmigan] on them, and a good many ducks and geese about. The former had nests, the latter none that I saw, and apparently on our arrival relinquished all vested rights in the property, swimming clean away in large flocks that never returned. They swim, of course, faster than an average boat can row, though I suppose they could be hunted down. Surveying over I found the others finished supper and sitting round a roaring fire, as there is abundance of driftwood on all these islands. They had not been egg hunting for a wonder or if so without success. I found lots, a few fresh, but mostly hard sat. This latter condition can soon be guessed without the necessity of breaking an egg, from the behaviour of the duck. We had, by the way, a ground seal hunt on the ice coming down. The brute lay very well on the ice and we got close to him. The captain, unfortunately, missed the beast, it being partially hidden by a snow hummock. I was on the stern, only a reserve, and had to stand up and fire over the others heads, as he jumped into the water. However, I gave him a bullet in the small portion of his body. I could see from the hummock, it had the desired effect of crippling his diving powers and after a long shot or two in the water, which they *will* fire, altho' if they hit, they know they can't get him, he came up quite close and we settled him, but he sank before we could get in the harpoon. Why, when they see this is invariably the case, they will persist in long shots I can't imagine. Had we not done so, we very likely might have got the beast, who was trying to get on the ice again. It is a great pity as they are very large and as valuable as a seahorse. This was one of the grey bearded seals. We got home about 4 or 5 a.m., some hours later than we need have done had we not had these provoking delays. Eaton has been travelling daily, right up to the glacier feet and, on the whole, has done excessively well in the way of plants in this bay.

July 25, Friday.

Turned out to breakfast. The owner and I afterwards landed on the island, as I wanted to survey it. This was done by lunch and the position of a shoal taken on which we had sounded a bit, painting &c is still going on, butchering, flensing and salting skins, preparing heads &c. We land again on the island in the afternoon to select timber for a davit. A good log is selected and some drift wood for the fires taken on board. In the evening P and I go for a goose hunt in the dinghy. We have a capital sail, but it comes on to blow a bit. We found lots of ducks but no geese. We dredged yesterday and also today. Tomorrow we are off back N and I may as well give a concise recapitulation of our proceedings, our "lark." We have killed two or three seals; 68 reindeer of which 65 were brought on board, of the others one being lost and two eaten by foxes; we have got seven white-fronted geese, new species to Spitzbergen, seen numerous King Ducks and evidence of their breeding, killed, seen red-throated divers and taken their eggs, killed a redpole, and could have killed more had they been required, found ptarmigan in summer plumage and found their nests and those of the snow-buntings, found and killed goslings, found young eider ducks, seen one ringed dotterel which I would have killed had I known that the little heap of feathers a rifle bullet would leave were necessary to establish its occurrence. Also found and kept a fish, genus *salmo*, length eight or nine inches, which I gave to Eaton, killed several Pomatorhine Skuas, found the breeding places of Mallemucks, found abundance of Tringa's (*maritima*) eggs, found several kinds of fungi, puff balls, and one or two other sorts, found several new plants, gentian, campanula, another ranunculus (?) growing in the water, fern on the limestone, a few shells, flat starfish similar to those the Swedes were so pleased at finding, a sort of anemone, a beast I believe never before known to exist in these seas; surveyed and attempted for an astronomical observation; discovered that the seawater here is remarkably fresh by repeated experiment, taken a few soundings and determined several good anchorages; and finally taken a few photos, and bearings and notes furnishing materials for a sketch on a large scale, to give the majority of the hills, valleys, lakes, rivers, and natural features should such be required for sporting or other purposes. We noted the character of the rocks &c and found one specimen containing apparently a trace of copper ore so that perhaps our proceedings may afford some interest to others, in addition to the intense enjoyment they have given us. Nor is this all. I have not touched on entomology, knowing nix on the subject. Moths have, however, been captured, both of the small kind, known to inhabit Spitzbergen, and a larger, I believe, entirely new sort; various kinds of sawflies, also new, I believe, a red *tramidium* of which we saw numbers; spiders and flies, some apparently ordinary housefly of commerce. As before, we also saw some small water flies, but I don't know if Eaton got any.

July 26, Saturday.

Dressed and on deck by 7.30 a.m. as I want to make a running survey. We get the log overboard

Andreas Umbreit

◀ **Figs. 9.21 and 9.21.1** The *Diana* visited Mosselbukta (Mosselbay, northern Wijdefjord) and the Swedish scientific wintering station there at that time of the Nordenskiöld expedition repeatedly during its 1873 voyage. The first time was for several days in mid-June when the three Swedish vessels there were still locked in solid winter sea ice and the Swedish expedition team was running short of provisions and accordingly welcomed a substantial food donation by Benjamin Leigh-Smith from the *Diana*. This image was taken by Chermside on a later visit, probably on 16 July, when the winter ice had left the bay. Unfortunately, the quality of the image is somewhat poor, probably taken from the ship with its movements, but in the picture, the wintering house Polhem and some of the scientific installations north of it are just about discernable in dark to the left of the middle of the photo. As Nordenskiöld largely relied on drawings for the documentation of his expedition, this photography by Chermside is a rare photographic account of the Polhem base at the end of its operation. The recent images by Andreas Umbreit show the Mosselbay area from a similar angle, though from shorter distance from the shore, on 5 July 2018, with the Swedish constructions collapsed. A closer view of the remains of the main Polhem building, taken in the year 2000, can be seen in Wråkberg (this volume, Fig. 2.6)

and then fire away at angles and bearings. This expresses my day, meals excepted, pretty well, until about dinnertime, at which hour we meet the long grey fog that is creeping down the straits. Up to now it has been sunny and clear, and the fiord pretty well free from ice. We are now nearly level with the northernmost of the three glaciers and ahead of us lies a great deal of heavy ice, floe pieces from some pack, presumably the northern, some four or five miles long, and one so blocking the channel that only a narrow passage is open in the total width of the bay, and that between it and the W land. As we stand across to this, we regretfully take our last look down the glorious sunny bay, where we have spent such halcyon days, and then as the chill mist comes on, we go in for the matter of fact consolation of a good dinner. We have sailed most of the way down, but are steaming now, off Grey Hook. It clears a bit and we make out the low land, then later on we see Mossel Bay, apparently deserted. As we come to Verlegen Hook, from the ice that was jammed in Wüde Fiord, there is evidence of a late gale and, as usual, the thick dark clouds are tearing and the remains of a gale blowing out of that ghastly funnel mouth of Hinlopen Strasse. We are soon in a heavy cross sea and this continues 'til about midnight ere which we had all turned in.

July 27, Sunday.

Soon after 2 a.m., the sea became calm. We knocked off steam and sailed along, sighting Low Island on the lee bow and Shoal Point. We stood off northward again under steam. Low Island is half covered with ice but several of the black, wicked rocks and reefs near it on this foul ground are visible. I get up two or three times and finally turn out about 9 a.m. The hard, heavy pack has, it seems, come down past the Seven Islands and stops us off completely from getting along in that direction. We explore the edge and bight after bight completely, until as about 8 or 9 a.m. it comes on foggy we stand off a bit. We get today as far N as the ice will let us, but then as it runs westward and it is not advisable to take or try to "take the pack" in thick weather, we proceed to dodge in a bight in which we find ourselves. This closes in at length and we stand S and W a bit as the ice trends that way. It certainly does not look pleasant to see the pack down so on the land, but it may have left a gap somewhere. About midnight it clears a bit and is pretty calm, and one time today it was rather rough and there has been a strong wind if not a gale blowing all day, it seemed quite strange to see the "fiddled" table again, and to have things taking charge. The temperature of the cabin, too, has fallen from above 50 to below 40, and one must resume some of one's left off clothing. We get bottom at 60 fathoms and take the temperatures. When it clears, I see the lie of the ice pretty well, and make out two of the Seven Islands. Turned in at 3.30 a.m.

July 28, Monday.

Calm enough, but not clear enough, so we continue to dodge. It has been decided to keep on at this until, if possible, a clear sunny day gives us a good look at the pack and the chances of success

Fig. 9.22 Another fine view of *Diana*

in forcing a way through it, in among the floes. It is when among the fields and floes that the winds acting on these large masses open lanes and channels thro' which ships can penetrate, and 'til we get a lead thro' them of some sort like that, we shall not I fear do much. The alternative is of course the straits again, but we ought to give this a fair chance first, so it is decided. We dodge, and we "ratch" in and out of the points and bights, and we sound, and then we do deep-sea dredging at 120 fathoms with a whale line, and a heavy dredge. The first trial is a failure, as we do not give enough line for the dredge to get to the bottom and a blank is the result. The second time we get a good haul at any rate, but I must confess I think it a snare. *Chez nous* [with us] there are three methods, and I must say the common domestic "swab" is, in my opinion, for ordinary purposes, the safest and best, that is in bottoms such as these and deep water. But for good results, as the books say, No. 1 and No. 2 give the best, when combined (mind I only speak of dredging conducted here, for I have no other experience, and the others I fancy but little) in the hands of inexperienced dredgers. First, let us take the swab [for] dredging. This is very simple. One of the long bundles or brooms of oakum used to clean the decks; the larger the better, then a deep-sea lead line, with its lead as sinker, the cavity in the sinker as usually neatly filled with clean wax or grease, to show the nature of the bottom. We also usually add a small tin, with a cranked or bent wire handle, and a little lead run in one side, at or near the bottom—this tin to act as a scoop, and bring up mud, stones, or gravel. Sometimes a small hoop net is added, with gauze, or fine meshes, and very likely if we take a deep sounding, one, two, or three deep-sea thermometers are set at the surface temperature and go down at 50 or 100 fathom intervals. Over it all goes, and line is paid out until the lead finds the bottom. Register the depth, pay out more line, and let the ship if she is not driving of her own accord; go ahead a bit to drag the swab, then haul in by winch or hand. Touchingly simply and applies to any depth. When it comes up you have your temperature maxima and minima to read off each thermometer, your lead to examine for the nature of bottom, and any fragments of the said bottom that may come up in your tin to examine and if necessary, preserve. Then there is the swab. This has been dragging along, with its floating hairlike fibres distended like sea weed or feelers, and will have picked up, doubtless, star fish, five fingers, alectos, echini, stones, sea weed, *et hoc genus omne* [and all that sort of thing]. Next the medium dredge: this is an ordinary oyster dredge, iron handle, with a bag formed of wire netting that can be if necessary supplemented or rather be covered partially or completely with sackcloth. The objection is that having only one long dragging flange or bar, if it goes down the wrong way or gets turned over by a current, it may come to utter grief. As yet it hasn't, but we have [not] tried it at any depth. Third, there is the heavy dredge: great double-handed iron frames, weighing I should say nearly 1cwt [112 lbs], with an open rigid mouth of two parallel bars or flanges from four inches to a foot apart, and the inside bag of which is lined with a coarse sort of cloth net, like Cash's towels. This will of course scrape along the bottom either way and cannot well go wrong, but it is cumbrous and heavy (even the lighter one of the two) and has to be let down in pomp with a whale line, and requires all hands to haul it in so that many and frequent hauls with it on one day are out of the question. It cannot well from the closeness of its net let anything out, but it can and does let a great many unnecessary substances in: if therefore any great rarity is at the bottom it ought for its own reputation to catch and retain him, and therefore prove its existence. More it cannot well do, for even in a muddy bottom it sinks so far in as soon to fill its capacious maw, with most interesting mineralogical specimens of the bottom, in such numbers that by the time he is with infinite difficulty hauled to the surface, he resembles a bag of ballast, and the aforesaid rarity's most intimate friends would have the greatest difficulty in recognising him. In other words, the bag gets full of stones, and all the animals are pounded to fragments or reduced to become amateur jellyfish. It is a bit clearer by 7 p.m. this evening and we stand in to and along the ice edge. It soon comes on thick again and begins to blow, and we pass the night knocking about.

July 29, Tuesday.

The weather still bad and the ice hemming us in. What a strange contrast to Wüde Bay. As I go on deck before breakfast, I find myself forcibly reminded of Merryat's opening morning in *Snarleyyow* [Frederick Marryat (1792–1848), British naval officer and writer of sea stories, penned *Snarleyyow, the Dog Fiend*, about a dog who pops up again and again no matter his ill treatment, in 1837], cold and just freezing, wet, and snowing, foggy and chill, nothing to show us where we are, lots of heavy ice about, and a gale from the Northwest. So, we run, and dodge, and ratch, and lie hove to. With the fore square sails aback everything is reefed, as there is a great deal of wind, but a remarkably modest amount of sea. When practicable we sound, and dredge with the swab and get some fine specimens of coral, which have been broken 'ere I write this. There has been a great worry and cleaning up in the cabin today: chief results utter inability to find anything one wants and the best photos of Wüde Bay—one by the way with a reflexion of the hills, I believe a really good negative—damaged. We have drifted before the drifting ice right before the mouth of Hinlopen Straits, which rejoice in their chronic gale, fog and wild turmoil of waters. Tonight, we try for Hecla Cove but can't get in. We have steam up; in fact, steaming anywhere we can get, as there is ice all round us, and we have apparently a remarkably good chance of getting beset, so let us soap our noses and keep our tails up! 11 p.m.: wind shifting, backing in fact to S.W., looks better. Any anxious friends or relatives who think we are starving here on salt junks of pork &c, if they do, make a tremendous error, our *petit* diners would make an alderman's mouth water: bottled beer holds out, and stout, and the owner has lots of unexceptionable dry "Carte Blanche" [champagne], soups, Fortnum and Masons [tea and biscuits], or Venison and other game as stock,

Rotis, Venison as often as we like (and a young reindeer is perhaps as good as any meat the world produces, the ground willow on which they feed, giving them a very delicate flavour). Then there are lots of ptarmigan, excellent birds and then a bird, facile princess, for eating considerations, the wild goose, to say nothing of loon and rotchie salmis, so that we may with justice turn up our noses at the guinea fowls, ruffs and reeves, fat quails, and skimpy ducklings, spring chickens, sodden pigeons, and tough old prairie hens, the present staple commodities of the London game-market. We saved ourselves from being carried on to Low Island or Shoal Point and that is about all, and late on this evening found our bolt shot, and that we were completely beset, so hooking on to a piece we knocked off steam. The mist lifting a little: we found ourselves very close to the shore, between Cape Foster and the first glacier on the western side of the straits. I saw two large skuas tonight of a different kind to the two we have met with up here; larger and I think the same as those that frequent the Norwegian coast.

July 30, Wednesday.

During the night, we drifted along with the ice that so far only encompassed us, and at one time were over a good way not far from Murchison Bay [Murchisonfjorden]. This, however, is still full of ice, and the navigation at any time must be very intricate from the number of islands it contains. Rooks, too, appear numerous, and altogether the prospect of that part of the coast of the North East land is about as ugly as one could desire. When drifting off the glacier we got up steam and managed to push thro' into Loome Bay. This was still free from ice and we were at anchor by breakfast time much in our old berth. Captain Mack's smackie is at anchor near us and the other three at intervals a mile or two down the bay, which is now free of ice for some miles; that is, to past Foot Island. In the course of the day, the heavy sea ice begins to appear, *creeping* in past Cape Fanshawe and then coming down towards us and past us, and long 'ere evening we are completely beset. The smackies have most been in difficulties, getting their anchors &c up, as they all took to dragging, and then there were their nets &c to be got out of the way. Eaton goes in the course of the day to visit the broken arm man; he has been wonderfully lucky and is doing capitally. Then when Captain Mack comes on board in the evening, we hear of our consumptive patient; we are indeed doctors, "*mirabile dictum*" [wonderful to relate], he is cured!!! and at his work again, only he would like some more medicine. The little *Diana's* medicine chest has certainly been of great use, and we are a sort of itinerant dispensary. Eaton is, of course, a surgeon and general practitioner, but we all have a turn, according to the nature of the ailment. We talk of course very grandiloquently about cases, "*ah yes, tightness of breath*" &c, "*I should say, so and so was the best thing*," but it is a remarkable fact that on retiring to make up our offhand prescriptions, we are, most of us, obliged to stimulate our memories by surreptitious reference, to one of the "Prescription Books" or the "Sea Captains medical guide" [*The Ship Captain's Medical Guide* was compiled by Harry Leach, the medical officer on the British hospital ship *Dreadnought*, and first published in 1868. It is now in its 22nd edition], so that our doctoring operations bear a strong family likeness, and much of our inspiration is drawn from the perusal of the list of ailments and their cures treated of in those eminent works on pathology. But our Swede or Norseman, it is he (the consumptive one) that has caused the new and strict regime of looking at prescriptions, and not only looking at them, but making them up, with scales and measuring glass, and finding how often the doses should be taken. It was perhaps rather happy go lucky before, [and based on] the "can't have too much of a good thing principle." Anyhow, when the consumptive one's medicine was made up, it was discovered on review by the owner that one of the ingredients was perhaps a trifle strong, for on close examination it turned out that a solution or tincture of one drop to 200 of water or something of that sort was the form it should be in, the quantity about two drams, by a trifling and unimportant oversight, a concentrated essence had been substituted for the tincture, and by another trivial inaccuracy owing to those

Figs. 9.23 and 9.23.1 From 15 to 27 July, the *Diana* explored 100 km long Wijdefjord over almost its complete length. Though this is one of the biggest fjords of the archipelago, it was (and is) rarely visited both nowadays and in nineteenth century. After the call at Mosselbukta, the next anchorage was in Dirksbukta (just south of Dirksodden) on the east side of the fjord on 17 July, viewing across Wijdefjord onto northern Andrée Land between Våtnedalen (valley almost to the left) and Gråhuken (Grey Hook, right, flat northern end). The corresponding modern picture by Andreas Umbreit (2017–08-24) shows about the same scenery, but from above Vassfarbukta, around 8 km further north, which means a different angle of view

cabalistic long-tailed threes that represent the symbols of apothecaries measures [that] 3.3 oz had been substituted for drams. For ordinary cases, perhaps the simplest way is to give something so ineffably nasty that the patient is only too willing to say he is alright, and the only question is what *is* sufficiently nasty. Our crew certainly have a taste for medicine; like the rats and mice of the advertisement, pills they eat greedily, castor oil is as honey, and Epsom salts a beverage, [and] looked on as a luxury, perhaps second only to grog. Black Draught [a nineteenth century purgative prescribed to constipated cattle and horses] certainly does choke them off. It is a great pity that their government does not construct a hospital for patients from these numerous Norwegian craft at, say, Hakluyt's Headland. I understand that the Swedes had some such intention, should the other nations give up their claim to Spitzbergen. Russia, however, refused. The establishment, which would have been ordinarily occupied only during the summer, was to have comprised an observatory. Captain Mack's yawl has been here ever since we left.

He, however, sent away a boat down the straits and soon after getting anxious from the appearance of the ice went after them in another. He found all well, and leaving them at Cape Torrell, returned. This morning, we heard a number of shots across the bay, and it proved to be his men. They could get their boat no farther than Auk Head [Alkefjellet], and leaving it there with some difficulty got over the high ground above Cape Fanshawe, as the glaciers were very bad going and the crevasses dangerous. Wahlberg's Island [Wahlbergøya] is now free from ice, and from the high mountain on it, the fast ice, which still holds all the land on both sides and completely closes the straits, can be seen extending indefinitely from Cape Torrell. It is a land floe of unusual thickness and there is an unpleasant probability of the straits being closed during the whole season. Not hopeful, as should all else fail, the exploration of the Kong Karls Land found last year would have been something. The boat had got five seahorses, a reindeer or two, and two bears. Captain Mack's vessel is English built and one of four belonging to a Bremen Haven Company, and strange as it may seem, with the little apparently to be picked up here, compared with the abundance in the endless sounds, bays, gulfs, and islands of the high latitudes to the West of Greenland. I suppose it pays as they are soon to have 36 vessels and six steamers or 36 in all. The steamers will, I suppose, all work on the principle of the one he is to command: they will have a large plant of boats, white whale nets, &c. After the Jan Mayen sealing in the early spring they will work up the western ice and then about these coasts picking up seals, seahorses &c, and then when the time for the white fish arrives will run all along these coasts from bay to bay like a coast guard vessel or revenue cutter, forming depots of provisions, men and nets all along. Then they will seal, and seahorse and white whale themselves, and then having picked up their men, plant, and blubber, bear up "for Norge." A very pretty theory and deserving of success and very sound on one point and that whereas at present, the erratic habits of the fish make it uncertain in which bay they ought to be, if there are depots in nearly all the best situations, they may rest happy, in expectation of their comrade's success even should their own be bad. The said white fish have visited this bay during our absence, but the fishery has not been successful, as they have been shooting at them to drive them from one another's nets. Who began it is not clear, but it is clear that it has been done, and 14 is the greatest score of fish any have got. It is lucky we never set the net for them as it is only a salmon net, and it seems they have been casually walking thro' a four-inch mesh net of material nearly as stout as sounding line. There has been a large whale in the bay for the past few days, a "finner" from accounts. We went on shore this evening for a walk. We also did some dredging in the course of the day, but our best trophy is a small thornback skate. The skipper of the broken arm man's smack tried smoking out a fox earth. They found and got four young: one is dead and one of the other three was presented by him to the owner. I hope he won't die like the Burgo's and Ptarmigan.

July 31, Thursday.

We are not by way of going far from the ship as our imprisonment may terminate at the turn of any tide or any change of wind. It is today and was yesterday lovely sunshine here, but in the straits the gale continues and the long streams of fog go pouring on. Today, however, we want to visit some rocks in which Capt. Mack thinks there are fossils, so after lunch we four started. We take up a stony valley towards the glacier so as to cross the river in the broad sandy swamp or bed below it. This river in size and importance ranks next to the Wüde Bay one: it is swift and deep, of the turbid colour of the old Tiber, or Anio, but though its course takes bold curves and sweeps, the bareness of the surroundings and the naked sterility of the slopes that dominate its reaches, render it dreary and ugly, the rushing water its only charm. We have about ¾ of a mile of fording to do over a broad sandy flat, with endless streams running across it, deepish for fording and icy cold, separated by patches of sand now hard, now soft, often wet quicksand, and with the hideous shivering surface, so well described in the "Moonstone" [Chermside is here apparently referring to the 1868 novel by Wilkie Collins, *The Moonstone*, an early detective novel about the inheritance of a diamond]. Then we take along the valley, the scene of our first unsuccessful beat for reindeer. Our walk was much the same, though we kept more in the valley. The scenery of this valley is very striking and the topography eminently peculiar. I do not admire it half so much as many of the valleys that have perhaps less remarkable features. First, there is the right hand as you ascend: this is closed for the whole length by an immense glacier, not so striking in pictorial effect, as strange and awful looking, in its high precipitous edges, in its huge ploughed up moraine, and in its sounding interior hollowed and cut into subglacial channels by the water, spouting out under great pressure from cracks and crevasses and fallings in high cascades and in its long serpentine length winding away in various branches between the hills and so completely filling the valleys between them like a huge parasite. But their summits appear on either side, a heavy cruel many-headed hydra, one of those that hold the whole interior in their giant folds and coils, keeping it, barren, dead, and cold. On the left are hills of considerable height and very bold profile. These rise above the bay, but are rather disconnected except on the sea side and of the isolated sugarloaf form behind, so that the valleys run in the most strange and complicated ways, and a watercourse standing at the back of them travels first round one and then another right inland, and has to go round four or five miles to reach the sea, from which it was not half a mile at starting. On arriving at the bay that had been our furthest point before we took home by the shore a walk I have already described. We found from numerous traces that the Norwegians have been up here fishing. The reputed fossils were not such, but a face of rock from which endless hard round nodes of stones protruded, at least of circular section, and of various lengths. We couldn't make them out but brought away a specimen. Most of the rock on this side is schistose, quartz, or quartzite, I think; in fact, it appears to be much like the Hecla Cover formation: there are the same marble-looking blocks of pink stone, and grey and green ones besides with white and yellow veins, respectively. When near the river that runs into the sea thro' a deep cleft was a very formidable obstacle, a friendly smack sent off a boat, and keeping close in shore where the water was too shallow, to float the heavy ice we rowed home and got on board by 9 p.m. We got a ptarmigan with a stone and were successful in getting a plant of the _____ genus, not known before, we believe, and a new fly.

August 1, Friday.

A fog came on last night before we turned in but was gone from our immediate vicinity this morning and it was lovely as ever. There is a good deal of mist about Ice Cape [Isrundingen], and in the straits the fog continues and the wind is in the same direction, but I fancy there is less of it. The ice is a good deal slacker and we might push out, but where to would be the question, as nothing can be seen outside. This inaction is very galling, but there is little to do here: on the one hand a glacier and on the other a shingly beach, and no game beyond that to be beat for as the Norsemen have driven it away. I am anxious to go to the Cape Fanshawe, but although I can get there now, my absence would be perhaps the cause of losing a good opportunity of getting out, and my return for some time in the present shifting state of the ice is extremely problematical. The ice is very heavy and the strain on the cable and anchor is very great, but everything has stood well. One reason for our not moving today is, we fancy, the inevitable Friday, tho' it is put down to the painting, of which a sort of jubilee is going on: everyone from James downwards is going about with a brush and vessels varying from jam pots to buckets full of paint, at the cross trees, on the rigging, right up where the topgallants should come, along the yards. The busy artists swarm and soon they get to the deck level and attack everything with fiendish glee: boats, booms, bowsprits and everything else with a "b," as Alice would say, such is the mania they don't even spare the stags skull and horns at the extreme end of the said bowsprit, the *Diana's* fitting emblem. We do as much photography as circumstances allow, but the rate at which the ice

Andreas Umbreit

◀ **Figs. 9.24, 9.24.1, and 9.24.2** By 21 August, *Diana* had reached East Fjord (Austfjorden), the innermost eastern branch of Wijdefjord, and anchored on its west side on the height of what they named Diana Island after their expedition vessel and nowadays called Bjørnneshol-men after a Norwegian trapper who had his main station on nearby Austfjordneset for seven years from 1924 onwards. Here, the Leigh Smith expedition spent five days on hunting and exploration, and Chermside took these photographs from Diana Island onto the glacier-polished rocky hills of Austfjorden and behind it the mountains from Rutherfjordfjellet (left on this figure, up to 1329 m) via the glacier Schrödingerbreen to Grossf-jellet (right on this figure, up to 1172 m), the tops hidden in a cloud layer starting at about 1100 m above the sea level. Some 12 km inland, behind these mountains, mount Perriertoppen (1712 m) is hidden as the second highest peak of the archipelago. The recent image by Andreas Umbreit was taken from the hilltop of Austfjorden on 6 June 2006, looking into easterly direction onto the same mountain scenery as the two images by Chermside, 133 years earlier. Both his and the recent pictures show remarkably little snow and ice coverage of this area despite the partly high elevations, which is due to the extremely low annual precipitation (meaning first of all very little snowfall) in this sheltered area of innermost Wijdefjord deep within the main island

Fig. 9.25 This image by Chermside probably taken around 17 July from on board the *Diana* as it approached its southernmost reach in Hinlopen Strait shows the northeastern side of Wilhelmøya between Moltkeneset and Tumlingodden (Thumb Point) from North East. For a modern view see Fig. 9.18.2

moves, the perpetual swinging, and the motion of the ship from the constant shocks preclude our desideration, a photo of her ladyship beset. So, on the whole, we loaf—as they say across the pond—talk and was idle as it is too fine for uninteresting cabin work at charts, angles, and tables of figures. We did dredge up a few shells and things by the way and shot a snowbird or two for someone who wanted skins.

August 2, Saturday.

Steam up and about 6 a.m. We started, and after pushing thro' a good deal of heavy ice succeeded in getting out of the straits. But beyond them the ice was very impracticable, and tho' slightly scattered and slack in all directions it is determined to make for Hecla Cove, for a look out—any from the mast head being precluded by the cold chill fog in which we are enveloped. Soon after lunch, we come to anchor in Hecla Cove, which is rejoicing in warm, bright sunshine. The whole of Treurenberg Bay has been apparently for some time emptied of its own land ice, but like Loome Bay is full of this heavy sea ice, which drifts and drives and swings always on the move owing to the strong tides and currents prevalent in all these bays. Near us anchored the *Sofia*, an old wooden Norwegian schooner, SS, though her single engine which is dreadfully dirty is of no power. She had made a good voyage having 86 seals and 46 white fish, but at present seems in rather difficulties, as her forefoot is much smashed up, and they have lightened her a bit to keep it out of water. She also has a foul anchor, dragged, I fancy, by the ice and a broken cable on it. These she succeeds in fishing up in the course of the afternoon. [This is not the same *Sofia* used by Nordenskiöld in 1868 in an attempt to demonstrate how close to the Pole one could force a steamship north of Svalbard. That expedition reached 81° 42′ N, but the ship was holed and the expedition forced to return south.] The owner and captain start to climb the high hill and we give our minds to photography, wet and dry plates, those from the ship unluckily from her constant swinging have to be instantaneous. The lookout people are back in time for a late dinner, after a longish walk and some stiff climbing. They were, however, rewarded by a view over and beyond the fog, embracing from the North Cape to the Norway Islands—the mystic Seven [Islands] being invisible and veiled in mist. This view, however, served to show that this miserable sea ice is jammed down tight on the whole coast, the only apparently slack ice lying N by W. The remainder of the sea prospect being simply ice close down on the land everywhere, and especially heavy along from Red Beach at the entrance of Wüde Bay. We shot a few birds this evening and got a small seal today, which was stated to be a freshwater seal. I don't know what it is, but it has a pretty coat.

August 3, Sunday.

After breakfast, Potter, Eaton, and I again climb the hill for a lookout. We saw a glimpse for a few minutes of the Seven Islands, or rather of one of them, but the outlook generally was much the same as that of yesterday: a good view down them, as well as into the interior of the country, which can briefly be described as consisting of ice, tho' from the sky there would appear to be a valley here and there. The only slackness visible in the ice, lay over Verlegen Hook, the same direction it showed in yesterday. The height of this hill by the mean of the two observations is 1,630 above the sea level, leaving a neatly done up memorandum in two metal cartridge cases. We effected a safe descent of the steep stone slopes that rise above the bay and get on board by 2.30 p.m., and by 3 p.m. were steaming out of the bay, for a fling at the pack. As we kept northwards, we got into the fog, and the ice became heavier and closer 'til at last it became evident no further progress was to be made, so we kept along westward, finding no opening. It is decided to make to Moffen Island in case there is any seahorse, and we get up steam, as the wind is fallen. Hitherto we have had a fair, strong wind, and a good deal of sea. We are now among the ice with a vengeance, nothing but heavy masses, hummocks, and large floe pieces to be seen in any direction and long before we turn in. Our progress has become very slow. Soon after midnight, we are all in our bunks but for some hours yet there is to be no sleep for us, and no rest for the Captain who is conning from the nest. This is really ice navigation among heavy ice and were we only making way northwards would be very satisfactory. The engine room bell is perpetually going, with an accompaniment from the rudder chains as the help goes hard over from side to side as occasion requires, *ting ting ting* (3) and the screw revolves slower with the help now one side, and then the other, we push through a winding channel, *ting*, on she goes again, to get way on to charge an obstructing mass, and finally, *ting ting*, the propeller stops and settles oneself for the coming shock. There it is, and she pumps right up on the piece and,

Figs. 9.26 and 9.26.1 In its first attempt of 1873 to go deeper south into Hinlopen Strait, *Diana* met heavy drift ice around the Foster islands (visited also by Parry in 1827, and named after a member of his expedition), combined with strong winds, which kept her in this area for some days around 7 July, close to Sværingen, the main island of these small rocky islands in the middle of Hinlopen Strait. An intended landing was not possible under these conditions, but Chermside took this image from on board the *Diana*, showing the south side of Sværingen. The corresponding recent picture by Andreas Umbreit shows the south side of Sværingen, too, from a slightly more eastern position, on 5 July 2011, without any drift ice around it. However, also nowadays drift ice is very much possible in these waters in July, varying strongly from year to year

slicing off, rocks tremendously. Again the bell and we go astern and charge again, no go and now we work slowly up to it, and getting her nose against it, crack on steam and slowly slew the piece round 'til a channel is open, and so twisting and turning to the port and starboard now going astern and now ahead with many heavy shocks that set everything ringing; we go on, until in the early morning the fog, having hidden the island again, we bring up a mile or two from it and hook on to a grounded piece.

August 4, Monday.

By 10 a.m., we have breakfasted but no island is to be seen. It has, however, been visible but unluckily the bearings were not taken. It is the general verdict that it is not a quarter of a mile off. This statement, however, we accept with reserve, for two reasons: first, the extremely hazy notions of ranges and distances we find to prevail uniformly among the crew on all occasions; and second, the fact that even in the fog we can see nearly a mile. So, Micawber-like, we wait for something to turn up, in this case the sunshine, or rather a clear atmosphere, as there is a good enough sun. The mist is local enough and one of the low-brooding clouds, like those we saw from Hecla Cove, and above it are the bright sun and glimpses of blue sky &c. The sun is duly shot from the ice with the aid of the artificial horizon for latitude and later on we get the sight for longitude in the same way. We also print a few small photos that are, to our partial minds, a fair success. We are in five fathom water, and during the day our grounded floe piece begins voyaging. There is a tremendous tide here so that all his *confreres* likewise voyage, and "a pretty considerable smart pace they travel," so steam is kept up and by its aid their dangers are avoided, though there are a good many occasional excitements and shifting of ice-hooks and warps &c. We have a good many visitors from Spitzbergen in the way of birds, snowies and dovekies and rotches, all come out to fish for their wives and little families. Indeed, for the past few weeks we have been much amused by their proceedings, in this respect in the bays. They have come long distances to fish in their favourite spots, and when they have fed themselves take in cargo in their pouches and necks to be unshipped at their homes in the various looneries. This takes some time and it is very amusing to watch them fishing and fishing, and keeping all they get without swallowing it. The loons generally grout in with as big a fish as they can get to bottle in the rest, and their holds are so full that his head or tail only which acts as a stopper is in their throat, the remainder protruding beyond their beak. Then they get under way and off they go. The quick, active, little "Rotchies" are especially industrious fishers, and having a large receptacle below the lower mandible of their beak lay in such a tremendous cargo that they find great difficulty in rising off the water. Often you can see them flap-flap, along with a great deal of splashing, but they can't manage and so are flapping on until they are either successful or have to lighten themselves of some of their cargo. Talking of bird's habits, I see Prof. Newton stating that it has been remarked by all observers, or nearly all, that it is a very unusual thing for snowbirds, that is Ivory Gulls, to settle on the water of their own accord and swim. I can only say that in our experience this is not the case. A good many are shot by the engineers for their skins and these are almost [all] killed either on the ice or in the water, tho' now and then we are informed that Jimmy has shot one on the wing (as a very great feat). In Wüde Bay, they settled and appeared to catch shrimps and I saw the Captain shoot three running and swimming about, and we ourselves shot some in Loome Bay as specimens of those we had shot that way, that we could witness to. We started in the fog this evening in a whaleboat to try and go to the island, but it is no wonder that we could not find it, as no one knew its bearings. Tho' our idea of them subsequently proved about right, we were now in some doubt of being able to get back, the ice having rendered our attempt as to where we supposed the island to be futile. However, after a shot or two we got an answer and got back on board alright.

Fig. 9.27 Cape Fanshawe, depicted by Chermside from ashore at Hingstsletta on the west side of the entrance of Lomfjord when the *Diana* visited this bay for the second time at the end of July, anchoring at almost the same position off Hingstsletta. Thick drift ice had entered the bay again, but snow cover on the slopes of Cape Fanshawe is less than during the first visit about 20 days earlier. For a modern view see Fig. 9.9.1

August 5, Tuesday.

Still fog and mist but, towards 2 p.m., it cleared and we made out the island at what the crew said was ¼ to ½ a mile and what we judged as one-and-a-half to two miles, and it proved to be nearly two miles. The large heavy ice, which had been stewing and grinding all day, causing us a good deal of inconvenience, was now pretty slack and we got to the island without much difficulty. It is a mass of low flat shingle only six feet above the sea level and has apparently been ploughed up by the ice grounding on shoal ground and carrying up stones before it. There is no rock to speak of in the island, only a few granite boulders, possibly brought by ice, and some decayed completely away to cracked masses of hard red clay. We started to walk round the island, one party going each way, and in about one-and-a-half hours met at the opposite points of a large natural bay or harbour which occupies the greater part of the interior, and which had still a great deal of land ice in it. The complete circuit of the island is about seven miles. We saw no seahorse but came across the "krengs" of some killed this year, apparently killed at the edge of the furthest inland portion of the lagoon or harbour, either on the shore or ice, or in the water here very shallow for some distance. We also saw, which was far more interesting, the bone-covered sites of what at one time must have been glorious engagements with large numbers of seahorse. There were three of these spots, white with immense quantities of bones of the animals evidently all killed at one time, and the largest of these was over an acre in extent. I am rather sorry I did not count the number of skulls. The seahorses, it appears, in August or September take to coming on shore, on shelves of rock, or on low flat shingles like this. Though I believe it is otherwise in other parts of Spitzbergen, shingle beaches are excessively common in all the bays on this coast, often extending as in Loome Bay for two or three miles at a stretch. On the chosen spots the seahorses assemble in large numbers, and it is, I believe, an open question whether they do not hibernate in a sort of way, remaining packed here like sheep in a pen during the whole winter. The fact of the spots selected being 200 or 300 yards inside a beach of steep shingle for such unwieldy beasts, in the cases we saw, seem to favour this supposition, and if they remain on them after the end of September, they would be unable to get into the water until the spring. This island is the scene of the attack on a large number of seahorses by the boats of Phipps' (afterwards Lord Mulgrave [Constantine John Phipps, 2nd Baron Mulgrave, leader of a Royal Navy expedition towards the North Pole in 1773]) expedition, depicted in the picture hanging in the dining room at the Admiralty. When a vessel is lucky enough to fall in with a number of these animals on shore, they attack with lances and rifles those next to the water, and their carcasses form a sufficient obstacle to the rest to prevent their reaching the water. The beasts too become ungovernably savage and attack and kill each other so that if the first onslaught is successful, large captures are usually affected without much trouble. This way of theirs of fighting each other is found very useful: when a boat succeeds in getting fast to two or three at once they not only often fight but the lines getting foul and they drown each other. Some years ago, the *John* of Greenock came up here from the western ice, in the days I fancy before the abundance of these Norwegian craft scared them away. She fell in with a large number in the water and on a rocky shelf, and the boat sent after them attacking those in the water. Those above came down in such numbers as to oblige her to put back for assistance. All hands then attacked them and with such success that she left Spitzbergen with 700 seahorses. The island is a great breeding place for eiders and they must have been visited this year many times. There are still thousands of nests, but on finding the eggs not fresh, we left the poor ducks in peace to a great extent, tho' we hunted a few and I killed one with a stone, and the other was caught alive and I believe it is still on board sitting on some eggs. There were also a great number of Arctic Terns breeding here and most of these are young. The opening of the bay is to the NW, not the N side as shown in the chart. The ice on it was thick, but very rotten, and having taken the floe in pursuit of a seal, I had great difficulty in getting off of it again. We put back to the ship between seven and eight, but though the island was still visible it was far from clear. Soon after midnight the ship began to take a very decided cant over to starboard and after making things secure I rushed on deck. There I found two huge floe pieces driving together with the tide, and that on the starboard hand had a long, submerged tongue, forming an inclined plane up, which the whole ship was pushed bodily by the other. This prevented the "nips" in which we were seriously jamming the ship, and steam being handy with the aid of the warp her head soon slewed, and we got off without much difficulty.

August 6, Wednesday.

It was not very clear. That is, you could not see over three miles or so even from aloft, not that I ever make that experiment, owing to the strong temptation I habitually feel on such occasions to hurl myself down. The ice was very thick but still, wherever the cracks of water among the pieces are of a ship's breadth, there, if well handled, she can go. So, it was determined to go somewhere, and rather after 10 a.m., we started. All day we were pressing on N and E wherever there was a chance at times getting into holes and patches of slack ice. We saw a great many seals, none however on the ice, and since they did not promptly shoot the hatch, we left those in the water alone. All day we go on turning and twisting and taking advantage of any holes in the ice shown by the sky. The ice only gets heavier and thicker, and steering is such hard work that the wheel is double-manned. Towards evening, these "circumstances over which we had no control" drove us down southward, and at last finding further perseverance useless we stopped for the night, or until any change of wind, tide, or anything should cause the ice to slacken. It was now a bit clearer and we could dimly make out the land. We were from three to four miles W of what is called Low Island, but it apparently consists of three islands. Parts of low coast land could also be made out in the intervals off fog.

The water at first was only ten fathoms, an hour or two ago it was 14. We dredged a bit but without much success. Nothing but ice can be seen anywhere or is denoted anywhere by the sky, with the exception of one very small pond of water a mile or two ahead of us, and a little water to the W of Low Island close to it and formed under the lee of the numerous grounded bergs lying there. The floe pieces all around are unusually large thick and heavy. There seems to be, by the sky, a little water about the mouth of Hinlopen *Strasse*.

August 7, Thursday.

Foggy and the land not in sight we have been driving a bit with the ice, deepening the water as we go, 10, 11, 14, 18, 22 fathoms, and then 28 and 30. We do a good deal of dredging, getting some corallines, an alecto, sea spiders, starfish of two or three sorts, some with 12 and 13 points, [and] any number of worms. We also take the temperature of the water below and fish up some from the bottom to test its specific gravity, as that at the top is extremely fresh. The land is not in sight all day and we wait for it to clear. A very large fine bear is descried quietly sitting on his haunches under a hummock at no great distance from the ship. We get up our rifles and as it seems possible to get over the ice, are just holding a council of how to surround, when one of the groups watching him, in the most thoughtless manner, gravely walks forward and sounds the seven bells. Brunie at once gets underway, and in spite of numerous burnt offerings has the bad taste to take himself off. Pursuit is useless as ice is impassable for a boat and one can walk over it only very slowly, owing to the gaps, channels, and hummocks, tho' after a bit the owner and I try it, but with no success. It clears up for a short time in the evening and we get a fair lookout seeing Shoal Point, Low Island, and the Capes forming the N and S promontories of Brandy Bay [Brennevinsfjorden]. We also seize the opportunity to photograph the ship. This lookout northward is indeed discouraging and we fear that the prevalence of westerly winds that have occurred all the season thro' render any considerable journey northward impossible. Nor will the ice it seems allow us to reach the Seven Islands again and explore the northern coast or seek the supposed land beyond, so it is determined to make again for the Straits and to reach Cape Torrell if possible.

August 8, Friday.

Early this morning steam is up, and the fog has again lifted, and we start for our destination. It is indeed wonderful what a ship can do with the aid of steam, and to any person unacquainted with ice navigation, the statement that we were going 'round Shoal Point and down the straits would appear an absurdity. Masses of ice of every sort and size in all directions, so completely appear to cover the sea that it looks like a solid irregular field of solid ice. Not a spoonful of water can you see in any direction from the deck level, beyond the distance of a hundred yards or two. Why, you would say, it is impossible for even a boat to put through that. Quite true, a boat cannot, but a ship can and by midday, that is, in less than eight hours, we are round Shoal Point (15 miles). The little *Diana* does indeed seem to be a remarkably handy vessel among ice and bores her away along marvellously. She stands the shock, too, of charging heavy pieces, at speed, with intent of smashing them, far better than from her seeming tenderness under canvas in a heavy wind. I should have imagined, a shock, a tremulous quiver through her whole frame, a rocking from side to side so that the keels of the boats hanging at the davits actually touch the ice at times, and it is all over and [yet] she is ready to go at it again or, insinuating her nose in the crack she has made, worm her way through. After passing Shoal Point, the ice begins to show more scattered, pools of water 100 yds across, increasing in size, until there are only streams of drift ice to be negotiated now and again. It is, for a wonder in the straits, a lovely clear calm day, the water alive with loon-kind fishing, and large numbers of seals, and these are, I think, from their appearance, harp seals or saddlebacks, tho' how they come here if so, I know not. They are not only too large for "dorothies" or for "common seals" which are the varieties of small seals we have met with here but are apparently gregarious as we see numerous small droves, some numbering a dozen. Now, those mentioned have never been hitherto found in droves, tho' it is possible they herd at certain seasons of the year. Captain Mack's smackie, looking very smart as she ought, being a Grimsby boat, is standing out with the slight breeze that has sprung up and he is much relieved to hear that he can get to Hecla Cove without much difficulty, as there is nothing but two or three streams of loose ice for him to negotiate, if he hugs the western land. Loome Bay is full of ice, which has stopped the fishery. It was with great difficulty that he managed to get out by towing, warping &c, and he has had to leave one of his boats. The others cannot get out with the exception of the *Johann* who left before he did. We steam on now in very open water, taking to the E of the Foster Islands, as heavy ice lies all along the land from Cape Fanshawe downwards on the Western side. Wahlenberg Bay is still almost completely filled with land ice, tho' in passing Murchison Bay today we were surprised at how much water [lay] inside it. We pass within a mile or two of Cape Selander and very striking is all the eastern coast from this point, along to Angelins [Angelinberget] and Black [Svartberget] mountains, high outward sloping cliffs of mountain limestone, with the markings of the strata and deposits curiously and plainly defined, the beds having an even but very gradual and gentle dip to the south. The faces of the cliffs are deeply seamed with rugged grooves and furrows running up and down them and considerably wider at the base than at the top. There are one or two "faults" in the beds, where when they have with their even and gradual slope slowly arrived at the water's edge, the same strata in continuation suddenly appear but at a considerably greater elevation. Crowning the limestone appears knobs, layers, and beds of dark hyperite and these show grandly but irregularly in bold eminences as Angelins and Black mountains. Some of the rugged and dark columnated seams of this rock look very majestic. Wahlberg Island bounds the other side of this channel and here again this rock assumes the same columnar form: dark basaltic looking pillars in half relief protruding from the face as at Alkrange. We speak a smack off here, for whom we have a letter, and then proceed on our way at a good pace, making way while the sun shines, at length we come abreast of Cape Torrell and rounding it, steam eastward, being brought up in about ten miles by the fast floe. We are here in very unknown waters, being beyond the southern mouth of the Straits, and I understand that this coast running from Cape Torrell to Cape Mohn is only delineated in the charts from observations from Cape Torrell and the distant Thumb Point the other side of the Straits, the principal observations having been those taken in 1871 by the owner and his Captain, [Erik Andreas] Ulve, a Norwegian. The said coast anyhow differed materially from its assigned form and the bearing of it is, we believe, too southerly, but until we determine the variation of the compass, it is hard to tell. The whole of the interior of the country that is visible from here consists of ice, mountains, valleys, cliffs, watersheds and watercourses, and even the coastline, all moulded on ice. The gigantic glacier that apparently extends from Cape Torrell to Cape Mohn is not however unbroken as shown in the chart: dark hills projecting above our anchorage here, and at a point further on, at which point a cape is formed by a long low spit of limestone or yellow shale jutting out from below the darker formation. Beyond this, the coastline is rather deeply extended, and then there trends away to what we take for Cape Mohn, a vertical wall of ice cliffs, decreasing a good deal in height towards that point. There are numerous islands about, with simply dark rocks; some of considerable height and with no attempts at beaches about them, all of the formation usual in these straits, their positions are tolerably well denoted, but they rejoice in the chart, in completely original and fancy sizes and shapes. There seems deep water round all or most of them, and we saw no sunken rocks. The bottom would also appear good for anchorage but none of them that is south of Wahlberg island seem to offer very sheltered positions. The fast ice consists of rotten floe, into which we do not penetrate to any great distance. Just as we are hooking on to the floe, a ground seal appeared and being in six or seven fathom water we shot him, but owing to a very strong tide or current could not get him. The ice extends from here in a broken and jagged line towards Thumb Point, but over on that side a great deal of heavy drift sea ice is jammed against the fast floe, and extends some miles northward, so that although there is an open strip of water beyond it, it is doubtful whether we could penetrate even across to the point. We are very anxious to ascend the dark hill that crops out of the glacier here, and rises above our present position, but the glory of the day has gone and it has come on thick and there is nothing to do but wait. It is considered expedient that the fox should have a chance of exercise, by being confined by a chain and collar, and leaving the little trap door of his tub open, and so like the Æsir of Valhalla we go after dinner to the "binding of Fenrir" [a wolf in Norse mythology who bites off the right hand of the god Týr (Tor)]. Tyr has been found and he emulated his prototype by getting his hand bitten, little wee foxie is not to be beguiled, he will bite any and everything offered to him, but declines holding on like a bull terrier to be dragged forth, and the hold is so small it is almost impossible to pull him out, with your hand; however, two of us at last get hold of him and he is successfully bound with an immense chain and collar; poor beastie, but it is the only one we have got and I suppose he will be happier so. We now turn in leaving orders to be called whenever it clears up.

August 9, Saturday.

This does not occur 'til between 6 and 7 a.m., and we all turn out and get breakfast as soon as possible. The morning was spent in ascending the hill and taking one or two photographs which, owing to it being a dull day and the ship and ice and everything in motion, were rather hard to take with any success. However, we did our "possible." The hill was 630ft above the sea level. It was of dark hyperite formation, with loose yellowish coloured shale, scattered about among the larger stones. There were a good many small fossils among this shale, specimens of which were secured. The top of the hill was very rugged and steep, and there were a good many birds breeding; apparently, snowbirds among the number. Towards Cape Mohn, the coast was clearly visible running as I have already described. Seaward were many large bergs projecting from the floe, the limit of which was not discernible. There seemed a good deal of water about Thumb Point and the ice looked practicable. Beyond this were visible capes Weyprecht and Payer and the numerous islands beyond them, [but] Kong Karls Land was not to be seen. During the day, however, I saw, as if by refraction, or mirage, which I forgot to mention, was the way we saw smack yesterday first of all, what to me undoubtedly seemed distant land. There were three distinct projecting mountains with apparent striae and patches of snow on them and of these I took the bearings. Finding that these talked with Petermann's chart of Kong Karls Land, I suppose we saw it; I say so as there were five or six of us. In the afternoon, we got up steam and take across towards Thumb Point, passing pretty close to some of the islands. When beyond them we get among the ice and this soon assumes a very rugged character, small bergs being very numerous. It is now a clear sunny lovely evening and the lights on the hills are magnificent changing from the warm sunny purple colour where the rays directly light the dark hyperite to the soft greyish yellow of the limestone scaurs with their plainly cut dark seams and furrows, and away into the soft blue haze on the distant glaciers and snow-clad hills. The sea itself is of a dark olive green here, and the sheen on the lighted parts of the ice forms from its very contrast to the intensity of the shadows a foreground as lovely as wonderful. When still about seven or eight miles from Thumb Point or perhaps rather less, our progress becomes so slow and our objective, if gained, is off such comparatively slight importance that we decided to give it up—at any rate for the present. We wait where we are among the ice for 20 min or half an hour to give time to get a couple of photos and then a "brunie" is descried, and I may as well say at once we have our usual luck with him. He is a long way off on the floe. "Oh, let's get to the floe then and take after him!" "You can't, sir, (this from the masthead), it is rotten and his hind legs keep falling through."

Andreas Umbreit

◄ **Figs. 9.28 and 9.28.1** Between 10 and 22 August, *Diana* reached its southernmost positions at the southern end of Hinlopenstretet (further south, the sea was completely blocked by very dense drift ice), between Bastianøyane and Vibebukta. There she did some scientific dredging of the sea floor as far as possible regarding ice conditions almost into Vibebukta, but with only very few landing possibilities. This image was taken by Chermside from on board the *Diana* in this period along the coast of Nordaustlandet (North East Land), and shows low Karl Alexander Island (Karl Alexanderøya) to the right and, partly hidden behind it, the bigger but equally low and rocky Franzøya with its western end to the left, seen from North, both islands situated south of western Ulvebukta, where *Diana* was struggling with the ice between these islands and the shore at Kapp Oetker around 17th and 18th of August. The corresponding recent image by Andreas Umbreit shows these two islands from north as well, but from a slightly more western position, taken on 7 July 2011, without any drift ice around them, which is not a given in early July even nowadays

However, as he is taking towards the loose ice, in which we could hunt him with the ship, we wait, but as luck has it, he spies a seal—the brunies we see always do get seals—then he stalks him, and when he gets near gets flat on the ice and drags himself along by his forelegs. He catches the poor brute, tosses it about and paws it 'til it is dead, and then after a short meal goes off again on the floe, and we have to give him up. We now steam back much as we have come and, passing between two islands, make for a shallow bay, almost a flat-shaped crescent just north of Cape Torrell, and of which that point forms the southern extremity. Here we come to anchor in eight fathom water about 1000 yards from the shore, as the owner wishes to explore the country here behind some black mountains or hills which with the valleys running from them, here relieve the monotony of the endless inland glacier and induce the belief that there is something to be seen thereabouts in land. The anchor goes down soon after 12 midnight.

August 10, Sunday.

There is no photograph of interest here to be taken, so I am off for a day's amusement, as there is no surveying on either. The owner and I go together as Potter is on the list, with a bad heel. We land after breakfast and take up the water-course between the two principal peaks; as we gain the top of the limestone formation, at the junction of that and the Hyperite (none of us can make out how to spell that word) we find a good many small fossils. Thence we ascend the left-hand hill and crossing a snow slope at the gully get on to the right-hand one. This has in parts steep rocky and almost precipitous sides but has a smooth rounded head, on which we erect a cairn and state in a memo therein deposited, that its height above the sea level is 870 feet, as well as other particulars referring to the visit of our illustrious selves. We get a round of angles from here and take the opportunity to munch our bread and cheese. Seaward we have a lovely view, the same islands, capes and headlands already described but from a different position; inland, all, strange to say, is not glacier and as the walking here is exceptionally good, except across the snow, of which the patches tho' wide are not over numerous, we are tempted to take onwards to a hill three miles or so inland. The geology of the country seems here plain enough, beds of mountain limestone underneath, above these the dark rock, we call hyperite, and above this shale, in some places light yellow, and in others dark grey, both very brittle and finely laminated, and the yellow very hard. Amongst the yellow shale were numerous circular discs of very regular shape, some eight or nine inches in diameter, and only half an inch or so in thickness. We found a sort of conglomerate fossil on the yellow shale, and a good one apparently "*mytilus edulis*," or the common [blue] muscle in the grey. The hyperite is here a good deal decayed and has on the hilltops been reduced, by snow and ice action, I suppose, to a sort of coarse powder formed of very small nodules scarcely any as big as a nut. These form a sort of soil, very good walking, out of which crop here and there larger lumps. There is a curious colouring matter in this, which makes your boots look as if you had been walking thro' a field of buttercups. It was a grand day for a walk, plenty of sun, but not too fierce and a nice lively fresh breeze. We at length gain the top of our hill, the highest for some way round (930 ft) and proceed to take some angles. Our expectations have been realised, and now below us at three or four miles distance stretches a bay, still full of fast ice. This is an arm running from a larger bay and this larger bay must be either a branch or the main fiord of the Wahlenberg Bay. No one, I suppose, ever dreamt of its existence here and I daresay even the hardy Norse adventurers are ignorant of it. The part above which we are is a fine landlocked bay with a mouth guarded by two bold rocky capes: one of the dark formation, and the other limestone with a dark crown, just Cape Selander with a difference. This bay bifurcates and we can only guess at the end of the further bifurcation by the glacier that closes it; that is, the water may run right up to it, or as in this indentation a plain of flat grey limestone may intervene. The main bay narrows opposite the dark and western point of the smaller one, but again expands beyond and turning southwards is on its far side bounded by ice cliffs, broken here and there by dark cliffs or rock, runs out of sight. The ice tho' dotted with blue pools looks still sound enough. Nothing more can be done except to explore the valley seawards as by this route we shall traverse another large inland valley on our homeward journey. The ground over which we have travelled today is frequented by reindeer, which are, however, not nearly so numerous as in Wüde Bay. They are, however, owing to its being later in the season, fatter and in more perfect coat. As it was much too far to carry them to the ship, we made the resolution not to shoot any except near home, but I am sorry to say it was broken by me at the head of this valley. In crossing some snow near a small plain, we were suddenly aware of two fine deer, and hoping to drive them down seawards, we waited. We lay down on the snow as they had to pass us, and I held the dog. They are, however, tamer than the Wüde Bay deer, and catching sight of us, insisted on coming to see, and so came back well. There they stood 70 or 80 yds from us, a fine stag and a hind, with pretty little horns as well, and hoping to drive them I let the dog go. Taking a sweep the brutes ran the wrong way, and the owner's attempts to cut them off were in vain, so I unslung my rifle and loaded and as they came to the stream, fired, chiefly to turn them, as being at a distance at which I could not distinguish the stag from the hind, I didn't expect to hit. It did turn them, the stag, so that he took back, and the hind so that she couldn't get out of the stream. An awful business I had, Rover got it by the nose, and I didn't want it hurt as I was in hopes of getting it to walk home, so there I was in the stream holding a struggling deer by the horns with one hand, and a yapping, twisting, squirming terrier with the other, until relieved by the owner. We tried to make the deer walk but it was no go, it couldn't or wouldn't, and so had to be knifed. The stag was very affectionate and wouldn't leave for a long time, but we didn't molest him. We now tried dragging the beast, first across a swamp to a snow slope, along that by the horns, then by the strap of my cartridge bag; the handkerchiefs were added as the strap cut our fingers, and then we harnessed ourselves by our slung rifles so as to get the pull on our shoulders but it was terrible work and the snow did not last for ever, tho' I felt any amount of labour was but just retribution on me, for killing the poor brute. Then I tried carrying it and after a short spell of that it was finally abandoned. I, for once, certainly realised Sinbad's feelings on getting rid of his ancient jockey. The stag, after being driven for some time, took away over the hills. We were now near the head of the series of valleys running into Wahlenberg Bay, and this ended in a most abrupt and peculiar manner. On the left were dark hills, with a good deal of snow on them, and the gully between up which the stag had toiled, completely so. On the right is a glacier, and in front joining the hills and glacier a dark hog-backed barrier or ridge or rocks. Towards this the ground we are on is falling. In fact, in this unnatural valley at first sight there is no way out; the ground falls in every direction and the water too. The latter explains it; having fallen as far as it can, it retires under the glacier, and so by some hidden valley, no outline or contour of which is even vaguely shown by the even ice surface above, into the armlet of

Wahlenberg Bay. It is such phenomena that bring home to one the idea of the enormous role glacial action must have played in the formation of the natural features of this world of ours, how it is thus the valleys, glens, and watercourses we know so well, and how once they were bare and cold, and in the rough like the land before one now. It is as if one saw with ones very eyes, if it is not actually the case, a country in an earlier stage of its formation, as if one saw it, like the rough block of marble in which as yet, under the hand of the *asbassatore's* [sp?] chisel, the image is but duly foreshadowed, and to which must be given in time all the fine detail and carving of the masters hand 'ere the statue is complete in its glory, as if the master touches were to be warmth and light, the sun under the omnipotent control of the master of the universe, the agent. Is there a revolving cycle of climates, is all to wear forever the same crude and unfinished air, or will in ages to come the sun have clothed these slopes with luxuriant forests, and these plains and valleys with a more luxuriant verdure, not so ephemeral and with such short faint glimpses of real summer? It is easy to imagine this—a land in the course of adaptation for man's inhabitation: there are the glaciers, scooping out valleys like gouge and chisel, and there are the plains already formed either by the heavy flat masses of ice or by the raised beaches. It would seem as if these mountains would be mountains and their endless streams but mountain torrents flowing into the fiords, the fiords to be empty of sea, with fine large plains and fertile valleys, watered by rivers running in the central channel and fed by the mountain streams, this would be the mountain country. The owner always says it is like Switzerland with the sea, in the valleys and, for aught we know, under the bosom of the deep is forming flatter and less hilly country. Anyhow, to give some reason for this perhaps vague dreaming, parts of this country seem more finished than others, in parts are visible the striae and marks of ancient glacial action, abrasion of stones &c as if it was over, then look at the difference of Wüde Bay, Ice Fiord, and other similar spots to this Nord Ost Land, still completely overpowered by ice. Look how some of the rudely chiselled ravines and hillsides are softening and as the rocks seem to feel the influence of weathering become of softer contour, while others again but at no greater elevation have knife-edged crests and sites at angles of 45 and 50°. Then there are the shingle beaches with the remains of whales killed perhaps 100 years ago, the dates indeed have been in some cases determined. The skeletons too heavy to drag above it must have lain in the water and now you find skeletons almost complete at six, eight, ten feet, above the sea level. At Loome Bay, we found a height of 30 or 40′ above the sea, a sort of alluvial soil of which endless shells formed a large ingredient. Then again today at a distance of two miles at least from the shore and at a height of 100 to 150′, the plain gradually sloping upward to the base of the hills ends and all the way up it we found whalebones, and masses of imbedded and decayed driftwood in far too great profusion, for any chance action to have brought them there. Then, too, what from the old Dutch charts are given as sounds and channels with the anchorages marked in them, but with existing glaciers and mountains shown in them, are now found to be valleys, some of their islands parts of continents &c.

Crossing this ridge, we come on a long flat sort of table land, the water from which also sneaks out at a corner with the help of a glacier, but this time into the straights, we came to another set of black rocks. These, from the ship, appear as a ridge of hills but, in reality, are only the revetting scarp or retaining wall of the terrace formed by the table land beyond. Below them two or three miles of gently sloping plains brought us to our starting point, and we were soon on board (8 or 9 p.m.). Eaton today made two journeys to the hills for fossils, the second after our return and got back before midnight, soon after which it began to blow.

End of Volume 2.

August 11, Monday.

About 6 a.m. this morning, a small mass of flat ice drifted across our bows and cable. The latter, that held so well in Loome Bay, must have got heavily strained, as under the momentum of this

Figs. 9.29, 9.29.1, and 9.29.2 These images were taken by Chermside on or about 17 and 18 August, when the *Diana* struggled with the ice conditions between Karl Alexander Island/Franz Island to the north and Kapp Oetker at the shore of Nordaustlandet east of Kapp Torell. Both images show neighbouring stretches of the coastline a bit east of Kapp Oetker. As a comparison, see the contemporary corresponding image by Andreas Umbreit (taken 4 July 2011), showing that the shore was far more covered by permanent snow accumulations back in 1873 than nowadays, even more so as the Chermside photographs were taken six weeks later in the season, close to the average minimal annual snow coverage. In the background behind the coast and the partly snow and ice-covered ascent, the gentle rise of the huge ice cap Vegafonna starts (hidden in low clouds on the Chermside images)

insignificant piece it parted, and away we drove. Steam was not up until we had drifted unpleasantly near the glacier that forms the projecting ice cape at the northern end of Augusta Bay [Augustabukta]. We then steamed back and anchored as near as possible to our former position. It was blowing too hard still for much to be done for an hour or two. After breakfast, however, the three whaleboats were lowered and dragging with grapnels, to hook the cable or flukes of the anchor, persevered in for some hours. After noon, we got underway and, with a large heavy warp out with numerous grapnels attached, proceeded to drag under steam. This manoeuvre was also unsuccessful, and as the next tide filled our beating ground with ice, dragging in this way had also to be relinquished. It was now blowing hard and we had to seek for shelter and came to anchor, in the northernmost of two little coves formed by the projecting point of a glacier, near the southern end of the bay. It is a great misfortune, this loss of our best bower, as we have no other really reliable anchor. The search will, if possible, be prosecuted tomorrow but our chances of success are small, as the bottom is densely covered here with oar weed and other marine plants and amongst these the anchor and cable must be buried. Had it only been smooth or as the sailors say small water, we should have had a good chance, as our old anchorage was very accurately determined whilst surveying, not only by bearings but by certain points, headlands, hills &c being in line, and we feel positive that the boats were kept plying as also the ship, almost exactly over the right ground. Then, too, we had soundings round our old position and knowing that the anchor was lost in 7¼ fathoms, the shoalest place for some distance round, which was also a clue. It comes on thick in the afternoon, and as it is blowing too hard, we do not go ashore, as the men have had a hard day's work and it is rather rough for the small boats. Neither does the shore present an attractive prospect: snow slopes and glaciers and a long low beach of yellow shale. We have been very unlucky in weather for "sights of the sun" or, as I suppose one should say, in refined phraseology regions, "astronomical observations," since we have been in these unexplored regions. But we got some good observations for Lat and Long, yesterday for the first time.

August 12, Tuesday.

At 4 a.m. this morning, heavy flat masses of ice appeared, bearing down on us at a great pace, with the swinging tide that runs, as is always the case in straits and narrows like this, with tremendous velocity. Steam is ordered to be got up at once and all hands are called to weigh the anchor, and although there were ten lbs of pressure of steam in the cylinder, at the time the order was given and the ice then more than a mile away, a great piece comes right across our bows before the engines can be got going. All hopes of getting the anchor up at present are over, and the only thing that can be done to humour the ice is to pay out cable. Slowly and surely the heavy mass presses on us and the cable now extends far underneath it, the piece tho' not of great thickness is of large surface, that is from two to three acres, and it gradually pushes us along, anchor and all. Now, indeed, we feel sadly the want of our guncotton: a few pounds just laid on its surface would crack it right across and with steam up we should soon have our anchor at the bows ready for "fishing." As it is, the strain on the cable is very great, so great that a strong hand at the compressor finds himself unable to move it. Gradually we go right round the ice point of the little bight I have spoken of and drive shoreward across the bight lying the other side of it, between it and Cape Torrell. Sixty fathoms of chain cable have now been paid out and we drive slowly but steadily before the mass. The ship is in a little sort of bay in it, and as it comes right on the shore and grounds, her stern is not 20 yards from shore and she is only kept from being driven ashore by the horns or promontories of this bay. Having grounded, then, the piece slews round a bit so that we come parallel to shore and all the rest of the bay being full of ice, it gets packed against the ice cape "C" and a comparative state of equilibrium ensues (see Sketch 9.7). The water here is luckily deep, a sounding close inshore giving five fathoms. It is a cold thick foggy morning, snowing hard, but we take a photo of the ship nearly driven ashore in a snowstorm, from a convenient point on our assailant, more perhaps for the sake of reminiscence, than as a work of art. This would perhaps be a remark equally applicable to all our photos. In a short time, we get the anchor up, and very glad we all are to see him, as though he is nothing out of the way in anchors, our only remaining reserve now is a very "wee kedgie." For the greater part of the day this charming weather continues. We can just see Cape Torrell and no more. As the tide turns all the ice sets along towards the glacier point, we unhook from the floe piece and steam out to go anywhere we can. It now gets a bit clearer; all beyond the islands the ice is very close and drifting hard, so we keep inside of them and make along the coast. We do not get as far as the other day, as it seems that what I then called fast floe is not so but merely a drifting piece miles and miles across, and it has moved a bit towards us along the coast. There is a narrow strip of water, rather full of ice between it and the land extending a long way along the coast. This floe is now, as it were, hung up between the land and Carl Island [Carlsenøya] and were a strong gale to smash it up. A great deal of the ice behind it, now hopelessly pressing against it, would be released and follow it out. Beyond the ice and towards Thumb Point, there are many strips of water extending eastward, the direction we should most like to go in, although we should be glad enough to penetrate in any direction *forward*, but the ice here is much in motion and the lanes of water soon get filled so that at present, 'til a better lead or one extending further shows itself, it is not worth risking the "nips" between the heavy floe piece and coming together, for the sake of penetrating four or five miles further into this mass of ice. We anchor to the floe, not quite so far E as before, and much nearer to Carl Island, and hope tomorrow to be able to climb the hill above Cape Atker [Chermside here is almost certainly referring to Kapp Payer, the easternmost point of Vestspitsbergen]. It is clear enough tonight and we make out Cape Mohn well enough, and the land all the way to it. From the masthead, those who are adventurous enough to go there, are rewarded with a view of dark distant land, either islands or the "Gillis Land" we are so anxious to explore. The wind has gone down and it is lovely sunlight, the ice is however too thick for a boat to reach the shore and the floe, as I have said, is not connected with it. Rough on the grouse, at home, today, I expect.

August 13, Wednesday.

At 5.30, I get up to assist the owner, who is already up, in getting observations to fix the position of the coast &c. It is a lovely morning and there is a good sun so that by breakfast time, our first set, for longitude, variation of compass &c, are completed. After that important meal, surveying commences. We get a good many angles with the theodolite, then a base is

Sketch 9.7 Grounded, 12 August

measured on the floe, but we unluckily find at the end of our operations that the brute has moved himself bodily some distance. Then there are more observations &c. The ice slacking off with the afternoon tide, in the direction of the island, we make a voyage there in the casey. It is deep water close up to the island, which is rocky and has no beach. The rocks are of the same clay formation as before, not unlike those of Staff and the "Giants Causeway" but not of such regular form. It is 1,500 yds long and the interior is a tolerably level plain of small round light-coloured stones; this plain shut in on almost every side by dark rocks. Among the stones were found a few fossils and what we believe to be a "madrapore" or fossil coral. The southern side of the island is the highest, precipitous cliffs rising here, sheer up from the water to about 200′. The bottom beneath can be clearly seen for a short distance, but the water is deep all along. It was very warm and sunny on these cliffs and we had a good lookout tho' there was a blue haze in the distance. Between us and Thumb Point at about a mile or a mile and a half distance lay King Fredericks Island [Chermside is here likely referring to Langeøya], rather longer than that we were on but only rising to a high point at the northwestern end. Thro' this sound the tide rushed like a millrace. The sound was almost completely full of ice, fighting to force its way out, and from our commanding lookout point we enjoyed a fine spectacle of the conflict: large flat masses 400 or 500 yards in length and breadth bearing down on where there was already a jam. Then, as they met it, how the edges scrunched up, as perhaps they lifted themselves bodily on to it. Here was an exemplification of "shearing stress" carried out with irresistible power. Then the jam would be clear and piece after piece would turn and slew and then getting free move smoothly on its way, and so on. With frequent stoppages the procession continued. The noise they made was not excessive, but there was a sort of harsh crepitation, slow and deliberate, as the edges crunched up under the resistless momentum of the large masses that sent a shudder thro' one, at the thought of a ship caught, "in the nips," and that formed an expression of power, in full accordance with the scene before us. The lanes we had seen running eastward last night and this morning had all filled in, and tho' a few patches of water showed towards Thumb Point, the ice in that direction was thick, close, and heavy. By the bye, we recognise an old friend (or many) coming along towards the race in a very high and mighty way, nothing attempted to stop *him*. He was a "Bergie Bit," an Ice Berg. Yesterday, after we had hooked onto the floe, we saw him coming up, much in the line we had come but nearer to the island; by the bye, he came abreast of us, and then up to the floe. Into this, he walked without stoppage in a marvellous way, like a mighty steam plough walking into a grassland, and he didn't bring up for a long time, not I expect until deserted by the tide. Today he returned; this time, however, his majestic promenade lay thro' the sound. He subsequently came back and completed the grand tour of the island, and now lies, aground I fancy, near its eastern extremity, at no great distance from the ship. There are a great many bergs in these waters, mostly in the offing, large enough to be mistaken for islands &c, and such as the crew say, are the famous ones that work down from the coast of Greenland to Labrador &c. The cliffs on which we sat were the first "Kitty Head" that we have come across. We have seen "Loom Heads" and "Rotchie Heads" and "Mallie Heads" and this one is inhabited by a large colony of Kittiwake Gulls. These small birds look beautiful as they fly screaming about the cliffs and settle on the ledges a few feet below that one can almost touch some of them. All the young seem hatched and are very nice-looking spotted callow little birds, and very clever not to tumble off a ledge six inches to a foot wide. Great fun it is, too, to watch the parents feeding them, especially if they arrive safe home after successfully eluding the chase of one of the numerous predatory Skuas ever on the watch to rob them of their prey. How they scream as the chase becomes hottest, and when after a long journey so nearly safely accomplished, they have in very sight of their nest, with one long last despairing shriek to yield the hardly earned morsel to their relentless foe. The Skua that we see here is the third species that inhabits

Spitzbergen. It is larger than the other two, and tho' common here, elsewhere comparatively a rare bird. We shot one today for a specimen. The ubiquitous Dovekie, of course, breeds in these cliffs, and beyond the "Kitty" part of the cliff and immediately adjoining it, there is a "Burgie head." Nor do these birds seem in any way to molest the Kitties, tho' I expect the Dovekies have rather a rough time of it. The "Burgo-master" proves certainly not half so much a king, as we understood he was. He is rather on the "live and let live principle" as far as the other gulls are concerned, and he is dreadfully afraid of himself, so to speak. That is, if there are nine or ten hungry Burgies round a "kreng," all the rest look on until the "cock of the walk" has had his feed and then another goes at it and so on, perhaps two at once if it is a big "kreng." But then they keep far apart, and are like two street boys in a row very much afraid of each other. Snowbirds from their numbers or their importunity can't or at any rate don't keep off, if the "kreng" be on the land or ice, and if in the water. The same is the case with the mallies. Still, they never dare attack him, nor have we ever seen the Skuas, which bother everything else out of their lives, ourselves included if we are near their nests. The inevitable reindeer horns are to be found on the island. They come over the floe I suppose, though with what object except to improve their topographical knowledge, it is impossible to imagine. By the time a survey of the island is complete, the ice between us and the ship is beginning to pack, so altho' there are several very pretty views of cliffs, bits of the plain with dark rocks rising along Thumb Point, Cape Torrell &c as backgrounds with these dark columnar rocks as foregrounds, there is no time to expose the dry plates we have with us and we make our return voyage. All day today it has been out of the question getting to the mainland, nor would the ice near the islands allow us to prosecute our intended soundings. We have driven slightly towards the island and a good deal to the W, tho' there is not much wind. The barometer has, however, been unusually low. I am very anxious to get to the island again tonight, but the ice does not slack sufficiently, in fact hardly at all, with the next tide.

August 14, Thursday.

The barometer up with a rush, and I wish it wouldn't, as it has brought no wind. A gale from the S.E. might be of some use but only a raw, chill morning, and absolutely nix to be seen over 200 yds from the ship, even over the ice that always makes things bright. I omitted by the way to say that the ice was up to vagaries last night and we got up steam, and got into a sort of dock, as piece after piece comes cruising along the floe edge, which we find ourselves in snug enough shelter; only the dock being in thin ice it would be no real protection against heavy floes. We are busy below all day, working out observations &c. We see the island once or twice, as thro' a fog dimly, and perceive that we are a good deal nearer to it, *voilà* and out. We do a great deal of dredging today with the heavy dredge, and with great success: "No anemones to be found in these seas, aint there?" We were delighted with our Wüde Bay one, then we got another in Loome Bay, who yet lives and for a long time has been our pride, a very fine large red lad. But today we have got them simply by scores of different sorts, too. Our old red friend is rather a sickly beast, he lives in a pickle bottle, and is continually getting out of sorts, and then he distends himself to such an awful size that we have been afraid of his going off by spontaneous combustion and disappearing altogether. Then when he has been coaxed back into health by fresh seawater and weed &c, he looks so blooming that our hopes of taking him back alive suddenly revive. Then he has another relapse. The misguided beast is sulking I am sure and refuses the tempting shrimps we offer him, and it was decided today that the next time he resumed his natural appearance, he should be treated to enough grog, consisting of alcohol and glycerine, to make him very dead drunk indeed. But the arrival of these have given him a reprieve. Corals and Corallines are a drag on the market, the forecastle is full of them, tho' of what value except to museums they can possibly be I can't imagine. Barnacles, altho' in England I heard we should not find them here, we have found all

Fig. 9.30 Washing day on board *Diana*. This image was taken near Low Island (Lågøya)

Fig. 9.31 Crew of the *Diana* towing the skins of a recently killed bear (left) and a walrus

along. There is a cluster of them in a tray, creaking in their attempts to make a little promenade, in the most unpleasant way, at the present moment. Then there are sponges, a good-sized one and very like the domestic sponge, and shrimps and prawns, in such numbers, that we ate them for tea tonight. Then there are shells and worms and other animals whose general characteristics are sliminess and creepy-crawliness, and I had nearly forgotten, a lot of very thin-legged crabs. The barometer is still going up tonight, and there is a light steady breeze from the S.E. Of the ice and its movements today, beyond those in our own immediate neighbourhood, we have no idea. There has been a "worry" at the engines, which had got slack, and they are again reported O.K.

August 15, Friday.

A nasty thick morning and raining hard. Macintoshes are the order of the day, for our morning promenade on the deck. Soon after breakfast, the owner and I take across the floe, as it seems as if we could get right across the floe to Cape Linderman [Lindemanodden], as the next cape shown on the chart is named. There are two shown in the said chart as lying between us and Cape Mohn, whereas in reality there is only one, our present objective. The floe proves very rotten and soon becomes intersected by numerous pools and channels in the ice, and after a good deal of wading, and a partial ducking for me, in pursuit of a wounded bird, we give it up and return. Some more of these bangtailed Skuas put in an appearance today. These are the Pomatorhine Skua, those that we took for this species before, being the long-tailed Skua. They are shy but a novel method of pursuing the *chasse* proves very successful. We have observed that Skuas always fly round and torment a dog as we have seen them do to foxes, with what object, unless from pure mischief, I can't make out. So, we take the bull pup onto the ice, and they come screaming and circling round him, and the pair thus fall an easy prey to the gun. In the afternoon, it seemed a bit clearer and the ice slackened at 1 p.m. at the turn of the tide. Towards the mainland, it is still impracticable, but we have drifted to within a mile and a half of the island in our moving dock. A heavy southeasterly wind has been blowing all night and is still doing so, and we take a long time getting over the distance as there is a heavy cross tide. Neither is the passage clear and a good deal of pushing across the ice has to be done, but the casey is very handy for such work. We land and scramble up the slippery ice slopes between the rocks and then separate to gauge our respective gaits, meeting at the summit in due time. By this time, luckily, my survey is complete, with the exception of one or two distant bearings, as the thick and blinding snowstorm comes down. It is so cutting that you can hardly look over the cliff edge, the face of which is fully exposed to the storm. The kitties and burgies seem all at home, sitting demurely huddled up on the ledges in great numbers. It is comfortable enough under the lee of the topmost rocks, but we soon see all chances of a lookout is at an end. The central plain of which I have spoken is in its upper portion 100 feet above the sea level and seems to have been once a sea beach, at no such very distant period either, as firmly embedded in the soil, we find the jaw and crown bones of a whale. We get safely on board, rather late for dinner. The evening was wet and snowy and brought no change for the better, and after dredging some sponges &c, I tried first to catch some of the small fish the loons catch so many of, and failing that tried for cod, should such be denizens of these waters. No go. So, I bethought me that if I could not catch the fish myself that we wanted for specimens, I must do as the predatory gulls and rob a loon. Here comes one in a desperate hurry alright, fish in his beak. Down he comes, poor bird, fish and all, but drops the fish and dives. A watchful malley secures the prey and ere I can take steps to prevent it, bolts it (the fish, not the loon). I have rowed to leave the malleys alone, so I give my gun to the engineer to shoot him, but the shot is too light, as he is not done for, tho' as he takes the ice and lies motionless, we think so, and go after him. He, however, takes himself off and so away goes our fish and we have nothing to do but pick up the loon, by this time defunct, and return.

It is perhaps a good opportunity for giving a rough sketch of our hours of feeding and general

Fig. 9.32 *Diana*, draped with dozens of slaughtered birds and at least one bear

habits. Between 7 a.m. and 8 a.m., a "worry" for potatoes or other provisions takes place in the "dining room," a sort of small after hold, the grating closing the aperture to which is just outside my door. This generally wakes me if not already awake. The steward then looks in at very short intervals, with many varied and hackneyed remarks about the lateness of the hour, and how we are putting back the men's dinners, &c, 'til soon after or as a rule, at any rate, always in time to wind the chronometer (8.15 Greenwich time, tho' now that we are in a Long[itude] some way E of Greenwich, this is generally done after breakfast) in sheer desperation I turn out. Then the owner is told that I am up, or if he is astir first, I am informed of it. The others astern are deaf to the steward's blandishments and I often have to resort to forcible measures. Breakfast between 9 and 10, then we pass the morning in our respective pursuits. There are of course readings of bar[ometer], therm[ometer], and all other observations, often altitudes of the sun taken before breakfast. After breakfast a promenade on deck and a pipe for the smokers. Then if we are not underway, we work out observations, or varnish photos, or dredge, or shoot in fact whatever turns up. Lunch at 12, then after that we are almost sure to take across the floe, or land, or sail or something, but if on board coffee at 3 p. m., then dinner at 5 p.m., tea at 8 p.m. and after that we tail off to roost as we feel inclined, tho' some of us are generally up 'til 12 a.m. or 1 a.m., but these habits are of course necessarily much broken thro', [as] a boat may start away for the shore at any time and then, if it is a boat expedition, bread and cheese (biscuit is called bread), water and a bottle of grog, and perhaps a joint of deer are bundled into her and off we go. It is wonderful, after the way one is accustomed at home to arrange and plan for an expedition, how simply and quickly we are ready here, rifles and ammunition up in a minute, stuff compasses, aneroids, notebooks &c into our pockets and in six or seven minutes all is ready. It is fair to say, I

Fig. 9.33 This image, with the hunting party on drift ice in the foreground, shows the Seven Islands (Sjuøyane) from south, with Parryøya to the left (mountains Øykollen and—with the big snow field—Gullberget, and then left, from the middle, western Phippsøya with Phippsøyhorga in the distance, right of the middle eastern Phippsøya with rampart-like Høgberget and Trollenykjen in front of it more to the right, and finally to the right of the image, western Martensøya). The photo appears to have been taken during the second visit to Sjuøyane, when the expedition explored also the area between Chermsideøya, Kapp Platen, and Sjuøyane. It also seems to reflect Chermside's difficulty in getting the crew to stay still in one place while he composed his images

am always last. If we are merely going to be put ashore, it is simpler still: two hard biscuits and a lump of cheese, and perhaps if we expect to be away a day, a haversack and a pocket flask, and all is ready. Then one can make two meals ashore and that carries one from 12 to 16 h. High tea is ready laid out for our return at any moment, and hot cocoa and tea are in the cabin as soon as ourselves, the said high tea being at any hour until it often practically becomes breakfast. Then, when we are on short expeditions, we are home usually for dinner and it is one of the greatest luxuries I have ever experienced. The way, on board this ship, things are always ready [as soon as] the boat has been seen coming, or ourselves if on land or the ice approaching the beach or ship, no waiting wet and weary as at an inn or to perform a toilette as at home, or order fresh things as after hunting, before you have time to change, and we are often so hungry that we don't. The soup is on the table, and a good dinner rewards a long tramp or a heavy pull. On the hillsides, there is always good water and we have never failing appetites. Occasionally, however, when we have boats crew away, we have a campfire and bivouac. Altho' of course we feel some disappointment, especially on the owners account, at the slight success that the unfavourable nature of the ice permits us to attain, in the way of exploration, the charm of this roving life is wonderful. No long weary journeys to be made as in land sport, locomotion provided for one, all the charm and more than the charm of ordinary yachting, combined with the same sort of pleasure as men get in hunting trips to the backwoods, but all the difficulties of travelling, conveying food, ammunition and change of clothing, as well as long journeys between favourable hunting grounds, and cheerless nights on the cold ground, removed by the luxuries of our floating home.

August 16, Saturday.

Gradually clearing up, and a strong breeze from the E.N.E. No change, however, in the ice: we are nearer to the land (the main) and further E, tide after tide, ebbs and flows, and a fair wind for smashing the ice blows, but this luckless floe holds on. Early this morning we got a large ground seal, in the water for a wonder, but then he was only wounded, and wasn't far from upsetting the dinghy when the harpoon was struck in. It gets bright and sunny after breakfast and the clouds lift off Cape Mohn, and we again make for the island as soon as the tide turns. After a little hauling over the ice, the casey is launched and we get up sail. There is a strong wind blowing, a "*sojer's wind*," fair both ways, and the casey behaves beautifully. Having no ballast we sit in the bottom and simply tear along. There is a good deal of ice in the road; now there is a stream of it ahead: down sail, to break the shock, in among it she charges; up sail, again to drive her through, in which manoeuvre the rudder goes adrift. It is soon recovered and we make a splendid passage to the island, and hauling her up on the ice, again climb to the lookout point. As a spot for the contemplation of nature, it is charming on a sunny day, and if it be clear equally so for a lookout, but it is emphatically *not* the place for a conversation. The birds make such a noise you can hardly hear yourself speak, and in the number of simultaneous orators, the vehemence of their language, and the sameness of their remarks, quite eclipse the august assembly at Versailles. Nor does the din diminish when two fine nearly full-grown young Burgo-masters are removed from the parental abode to form an addition to the *Diana's* menagerie. The Kitties were too young to take, as our former experience has shown us. The lookout is not very cheerful, the ice eastward is not practicable and that towards Thumb Point, tho' broken, very close packed. It is decided, however, to make a move at the turn of this evenings tide. After scrambling all over the place with the "cambra," as the Captain calls it, to expose dry plates, we again embarked and went dancing along gloriously and after a splendid sail got on board about 2.30 p.m. All day there have been crowds of snowbirds about the seals' "kreng" and the usual bombardment has been going on, with comparatively small results compared with the number of cartridges, with which the sportsmen were provided. This evening it has been coming on thick and cloudy and when, between 6 and 7 p.m., we steamed out of our icy dock, it was raining hard. Very rough on the ice this rain. I only hope it will melt it and smash it up as it ought. By the bye, it came on in a way that I have never seen it do in Spitzbergen. We were going along like the sketch of the Rhine Steamer in Jones, Brown and Robinson [navigating without regard for a storm, from a scene in Doyle's *The Foreign Tour of Messrs. Brown, Jones, and Robinson*], with but very dim glimpses of the islands we passed, all soaking at any rate in appearance, but protected by Macintoshes or oilskins. After a bit, it cleared in some directions and there was a beautiful sky. These great dark indigo storms, for there were more than one, some to windward, too, rolling along, and hiding all they touched. Other hills [such] as Thumb Point are half enveloped in clouds and here and there bright patches of sunlight on the distant glaciers and snowclad hills. In the sky were also red-coloured clouds like thunderclouds that cast a rather lurid but rose-coloured glare all over the floe and ice amongst which we are steaming. On one or two glaciers, too, there shows this soft warm tint, and away N.E. there is a sharp cold steel blue sky, with beautiful steel grey clouds clearly defined, that throw the same blue grey cold tint on the glacier below them (the long one running into Wahlenberg Bay). It is one of the finest skies we have ever seen. Then it gradually overcasts or rather another rainsquall blots it out from us, and the pitiless storm is on us again. There seems somehow something fantastic and unnatural in this downpour, so thick and steady, that we can no longer see Thumb Point among these Arctic solitudes. All the way across we have been navigating among thick packed ice, and as far as one can see the Straits are so full of it, that it would be no easy matter to get out, even were we so inclined. When about three quarters of the way across to Thumb Point, we make out open water to the eastward very clearly, and the little *Diana* bravely strains every nerve to gain it, one of the first obstacles is a neck of solid unbroken but presumably rotten ice. At this she charges again and again and in half an hour or so succeeds in getting thro'. Then we go on among very close drift ice for some time, until at length only another and a narrower neck separates us from the open water. A narrow channel filled with broken and cracked masses of ice soon shows thro' it. Behind us and we are in the open water. We can now see for seven miles ahead, and there is not any ice in sight. The swell, too, that is on here is very encouraging as it seems to imply a considerable extent of open water eastward, and as she steams gaily on against a head wind, dancing over the waves, our spirits rise immensely. It is wonderful here when such difficulties meet us at every turn how great an effect the progress of a mile or two in the right direction produces on all of whatever temperament. The motion and pace, too, prove so exhilarating after the long monotonous pushing among ice, how one fidgets up and down when taking along the pack edge, and look at the compass every five minutes to see how the ice allows us to steer. The storm squalls are succeeded by dense fog, and as it is cold tonight, it is soon frozen, and masts, spars, rigging, soon even ourselves covered with a delicate white woolly covering. Continually the case up here, if on these occasions the yards require shifting, or if in fact anything is moved much above the deck level, a perfect hailstorm of icy fragments descends on the deck. This is preferable to the subsequent showers that the action of the sun produces when there is a constant fall of black tarry drops that seem by instinct to time their descent so as to alight well inside one's coat collar and trickle down one's neck so that the beauty of a ship so covered is soon lost sight of in the foretaste of subsequent discomfort. After a couple of hours steaming, we come apparently across the floe, and tho' progress is still practicable, it is considered advisable to hook on and wait in hopes of its clearing up. We hook on and sound 17 fathoms!!! in the middle of this ocean!!, a ludicrous scene indeed. All the water around the roller for deep sea, the owner and the Captain and myself deeply interested and setting the thermometers, '*Captain do you think 100 fathoms will be enough?*' &c &c and as the line runs out and finds bottom at 17 fathoms we burst into a simultaneous shout of laughter. After some hours it appeared that our floe was loose, and carrying us away to leeward, so in the early morning we again steamed on. There was not much sleep done tonight as may be imagined, so we were lying down in our clothes and pervading the deck at uncertain intervals. Then as the latest from above tumbled down the hatchway and removed his Macintosh, for at intervals it rained hard, eager voices, from the depths of blankets demanded the news, prospects of clearing &c. We were still steaming E by S and it was not for nearly two hours that we were brought up by what certainly seems remarkably like a land floe of fast ice.

August 17, Sunday.

A thick foggy morning and we see nix beyond 200 or 300 yds. Leeward there is a stout, fast floe of land ice, with pools and channels but not great apparent desire to become disintegrated and go on a voyage of discovery in detail. We have seen a great many seals about here in the water mostly harp seals or "saddlebacks" in droves and schools and these seem to be exclusively composed of females. We shot and secured one this morning. As it does not clear in the afternoon, I go on a cruising expedition in the dinghy northwards as that is towards the land near Cape Mohn. After a mile or so, I am picked up by the ship and we steam on down some few miles. We have hitherto been in soundings of 22 fathoms, the bottom very flat and muddy. We now get casts of 14, 17, and 15 fathoms, but a little further on to the N.W., deepens the water to 25. The floe retains an unchanged appearance and heavy fog and a soft Scotch mist continues. In the evening, the mate, engineer and myself start in the casey to row back to our last position, to see if the seals "kreng" has brought up a bear. Unfortunately, the pull is far longer than we expected and the strong tide that brought down the ship is now against us. The fog is intense and the birds &c loom enormous thro' it. Cheerless work enough as point after point that only appear 150 yds or so off is overtaken and past. We shoot three or four of these long-tailed Skuas as there is still a demand for them, and after a spell of three to four hours enlivened much by James' anecdotes we arrive at the spot where the ship was hooked on this morning. Wonderful are the old mate's experiences, and extensive his information and by pertinent questions and abstaining from chaff, one can hear a great deal of interest. How many years he has been at sea we can't get out of him, but he was fishing in Greenland in 1827 and was on board one of the whaling vessels that Parry and Ross's memorable voyage in that year, met off Prince Charlie's Foreland. He has been in his time a very celebrated harpooneer and many is the fish he has struck and killed. There is one especially memorable occasion of which he is spoken of as the hero. A fish was struck and killed far from the ship and a chill dark October night was bringing down at once, fog and frost and gale. All the other boats abandoned her, but James hung on and under the lee of the fish successfully rode out the gale. Then, too, he can tell you about the sealing, when flintlocks were the first weapon of the day, and how when the guns became so foul that the balls could not be rammed home, wads of blubber rammed down as a lubricant overcame the difficulty. However, here we are just as we had given it up at our late position, but alas our "kreng" is gone, a piece of ice has broken him off and carried him away before the wind, possibly a bear with him, as had such come up wind to him, no tracks on the floe as is the case would be visible. A few bad potatoes &c alone show that this is the wished for spot. We haul out our boat to empty her and, after great difficulty as the lights are damp, light our pipes. The strong breeze that has been blowing right ahead as we came up, and which we so calculated on to carry us back over the heavy tide, lately turned and now against us, has fallen completely and there is nothing for it but to pull. We see a few seals, one or two being large ground seals, but they are playing like porpoises, only showing for a moment and at some distance too and we can't get a good shot. We get back between 2 and 3 a.m.

August 18, Monday.

Not a glimpse of anything but fog. I omitted, by the way, to say that about 4 p.m. yesterday, it lifted for a few minutes and we saw the south coast of the N.E. land and the glacier in parts, but dimly however. Today it is decided to try back, for these reasons. On first striking this floe that completely bars our direct progress to Kong Karls Land, one of two routes had to be chosen, right or left. We chose the left. Now we have followed this road along for some miles and we find that our course has lain as it were on the circumference of a curve getting continually more and more westward. We cannot be far from the land but if a channel is open right up to it where shall we be, but a few miles from our starting point. But the other side of the moving floes held us rigidly at Karl Alexander Island [Karl Alexanderøya]. It is true that once near the land, a passage may open between it and the fast ice, but even if it were not, it would probably lie so close under the glacier that the danger of being overwhelmed by falling icebergs would be very great, and it is, I understand, the height of the iceberg season. The supposition as to our position and the proximity of the floes behind which we were before is borne out by two facts: one, our latitude secured from a watery and insipid sun who showed thro' the fog at noon for a few minutes like the circle faintly reflected on a sheet, from the lens of a magic lantern, whose lamp, as is almost invariably the case, refuses to burn properly; second, that there is ice at no great distance on our port hand, like that on the starboard hand land floe, but unlike the latter moveable and carried about by wind and tide. The present wind, however, amply suffices to keep it at arm's length and frustrates any delicate little attentions in the way of "nips" and it may be desirous of paying her ladyship. Now trying back we shall pass the first point we struck the floe at and following it along discover whether this is merely a scooped out bay in the floe, a semicircle at or about whose most projecting point we fell in with the floe, or whether there is elsewhere a lead and the fast ice broken so that we can penetrate through, or even possibly the whole ice edge to trend S and E. *Voyons* [see], tho' we fear that the semicircle or rather segment of a circle theory is likely to be nearest to the mark. If so, the soonest we know the worst and find the affair over the better, more time will remain for other little devices. We should then have proved that fast ice extended from a point on the lowland of the S coast of Nord Ost Land to either Thumb Point or the "Bastion Islands" [Bastionøyane], in a curve say the segment of a circle, the extreme point of which curve projected some miles further eastward than either of its points *d'appui* [support] on the land, and the great Straits question would be settled, for this year, at any rate. From the appearance of the fast floe, however, this end of the straits would appear to have been open last year, as it is a new floe probably one year's ice, two at most. Nor is this the only evidence of the state of the sea: here and there from the floe spring up large icebergs from its level surface. There is one inside the floe now, at about ¼ mile from the ship, dimly showing at intervals through the fog. It is allowed to be a fine berg even by the "straits connoisseurs." He is perhaps not much over 100 feet high from the ice level, but long and of bold outline, forming a sort of miniature little mountain range. Bergs are very numerous here but, of course, we only see those near the floe edge owing to the fog. They are mostly aground as can be seen from the sides of them washed by the tide, and must have come into their present positions, after being carried about by unusually high spring tides. Some of the smaller bergs indeed do not seem aground, as these latter which a floe breaks off, they cruise with it, whereas the grounded ones remain and the floe coming against them is torn away as it were 'til the main body gets free, so that except in fast ice, there is generally water under the lee of a big berg. The parent glacier of all these bergs must be, I think, the large one I have described as extending all along to Cape Mohn, or those on the E coast which connects that Cape and Cape Leigh Smith, and the water close in shore there must if so, be deep. So, we steam back to within about two miles of our first point of meeting with the floe. Not that this is actually so easy to do, as it seems to write, when all is a blank monotonous space of dull grey water, and dreary looking low flat ice, dotted with dark pools and cracks, each projecting point and bight of the ice like the last, and the prospect always narrowed to a hundred yards or so, landmarks are not easily selected, and even after selection mistakes readily occur where all have such a strong family likeness. Near us, however, is a berg that we noted in the casey the other night and so we know well enough where we are. There are a good many seals about and we have also seen a seahorse or two. Of loons, there is an infinity and thinking that like the partridges at home their time for good condition ought to be nearly ripe, the Captain and I go after some in the dinghy. They are so tame that I leave the gunning to the Captain. You can hit them with an oar and I am sure I could capture any amount with a landing net. The Captain is still greatly in difficulties with the safety spring and the left barrel and it makes one "*rire à gorge déployée*" [laugh uproariously] to see him, as I have done, point his piece at a flock of loons, and succeed in getting it from full to half cock seven times, without any further result, the loons being this time all innocently and inquiringly swimming round the boat, and too scattered and too close for more than one to be got at a shot, and the Captains face presenting a killing picture of the height of mortification and good humoured vexation. However, after several discharges of the right barrel, and triumphant one of the left, which was so unexpected as to add nothing to the bag, we got 13 or 14 brace and returned on board, and the keen air, albeit fog, and exercise of a pull always put one in good spirits. People have a habit of performing various manoeuvres and repeating little sayings by way of consolation under depressing circumstances, which I never could understand, tho' air and exercise always have that effect on me. One of my greatest friends, whenever anything turned out an unusual "grind," as he expressed it, invariably derived intense cheerfulness from his repeated assurance: "Never mind it all goes in the days work," an enviable style of philosophy.

August 19, Tuesday.

Fog and mist as before, ditto ice and icebergs. After breakfast we follow the floe edge along for an hour or two (11 knots by the patent log) and again hook on. The floe is here firm and solid-looking and we were brought up by rather an unfortunate occurrence as we steamed past this point. There was some large dark mass 150 yards into the floe, looming thro' the fog, and after a little consideration, the ship was turned and we came back to look at it. We ran nose on into the floe, close to it, and it was not 'til a hand was descending from the chains to approach and examine it that the mass assumed the form of two living seahorses, which on waking disappeared with a plunge into the crack near which they were lying, an awful disappointment as harpooning seahorse on the ice with a hand harpoon and holding on by main force with several hands would have afforded the most splendid sport one can imagine. After a bit, the owner and myself took across the floe to a berg that lay about half a mile in. The floe, tho' very solid, was intersected by many cracks, some mere cracks that had never opened, others that had opened and come together again. The pressure to which they had been subjected clearly denoted by the line of heaped snow and powdered fragments of ice running along their junction, which they covered, and others again open with channels varying in width from one or two feet to nine or ten yards. These, however, by no means appear here to indicate the approaching dissolution of the floe, the pieces of which are finally wedged together, and these are doubtless only the results of the rise and fall of tide. After a due amount of wading and the usual wetting over one's boots, we arrived at the berg rising out of the flat surrounding plain, covered with dykes and numerous inundations like a citadel in the Low Countries. It was about 120 yds in diameter and the *enceinte* [pregnancy] was circular surrounded by a wet ditch formed by the rise and fall of the tide and due to the fact of the berg itself being aground. Nor were beaches wanting and their debris, in the shape of large masses of ice, some blue, some brown and discoloured in many places, choked the *fosse* [pit], but a rugged and scarcely practicable road lay over them, piled up all along the foot of the citadel. Mined lately above the ditch and following the ordinary principles of defence, rose a smooth parapet of ice rugged enough and full of hills as we subsequently found to our cost, but beautifully smooth looking from its treacherous covering of snow. The rationale of this, I fancy, was that the floe had risen with spring tides and in falling broken off these hummocky morsels leaving them on the berg, or else at some time driven against it by wind or a swell from beneath, the edges of the floe being piled up as they broke as in the ordinary formation of hummocks, these at first of crude blue blocks of jagged ice become covered with a winters snow and when the mass has become welded by a summer's sun assume their usual soft round appearance. Another low parapet of the same sort, outside the ditch forming that of the covered way, the latter trodden out by the garrison and from the traces well patrolled it must be, too, by large-footed bears who retired on the approach of the enemy so that the place fell without a shot being fired. Choosing as good place as we could, we crossed the ditch, the water of which was very clear and showed as the submerged mass of the ice berg stretching far away into the depths below, the parapet of ice that rose abruptly beyond was snow-covered and the snow concealing the numerous holes among the sharpened blocks beneath, and few harlequins could have executed the great disappearing trick, with more *éclat* [style] than we did, many times too, during the five minutes it took us to cross it and the *terre-plein* [median] beyond. Over the parapet, we had to descend again so that all view for a time was confined to the scenery within the berg. Scenery is really not too fine a word. There were valleys and hills and ridges and rocks and watercourses running with the water and miniature cascades, and there were miniature lakes too of a clear blue and some of a considerable depth one of 20 or 50 yds length and ten or 12 breadth winding along and by its appearance surrounded by snowy slopes resembling a lake not a mere pool. The water in these iceberg pools, at least in all I have seen, is of an intense blue, like that of the Lake of Geneva. This iceberg had once been, we felt sure, the other way up, but had thought it fair to give the lower part a chance and we only hoped it wouldn't suddenly change its mind again. Otherwise, whence it's red clay in the valleys, whence the pebbles and shale and the boulders, the larger rocky fragments, it may have got these as its portion from its parent glacier, but we thought not. Had it been fine we hoped to get views with dry plates, to illustrate what we saw, but the fog put this out of the question. The peaks were very sharp and icy, but we found an easy enough ascent. The top was from 80′ to 100′ above the ice level. Unluckily the view was very limited. Oh!, for a fine day for a look out from a point like this, we should see a floe, I fear, stretching right away E. But a sunlight lighting up the countless high bergs that rise from its surface would be a glorious spectacle, some thousands, I fancy, one would see at once. Gillis Land we ought to see plainly enough from here and would of course have a splendid view of the gigantic glacier near Cape Mohn and should probably run in close to it. As it is, we can make out the ship, now and again only, tho' distant a bare half mile, while on the summit there appear below us on one of the channels leading from the main *fosse* an old and young seahorse who come up to blow. We do not shoot as we are in hopes of their taking the ice. They however go down, and as after a reasonable time there seems no chance of their reappearance, we descend from the top of the berg, and completing the outside circuit return to the ship. Dredging has been going on, but the level bottom of soft mud, does not seem very productive. When the ship is driving at all, the dredges are worked from the stern, but when as of now we are hooked onto a fast floe, a boat is used, and by an ingenious method first adopted by old "James," the mate, and now always put in practice when there is any drift ice about to render it possible, a great deal of labour is saved, and a small boat like one of the dinghies can be used even with the heavier dredges. The old fox was once away to dredge, as all the crew take to it and seem much interested in it, he was by himself with one of the oyster dredges and we thought he would find he had got more work than he had bargained for, and so he seemed to think too, for he quietly ran his boat's nose against a heavy piece of drift ice that came by, his way, and making fast unshipped his paddles, and allowed the ice to dredge for him, and when it had been down long enough, quietly hauled up his dredge, and casting off pulled back to the ship. Later on in the day we still steam along the floe, leading us now S.W. by W.; it now comes on exceedingly dark and thick, and we come on a berg quite unexpectedly projecting from the floe edge and not visible 'til we are almost foul of it. In a few more minutes, we come across another aground in open water, and we have to put the helm hard over and reverse the engines to prevent ourselves running full 'til against it, poor thing it might be injured. This one is not so large as his neighbour, the one we have just past, this latter rising with vertical and overhanging sides so that it seemed like being under a great cliff. He seemed to be about ten feet or so, higher than the masthead, and so towered tremendously above the deck. It seems now time to hook on to the floe again and as she comes round still steaming ahead, something dark appears in front, and in a second over goes the rudder and as she comes round sharp unavoidably running her nose hard against the floe, and bringing herself up with a round turn, her jibboom just misses another berg, that one never saw 'til it was close off the bowsprit. We gave it a little wider berth and hook on to the floe, and presently the owner, Skipper, and I go for a walk across it, eastwards. This is a thick heavy floe, rough and hummocky, too, and with the cracks that have existed when it was in the state of the one we saw yesterday frozen over and healed. It is ice of two, at least, if not three years standing and is the third sort of fast floe we have come across while following this ice edge, the other two tho' probably last year's ice being of different ages. My surmise of yesterday therefore that the Straits were open last year is by no means proven. However, this floe, we are now on, may have been a drifting one that got frozen in. It is where it touches the open water of no great width (a mile or two) we find and surrounded on both sides by bay ice of last year's formation, we think. This may, however, be equally the case if it was a fast not a drifting floe, as this would then be a projecting point, that escaped breaking off, and which became subsequently enveloped by the new ice. As we penetrate in, the old ice does seem to get broader, but we do not however see enough to demonstrate anything clearly, and it is not 'til an hour or two later when there is a slight lift that we make out the probability from the look of the fast ice, of a band extending across the straits between this and Kong Karls Land, having remained unbroken all last season. Meanwhile, we continued our promenade and penetrate about a mile into the floe, beyond which point, tho' solid enough, the pools were so numerous as to offer little inducement for further exploration. I forget what is the most generally accepted theory as to the blue colour of the water and the sky. But their pools are blue, intense blue, and when they dry up leave a sort of blue residue like starch. Some pools are brown, some greenish brown and stagnant looking, and you often meet deep green pools. The latter are usually full of salt or brackish water, which has welled up or been forced thro' holes in the floe, tho' we have found fresh green pools. The brown ones are mostly turned in, and by the melting of discoloured ice or as is often the case become dirty from the number of birds that frequent them. Green pools have been found both fresh and salt with their colour due to a plant (Conferrice) [?] growing in them, and brown ones with diatoms as the colouring agent. Here, however, the blue pools are fresh or brackish, and those near the floe edge have lost their blue colour and are of a dull grey just like the seawater here and many naturally enough as they are full of it. But it is hard to account for some that are of this grey colour being fresh, or nearly so, unless the salt is deposited or water sucked up from below thro' the pores, and its salt filtered out in the process. Old ice undoubtedly becomes brackish, especially when rotten, as it is then very porous. All sea ice is indeed far more porous than fresh water ice, and the large blue and black lumps, fragments of bergs and glaciers, that one sees floating about, even at the western ice are for this reason prized by the whalers for watering purposes. Making a detour to avoid retracing our steps we soon get back to the ship. This fog is very annoying for sporting as well as other reasons, as tho' we come across endless bear tracks, we are not likely to see one, unless we tumble right on it. There are seahorses about, too, and we see drove after drove of "saddles," these pass sometimes near the ship, but seldom give a shot as they are on the passage, all going N.E., school after school go in that direction but none return. The droves number from 10 to 20 or 30 as a rule and seem all females or the young of the year, none of the males with their black faces and grey hoods. They go at a great pace, coming up like porpoises, and at about the same intervals, and frequently leaping right out of the water like a salmon. We shot two tonight and got one of them. About 10 p.m., it began to clear to the west and south. We could see in half an hour a high but cloud-covered hill which we think is Thumb Point, some of the islands up the Straits, and the Bastion Islands. This is the only direction in which we see far countless bergs are in sight, some down towards the islands in large clusters. They seem, however, none, that is of the larger ones, to project westward much beyond the line in which we lie, as if a bank existed on which as they drifted from the E, they grounded, and beyond which none of the larger could pass. I try a photo of one, but it is already darkening and by midnight, it is as thick as ever. I waited up 'til 3.30 a.m. in hopes of another lift, as I yearn to climb a neighbouring berg for a look out, but no lift came.

August 20, Wednesday.

Still moving in darkness, we determine to seek some open water that last night showed in a bight near the islands. The ice here gave promise of seahorse and beyond the islands, there *might* be a passage. We soon leave the floe edge to take across the mouth of a deep bight and once left, although we took the bearings last night, he is not so easy to find again, as the numerous bergs, prevent our keeping a very straight course.

Today, one can see about half a mile ahead. By degrees, they become more and more numerous and we find ourselves at last in water where they swarm and there are simply hundreds of them, few of large size. We hook on to one but soon find he is not aground. It is all very pretty but rather dangerous being amongst this giant ice, in a fog, in 35 fathoms to 40 of water. Many of the larger bergs are firmly aground, but none that we can conveniently hook on to, the lesser ones all big enough to crush a ship, careening gaily about in every direction. It is blowing a gale from the S. E., but these fellows especially the low ones think nothing of coming to windward with a good tide under them, so we unhook and take back to search for the floe again. These must be evidently an "*eisted föd*" [Chermside here is apparently using a Norwegian sailor's colloquialism for 'birthplace of ice'], or some wonderful convention of bergs, as on our way back we meet a great many of the smaller ones. Those who can float in 30 to 40 fathoms of water hurry up to join the assembly. If they are all beating about when the frost sets in, they will get beset like a ship and become prisoners in a floe, their inevitable fate somewhere or another. Some of the large grounded bergs look as if they had been where they are for years, and very likely grounded when the channel was far deeper than it is now the case. We had to steam some hours before we got back, making at times very little way against the gale, and regained the floe about a mile to the N.W. of our old position being now rather beyond the berg we so nearly ran down!!!

August 21, Thursday.

The fog continues, and a very strong but shifty wind blowing during part of the day from the W so as to find us on a lee floe. The result of these gales has been to break away ice from the floe edge so that there is a good deal of drift ice about, none however very heavy except the bergs, of which a good many of the smaller ones who can float in these soundings are cruising about like privateers. The seals seem to have been looking about for a way out and to have failed as the schools are returning, still playing in their old restless unsatisfactory manner. It blows too hard to go hunting in the boats and the rain is very heavy coming down in perfect squalls at times. The anemones profit by the forced inactivity and are sorted into separate bottles &c. Towards night, the sky became thinner by slow degrees.

August 22, Friday.

Soon after 12 midnight, it began to clear and we were on the lookout for a complete lift. By 3.30 a.m., it was clear and we got ready to start which we did soon after 4 a.m., the lookout being by this time completed and the necessary bearings &c taken. We were some miles to the E of Thumb Point and beyond the group of Bastion Islands. One of these projects far more than is shown on the chart and we must have been very close to it in the fog two days ago. There is drift ice inside of these islands, but the passage appears rocky and dangerous and the abutment of the arch formed in plan by the ice as we had surmised lies on the land between these islands and Cape Payer. There is evidently no way round this way, as all the horizon S and E is bounded by heavy floe of a hummocky nature like sea ice and this water indeed may be considered open sea, as it lies beyond Barents Land [Barentsøya] and south of Kong Karls Land. We can see right into Einhorn or Unicorn Bay [Einhyrningbukta] and along Barents Land as far as Cape Barth [Kapp Barth], Helis Sound [Heleysundet], leading from this bay into the strait or fiord, appears blocked by grounded ice. One berg lying here is seemingly over a mile in length but, at this distance, it is impossible to be certain of its being a single berg. The land is here very vaguely laid down, tho' it has, I believe, been circumnavigated by Norse smacks. As the gales may have much altered the condition of the ice towards Cape Mohn, as well as to get a view of that land hitherto unvisited and only surveyed from a distance, we again take along the ice, steering a course, however, as we know its lie so well. Between 9 and 10, we hook on to the floe, just as our second edition of breakfast is over. As we are now about two miles from the shore in Vibe Bay [Vibebukta], fast ice lying inshore of us, we are by no means so far eastward as we were an hour or two ago, as the ice lying on the curve we supposed has brought us back westward continually. The wind is very variable and squally and at one time this morning blew with almost the force of a hurricane, making the vessel roll and lie over a great deal, altho' in the lee of the ice and not carrying a stitch of canvass. During one squall from dead ahead we hardly moved at all during half an hour steaming, tide too in our favour. I have called this Vibe Bay as that is the most easterly of two shallow bays in the chart, called Ulve [Ulvebukta] and Vibe Bay. In reality, but one exists, so I fear that one of these eminent explorers must transfer his patronymic to some of the other unnamed features of the country, sufficiently numerous indeed to make it no very difficult matter. The fast ice at the northern end of the arch, as I have called it, springs from a point three or three-and-a-quarter miles N.W. from us and above this five or six miles in land rises a dark rocky crag, which we determine to ascend. The owner and I and two hands start in the casey for this purpose, and after the lull at starting soon got a squally wind which just allowed us to sail in the desired direction and slumming along among drift ice in such a crank boat with no ballast made our journey exciting enough. She was soon hauled up on grounded ice and we started our way that lay across low land intersected by numerous streams and covered with fresh water pools from one to three or four hundred yards in length. This low land was composed of yellow laminated shale that is very flat and sharp, a great many of the pieces being stuck edgeways into the ground like panes of glass, either vertical or at various angles; strips like this often occurring for half a mile or more, and being by no means pleasant to walk over. This land was absolutely barren, with the exception of one of the yellowish white [ILLEGIBLE] of the country, among the stones here and there at rare intervals. There were numerous fossils, all of the ammonite, snail, or shell species, some of the latter as the scallops much resembling those we dredge up in these waters ourselves. The specimens were not over good but on our return journey we picked up a few. This barren and monotonous beach rises slowly towards the dark hill and at its foot has attained the elevation of 300 feet above the sea. In a distance of from five to six miles, near its highest elevations and on top of one or two of the numerous ridges, which appear to have been once sea beaches the stones are more rounded, and the ridges themselves would appear to have been subject to ice pressure, at the time of their formation, as they much resemble those pushed up by grounded ice at the present day. I much doubt whether this shale is limestone tho' the fracture is not unlike that of some of the limestone we meet with, there being two sorts: the upper and lower mountain limestone, with hyperite usually between them. This hyperite is, however, a rock of no definite age, being, I believe, a sedimentary deposit of volcanic ashes, much of it that we have found appears to contain small garnets. There were on the shale many boulders of granite, some of considerable size, the red granite boulders like that at Walden Island occurring more frequently than the grey. These must have been transported here by the ice. The dark crag is of hyperite and rises in front abruptly, almost precipitously, but by the aid of a small friendly glacier, which on his western side led nearly to the top, we made an easy ascent. The height is 750′ above the sea level, and we had a very good look out. We could see and sketch the coast to Cape Mohn clearly enough and away off it lie four islands apparently rather E of it, three of these appear to be but rocks and the fourth not over a mile long but of this one cannot be certain. Seaward of them all lies a grounded berg, higher than any of them and of larger extent apparently. All the view that we had as we steamed down this morning was now before us, but we could see further and it lay as it were mapped out underneath us. One great reason for the straits remaining blocked with fast ice is the number of grounded bergs lying to the eastward. These extended pretty well from Cape Mohn right across and about the centre. There is a large reef or chain of them so closely packed as to exactly resemble a mountain range. This chain is to the eye perfectly continuous and extends for some miles, and a bank is being doubtless pushed up by these bergs that will form either shoals or islands if it is not destined to close the straits altogether. As we steamed past the bergs this morning in the bright fitful sunlight of a wild windy morning the chain bore a striking resemblance to a mountain range with the snow-clad fantastic, acuminated [pointed] peaks so common in this country, and one can't help thinking that the range of peaks in this direction given in the new English Chart as Swedish Foreland [Svenskøya], that is, King Karls Land, owes its origin to these bergs being mistaken for land, perhaps in less favourable weather or from more distant lookout, as the land discovered last year bears little resemblance to that I have spoken of. We had little doubt that we could see it as from the distance we ought, but the ice seems indeed hopeless, there *may* be heavy winds coming that would drive away this floe, over which we can't see, or it may last the season out, but we can't wait for the chance any longer, so we take a last look at its dim blue outline and complete our round of angles and then eat our lunch. There is no temptation to take us further inland, one hill indeed mostly snow-clad rises just behind us but we can see as much as we want of him without climbing him, and behind him rises with its slow monotonous slope the endless glacier of which the land in the interior solely consists or rather by which it is hidden. We are, as it were, in a bay of the glacier in which rise these hills, as he curves round on both sides meeting the shore in four or five miles in each direction. All the way up we had found decayed driftwood, and numerous whalebones, firmly embedded in the stones. The crag of dark rock rises above this low land in *all* directions. It is in fact perfectly insulated, just as if the sea were to leave one of the dark islets in the straits, barren as it is, its mosses and lichens and few scattered plants (the willow amongst them) make it the most fertile spot in view on this crude land. Even the ubiquitous reindeer who seems by his horns to travel everywhere, I suppose when the snow is deep, as I am sure the shale offers no inducements for travelling, seems seldom to pass this way as today we only found two old horns. We descend the rocks at the back, then are at great breeding place for snowbirds we discovered, and the ledges on the cliffs are very easily accessible, but the season is much too late for eggs—only one nest of eggs this bird has I believe been taken—and here it is a great opening for enterprising collectors. To save our feet we followed a low flat glacier most of the way back, which has the same gentle declivity as the shale. There was however much water under the snow, and endless rills and streamlets, so that we got dreadfully wet. We started home with a gentle but fair breeze, and when within one and a half miles of the ship saw two whaleboats lowered away, after seahorse we guessed, as we had when steaming down this morning, seen a large school in the water, blowing and rolling along, at a pace that rendered pursuit hopeless. Our conjectures were right enough and we could soon make out the seahorse. There was not one but several schools, in fact some hundreds of them. When first seen from the ship, they had been unmolested in hopes of their taking the land or ice. When, however, they worked right up to her in the water and turned it was time to be doing and a great hunt we had but with no success. Like the saddles they came up and down so quick that tho' the boats were among them they could never get a sight on any of them or anything but a flying shot as they rolled like porpoises. It was a great pity that they did not try first to round them with rifles and then harpoon the wounded, each separate drove took such a long time that pursuit became useless and embarrassing. By the time they neared us they were so scared that we got but a long shot or two. We, however, wounded one in the head, but from the confusion of different ones appearing in different directions lost sight of him. We killed a ground seal on the ice today, after the seahorse hunt, but the ice breaking he unluckily sank. It is decided to give up Hinlopen Straits soon after dinner, so we get up steam and make a clean run to Cape Torrell. The drift ice was very light, and we were luckily enabled on this account, to run a base with the patent log. Before midnight we were at anchor in the little bight round Cape Torrell, in which we were nearly forced ashore when here before, but now free from ice, as we want to have another try for our anchor tomorrow.

August 23, Saturday.

The early part of the day spent in dragging under steam. Results: immense quantities of seaweed (by the way, a lot of dredging was done yesterday in Vibe Bay) and a broken grapnel. The said grapnel hooked *something* ere it broke, and we buoyed the place and tried with boats, then, after a last unsuccessful feel, try under steam. Away we went up the straits, earnestly hoping that the late winds we have experienced will have prevailed elsewhere and have cleared off the ice somewhat from the northwestern coast. From Augusta Bay 'til Cape Selander there is but little ice, and we again have the satisfaction of running a distance. Near Fosters Islands, we meet heavy ice, all sorts of ice "from Thumb Point bound for ___" [and] we make across to Cape Fanshawe, as should there be any smacks in Loome Bay in difficulties, the owner means to tow them out. Such a strange assemblage of ice, as we make our way thro', quite antiquated and old fashioned some of it, fragments and remnants venerable looking enough of bergs emigrating westward ho. From Thumb Point, some indeed looked as if they might count their age by centuries, old and dirty, of green brown and every conceivable hue in parts, decayed and hollow and very picturesque, sometimes assembling vast teeth lying fangs upwards. One was hollowed or honeycombed in an exceptionally peculiar manner, its face having a very vertical section as is often the case from their splitting right along one plane. The noise caused by such a fracture, occasioned by frost or concussion, is very sharp as well as loud, and we have during the past few days become quite accustomed to it, as well as to the clean blue appearance of fractures often almost mechanically precise. Off Cape Selander, we came first among the "loons of Alkrange." I suppose I should have put capital "L," as they are such an ancient and well-known family in Spitzbergen. All the young loons were on the water and there were simply thousands of them, generally one to each old bird, jolly little fellows not speckled but feathered properly just like the old birds. There is one smackie in the bay, but he seems all right and makes no sign, and it proves that there is nothing to stop him as the side under the glacier is nearly free from ice, and without a wind, at present favourable, a couple of tides with his boats towing ahead would at any time take him out. As we emerge from the straits about 10 p.m., the sea without is so clear of ice that we determine to stand on, without putting in for a lookout at Hecla Cove. It comes on to blow strong from the E and we knock off steam. We do not, however, [and] have to take our boats in from the davits. Things look more promising and we keep on N seeking the expected ice edge.

August 24, Sunday.

We have kept on so far, looking for the ice, that the islands lie but little to the N of us, in fact nearly E, and as the wind has gone down, we steam towards them as if free from ice they form a good base and lookout point for operations northward. It is a lovely morning, with a long soft smooth swell on the sea. How strange it is to steam along in open water with but a morsel of drift ice here and there, when ten days ago, scarcely a drop of blue water was to be seen all over the surface of the sea. After an hour or two's steaming, however, we make out the pack edge or "*Eis Kaute*," as the Norse say, running W from the Seven Islands. But there seems a deep bight to the northward somewhere here and we hope to get some idea of its extent from one of the mountain tops of the islands should it only keep clear. People say that the seals go N from the western ice, but where to they don't know. We know that at any rate they pass this way, if this is not the end of their journey. There are simply thousands of them in schools and droves of various sizes, and they seem only feeding and playing, many indeed are the young of the year. I always understood—perhaps I formed an idea erroneously as I can't remember when it was—that seals swam on their backs, as their natural mode of progression, but except when feeding or playing we have never observed them do so. About noon, we came among the ice to the W of Walden Island. Heavy drift sea ice is so abundant and all the sounds and channels are so blocked with it, that we can gain no anchorage among the islands. We knock off steam when near Nelson's

Andreas Umbreit

◀ **Figs. 9.34 and 9.34.1** At the end of August 1873, *Diana* approached the Sjuøyane again, this time with open water west of the islands. This image by Chermside shows the west side of Phippsøya (left, to the north) and Parryøya (right, to the south), with the southernmost tip of Martensøya looming behind to the far left, seen from southwest. The corresponding contemporary image by Andreas Umbreit was taken (2016–07-02) from south-west, as well, but from a closer position, therefore showing only the southwestern coastline of Parryøya (with small Nelsonøya (Nelson Island) in front of it, left of the middle of the picture), while Phippsøya and Martensøya (except of the snow-covered top of Solvberget in the background) are hidden behind

Island for about half an hour to enable me to try and take a photo. No great success as the day is now dull and overcast, and the rock dark, a bad colour for the instantaneous photo that the swell and the strong tide that is running necessitate. Then we steam on thro' heavy ice to the eastward as there is seemingly the best chance in that direction, and opportunities of exploring the N coast but seldom occur up here, I believe. Presently, ahead of us appears what we at first take for a square-rigged ship, the topmast of which surmounted by a flag is being continually hauled up and down. It is an illusion and a very striking one, the mast &c showing very plainly. It is, however, simply mirage. Later on, a great many more curious effects are produced by refraction, some of the most common but most curious being the appearance of small bergs, the refracted and inverted image of which appears above the real berg, with an appearance however equally real and solid. Islands, too, and rocks and capes appear in fantastic and distorted positions and a dark rocky island shows clearly to the W of us in water we have sailed over and where we know is no island to be. Never before up here have we seen refraction so clear or substantial, our glasses often not showing the illusion, as usually happens, and this too in the form of comparatively near objects. As the sun comes out again in the evening, we see a great many ground seals on the ice but as, of course, we only hunt when at a standstill as regards exploration. We don't go out of our way for them. A boat hunting amongst this ice might make a large bag. A few "saddles" we see, too, but unless those we have seen were "to take the ice" *en masse* as at the western ice, pursuit is useless, and you don't even get a shot. We kill two of the "bearded seals" or "*ogduks*." Grim beasts they look, as they look at you from the water. One of these was only wounded and getting fast to here in the dinghy, we had great sport. We now make for Cape Platen and when about ten miles to the westward of it, the ice becomes so heavy as to bar further progress, so we stop perforce and wait to see if it will open at all to the eastward. I take a distant view of the Seven Islands, not improved by the behaviour of the group in the foreground. The crew indeed are one and all excessively pleased at being included in a photograph but neither precept or example can induce them to assume natural attitudes or even to remain steady. We are seven- or eight-miles S of Martin's Island, ten W of the North Cape and ten E of Cape Platen. The ice closing instead of opening we soon find ourselves beset at any rate for this tide if not longer.

August 25, Monday.

The ice this morning in all directions is hopelessly tight, but we are lying in a small pool of water about 100 yds long or 40 or 50 wide. Two or three of these pools being left here and there where the floes and smaller, masses are jammed by the nips, we manage to get into this one this morning after a good deal of steaming and any chance of egress for the present is stopped as at either end it is closed tight by the point ends of two floes nipping hard, considerable hummocks being gradually thrown up at the points of the greatest pressure. Just before breakfast, a sea-horse comes up close astern and is wounded, and an exciting hunt takes place across the ice to the different holes at which he comes up, but all trace of him is lost. We find that since last night the ice has drifted E about a half mile, as well as having moved us somewhat nearer to the N coast. We are nearly in line between the N. Cape and Cape Platen. These are 20 miles distant from each

Sjuøyane - from left: Vesle Tavleøya, Tavleøya, northern
part of Phippsøya, the steep cliffs at Høgneset of Parryøya
seen from South. 2016-07-02,
Andreas Umbreit, www.terrapolaris.com

◄ **Figs. 9.35 and 9.35.1** Between 15 and 29 June, Leigh-Smith tried to reach the Sjuøyane for the first time during the 1873 voyage, but the *Diana* was stopped by dense sea ice off the west side of the islands, with only small Tabular Island (Tavleøya) accessible without ice. This image taken by Chermside shows a view from the *Diana* to high Høgneset as the westernmost steep cliff of Parryøya to the right and behind it the western coastline northwards of northern Phippsøya, then Tavleøya, and furthest to the left in the distance Little Tabular Island (Vesle Tavleøya), with Ross Island as the northernmost point of the Spitsbergen archipelago hidden just behind it. The contemporary matching image by Andreas Umbreit shows almost the same perspective, taken on 2 July 2016, with no drift ice—and this is not common so early in the season even today

other and the bay between them is very deep, 15 to 20 miles, and from its shores branch off numerous bays and fiords that penetrate still further into the land. It also contains several islands. The shore and hills are tolerably free from snow and I have no doubt reindeer are to be found in small numbers all along this part of the coast. The western side of this bay is, like the Seven Islands, of primary formation of rock, granite and gneiss, and the eastern of layers of sedimentary strata, limestone and clay slate like the Hecla Hook formation. Some of the distant hills were, I think, of the next lower formation from their colour, which is of the magnesium limestone of Hinlopen Strait, but this is of course only surmise. These hills were those that rise on the east side of Lordragers Fiord [Zordragerfjorden], a long deep fiord running far inland just to the W of Cape Platen, its other promontory, Cape Wrede [Kapp Wrede], being a dark striking headland. While taking observations on the ice for the variation and dip of the magnetic needle, we became aware of a snorting and roaring noise, and soon became positive that this was our wounded seahorse. We soon got scattered in pursuit as tho' the noise seemed close at hand it was in reality at some distance and opinions varied so much as to direction that we each went [in] our own gait. It was tremendously rugged scrambling work travelling over this ice as it was not solid floe, all the land floe having broken up, and when yesterday we rounded the southern point of the pack near Castren's Island [Castrénøyane] it appeared loose and accordingly when four or five miles further E we "took the pack," as it is termed in technical phraseology. Now it had jammed and tho' in places there were floes, this ice among which the seahorse was consisted merely of small floating masses tightly jammed. After some time, I found myself standing on a high hummock above a small flat level ice plain, which was really a piece of land floe some ten or 12 acres in extent, surrounded on all sides by the heavy hummocky sea ice and at a hole in this (it was full of holes) appeared the seahorse. Then he drew himself out with difficulty on to the ice, and as he there continued his roarings, one was forcibly reminded of "Philip Slingsby's Snapping Turtle" [from *The Fight with the Snapping Turtle*, by Scottish poet William Edmonstoune Aytoun (1813–1865)]. It is not exactly the thing to relieve one's feelings by a view [ILLEGIBLE], and a cap elevated on the muzzle of ones rifle is an unsatisfactory tho' efficient substitute. *Forrard on*, the Captain and his ally with a harpoon have a view, too, and are close on him, and I tear recklessly across the rotten floe, the pace too good to let one through, tho' "*As I pass beneath my feet I hear its surface crack*." Now we are up to him and the poor brute faces round on his enemies. Rover, as soon as he has satisfied himself as to the Captain's identity (he was with me, and always when away from the ship, attacks any but his own party), boldly attacks the seahorse. Our object now was to get fast with a harpoon before firing as with even a mortal shot the brute would get into the water and sink, but his hide is so tough and the harpoon so blunt the combined strength of two of his assailants utterly fails to drive it in. Poor brute, one does pity him, as unable to reach us he strikes his tusks savagely into the ice, then he takes the water, and as he rounds his back we get in the harpoon and hold on. Up comes his head and a bullet well placed finishes him. Poor brute, he was too sore wounded before to make a good fight. Detachments are arriving and a terrible pull five of us have to get him on to a pretty safe-looking piece of ice, and as more hands arrive two of them get a complete ducking. Something else comes up in a hole as the flensing is going on, seal or seahorse, and the owner and I go to look for it, and while casting about are aware of a "Brunie" coming towards us. We lie down and on he comes, having got scent of the seahorse kreng. Nor does he seem to take much notice of the noise going on at it. Stopping every now and then, and then continuing his advance in a very dignified way. The other two guns are in ambush also. He sees Rover but appears not a whit disconcerted. At 80 to 100 yds from us, he stops and surveys the scene, and would have doubtless gone on to the "kreng" had the men kept steady, but the temptation of a shot is great and we give him a volley, two bullets taking effect. He runs a few yards and then falls forward into a pool of water and is soon out of his agony. Rover, however, boldly attacked his head while he was still growling and snarling. He proves to be a very "bonny beastie" and is as large a bear I think as I have seen. We have about ¾ of a mile to travel to the ship and get back at once, and as the hands arrive with the skins, take a photo or two [see Fig. 9.31]. It is a lovely sunny day today, tho' the ice lies as tight as ever, no wind in fact, but a gale from the S or SE can do us much good, and strange to say we find but little tide here. It is in fact a very landlocked position, as all across the mouth of the bay I have spoken of but at seven or eight miles distance stretches the line of the Seven Islands, forming as it were a high distant breakwater. There are a great many young birds about, kittiwakes, snowbirds, and dovekies. The kitties are very pretty, black bills and legs, black marks on their heads, and stripes on their wings. They congregate on the ice in large numbers like the swallows at home, as if assembling for their migration. The snowbirds, that is the young, have the greater part of their feathers tipped with black, but it is the dovekies that one can recognise by their shape alone, altho' in mature summer plumage completely black except the wing patches, with red bills and legs these are of a soft grey with white breasts, and the white patches on the wings and with dark bills and legs. The old birds, too, are speckled and we are inclined to think that they assume a winter plumage of grey as many that we now see are very speckled. P, whose stiff knee is getting better, comes for a walk with me after dinner. We take along the floe that was on the starboard bow. This floe is of some miles extent, five or six long by one or two broad, and covered with small blue lakes. There are one or two ground seals in view, but we fail to stalk them as their hearing is wonderfully quick, and when after a long crawl on the snow, you at length peer round a hummock you are often surprised to find that your quarry has taken himself off. There was a very powerful refraction tonight both on parts of the land and almost over the whole extent of ice horizon, the latter indeed appeared completely bounded by a high range of ice cliffs, owing to the inverted image of the hummocks.

A breeze springs up from the E tonight and a pretty sharp frost sets in.

August 26, Tuesday.

A cold fine morning, with a strong easterly breeze. We have driven some distance W with the ice, but our attempts to get through it at the turn of the tides utterly fails. The ice indeed is all under strong pressure and when broken soon closes in, nor is there any water in which the loose pieces can play. The owner and I take across to the "kreng," now one-and-a-half to two miles distant, as besides moving 100 yards or so this morning the ship was steaming ahead or trying to do. So, during the whole time of our walk last night, and by the dint of our efforts during the past two days we have made perhaps nearly a mile of way getting into another of the small lakes between the nipped ends of floes. Our walk lies across the large floe I spoke of last night, along the edge of which we hope gradually to make our way out with the ship and firm as it is, we do not see how any of these boat sledge expeditions could hope for easy travelling except in early spring, when their sufferings from the cold would be very great. In those that I have read or heard of, 12 to 20 miles a day are generally assumed as the *average* distance in the *desired* direction, not merely as the distance to be travelled, and then when soft snow is met with and these endless pools necessitating such frequent and often long detours, these seem to be

Figs. 9.36 and 9.36.1 Southern Phippsøya in Sjuøyane, with Granittberget and Hytteberget, seen from west during the first 1873 attempt to reach the islands in second part of June, with the hindering dense drift ice visible in the foreground. A similar perspective (from a slightly more southerly position) was depicted as comparison by Andreas Umbreit on 2 July 2016

regarded rather in the light of unlooked for causes of failure. But how aught else could be expected at a time when the sun's power is sufficient to make the Arctic climate pleasant and fit for men to endure such constant exposure, I can hardly understand. At best were such an expedition to start in early spring, their average rate of travel could never hold for the *return* journey even suppose their strength to remain unimpaired, which I believe under the most favourable circumstances of travel is never the case and a good margin should be allowed for the stress of weather, as heavy gales and snowstorms are then very frequent. At last, we discovered our krengs and what a change in their surroundings of the thin fields of ice on which they lay yesterday. Hardly a recognisable morsel was left, like the chamber of torture whose walls closed in on the prisoner, the surrounding masses had come in and nothing now remained but piled up fragments of ice and snow, fresh cut blocks lying on each other in every conceivable position. One kreng which should have been 500 yds nearer to us than the other was now the further, the other on its slab of ice raised bodily out of the water and resting between two of the heavy masses that had been the active agent in this nice little grave. The numerous footsteps led any and every way and ended abruptly. Continually it was like a child's puzzle when the pieces have been shuffled up, and considering our own position, we regarded it with perhaps more interest than such an everyday occurrence here merits. On arrival at the ship, things remained *in status quo*. We however have another crack at the floe, and coming back to our old position worsted, give it up for the night.

August 27, Wednesday.

Early this morning we found our lake closing in and, as the floes which barred our way seemed slewing, we succeeded in smashing the point ends as they opened, and making our way a short distance into another long irregular dock. The morning was fine with a light breeze from the E. A sharpish frost still continues. About two to three miles inshore of us lies the pack edge, and open water seems to extend all along by Castren's Island to the N Cape. We are enjoying lovely weather and the chief annoyance of our besetment lies in the unpleasant feeling that you are a prisoner and cannot do what you like, as well as in the time that we waste, precious now that the season has so far advanced. This latter reason, combined with the frost sharp enough to freeze over the sea water in some places to an inch thickness, makes some of the crew rather nervous, altho' there has been no week all thro' the summer when a scum has not been forming some nights on the surface of the water and often has it increased twofold the labour of a long pull back to the ship. This was especially the case in Hinlopen Straits, a strange fact when one remembers the extreme average saltiness of the water we met with there. It, however, has an excellent effect, as they work with a will to get her along, whenever it is considered advisable to try. Seriously speaking, there is not, we fancy, a chance of this ice holding tight until the frost permanently sets in, and therefore we see that chance. We can afford to laugh, tho' we shall all be glad enough to get out. But lying here in fine weather and knowing that this month and September are generally the openest along this coast, one cannot really share such feelings. This afternoon the ice seems slacker and our present dock rather closing in. These docks are certainly grand inventions, and it is a great consolation to be lying in one, well protected at any rate for the moment, when all around, as has been the case for the last three days, the cracking and crunching caused by the nips and the noise of the hummocks and "bergie bits" as they split is going on. The owner and I have been pioneering all along the edge of this large floe, and a very little slackening would cause a small and twisting lane to lead us along its edge, at any rate half the way to the open water. We stalked a very large ground seal, but he was lying on new ice, so thin that when shot, his death throws broke it, and we lost him, chiefly thro' want of a club. On our return, the attack is again recommended and our slow, and circuitous advance, certainly does somewhat resemble the tedious operations of a siege. This time we go at it with a will, charging, warping, blasting, towing pieces aside, and removing broken fragments or punting out on them into the open water. All only serve to advance us 50 or 100 yds and we get into a very small dock indeed, so small that when the point ends close behind us, ice touches us almost all round, and but a few inches of water keep the ships side from the pressing ice. The nips are on in force indeed tonight, and the pieces are jamming and squeezing each other like a very ill-behaved crowd. The weather is bright sharp and clear and there is a very clear sky, with lovely colours in it. The thermometer is 7° below the freezing point. We smash the ice on one of the freshwater pools on the floe tonight and fill up our water tanks. Breeze from the S.E.

Figs. 9.37 and 9.37.1 Tryggve Granfjellet as the northern end of Phippsøya (right) and Tavleøya (left) depicted by Chermside in the second part of June 1873 from a southerly direction with the dense drift ice blocking the west side of Sjuøyane in the foreground. A matching recent perspective by Andreas Umbreit shows the same scenery on 2 July 2016

August 28, Thursday.

"*Great is Diana of (Dundee)*"!!

Early this morning we got two seals in the pool we left yesterday. One of them had had a narrow escape from a brunie, losing a hind flipper and carrying a very clear impression of his enemy's claws on his coat as well. We had during the night drifted some little distance to the E, but then we drifted with the ice and the open water was still visible from the masthead at about the same distance. We had already found that smashing point ends and masses of ice so as to let, *la petite*, get her nose in, was of remarkably little use. When the floes &c were under pressure and insisted on closing for themselves the gaps that we make, with a force that made their authority indisputable, that was a complete defeat for steam, [a] mere waste of the "puckle coals" [an apparent reference to the Parkfield Colliery located near Pucklechurch, England, from 1851 to 1936]. We were also aware of what immense masses a ship can slew by steadily steaming against them for 10 or 15 min, and among what heavy ice she can force a passage, when there is any play for the ice, any capability of its being induced to move so as to leave a passage the width of the ship. We now have to find what resolute organised efforts can do when the ice is tight jammed with no play, but *not* under a heavy

Figs. 9.38 and 9.38.1 Høgneset as the westernmost steep cliff of Parryøya, seen by Chermside from the *Diana* across the hindering thick drift ice from North during the first attempt to reach Sjuøyane during the second part of June 1873. A similar recent perspective by Andreas Umbreit from 7 August 2013 shows the same area, then free of ice. Nelsonøya is situated just southwest of Parryøya

pressure only so in places, and this is what from its appearance, it now is. By 9 a.m., steam is up and we commence, and hard work it proves for all hands, hardest for the engines perhaps, a tremendous test for even such good machinery as this has proved to be. With very heavy warping we get out of our first dock, for there was no room for a run, then we push thro' a little water, heavy work tho' as it is full of drift ice, and then we come at the next barrier with a rush. I may as well describe our modes of attacking obstacles, as these differ according to circumstances. There are almost always several hands on the ice, and when there is any loose ice these push the pieces aside or, getting on them, punt themselves along with their boathooks and then ferry them into any corner of water out of the way that they can. In the meantime her ladyship has backed astern as far as the dimensions of the pool or dock will allow her, a distance sometimes of as much as

200 yards, then she comes ahead full steam and jumping right on to the ice succeeds in cracking it. There she comes again and again, and several large pieces are by this time smashed. On to these men jump with drills and, boring holes, ice hooks and warps are made fast from the ship and they are towed out of the way. The boathook men (or numbers as we say in drill, removing the smaller pieces) if the floes be nipping fast little is gained, but if not, the passage is slow and gradually cleared, and at last only two meeting points remain. The dinghy perhaps picks up men floating hopelessly about the pool, and other scramble up into the chains and so all get on board, not to lose their passage as this time she is going slap thro' and will be rewarded the other side by 100 yards or so of better going. We are backing astern to prepare for a charge and now is the time to help her impetus by "Sallying" or "Sollying," as it is called. *Ting* goes the bell and she starts ahead. "*Give her a roll boys.*" All hands at once range themselves on one side of the deck: "*O—ver,*" from the mate on the bridge, and we all rush across. "O—ver" again and so on 'til by the time her head is straight at the fence and the Captain shouts "steady" from the nest. She is going at full speed and rolling almost to the rail. A crash and we are thro' and looking forward to the next attack. When the ice is too thick to break by repeated charges, I am often at work blasting to crack it, once or twice with great success. At last, the cracks are made and soon widen to a few inches; still we fail to tow the pieces out. Another blast and one gets loose, and then tho' tedious the work goes ahead well enough. The best way is, if possible, to explode the charge, just as she comes up with a rush and a sally. At times, however, no room can be got to charge, or a direct charge cannot be gained at the right point or, oftener still, the ice is too thick, or if smashed no room for towing away the pieces can be got. Blasting is then of little use, unless the passage to be broken be narrow as then she may ride down the fragments, and the only apparent method is to force the two floes apart. Warping is now resorted to. A hole is drilled ahead of the ship and an ice anchor made fast to this. The warp is led in from this thro' a gangway aft and forward to the capstan. She now has her nose jammed firmly in between the two masses to be forced apart, and steams ahead full speed, and at the same time all hands heave on the capstan, and then the helm goes hard over and as she swings the men take in the slack heaving a pawl or two on the capstan [a pawl is a set of short bars that engage with whelps to prevent the capstan from recoiling]. When no more can be gained over goes the helm the other way, and with the immense leverage gained by the whole length of the ship, the steam of course besides the pushing ahead giving her the power of swinging and thus the leverage, more slack is gained and so we force our way thro' inch by inch. When there is a heavy pull on the capstan, they are perpetually singing out, "*Heave and break the warp*" &c, and well is the injunction carried out. Yesterday we broke a warp and twice today new warp that had never done service. Often, we are an hour at a place like this but at last the floes yield or we get our chance for a rush, tho' sometimes after charging, and blasting and warping have all been tried we have to give it up and try another place. Up 'til today we have always had finally to give up, but now seems our chance and we stick to it. At many places, a way has to be broken by sallying and charging and blasting as well, before the point of heavy pressure can be gained and warping commenced, and a grand sight it is, as one stands at work on the ice, to watch the ship rush bravely at the obstacle at full speed, like a troop charging, and how satisfactory as her bow leaps out of the water onto the mass, 'ere she falls back to feel it groan and quiver and crack under your feet, a dull dead muffled sound perhaps and a new dark crack or two telling you that your blast has also done its work. Steam can indeed do wonders, and I fancy guncotton would by no means form an inconsiderable ally to it in these besetments and might even lessen the dangers of the dreaded Melville's Bay of the whalers [Melville Bay, in Northwest Greenland, was the sight of several nineteenth century tragedies involving whaling vessels trapped by ice]. By 7.30 p.m., the last heavy

point is smashed and we are soon in fairly open water which extends past the North Cape. Thro' this we steam passing very close to Castren's Islands. Dark mountainous islands they are, the sharp hog-backed ridge that forms the backbone of the larger rising from the sea to an elevation of between 800 and 900 feet. It is a glorious sunny evening, and many of these wondrous effects of refraction are very clearly visible, Walden Island indeed rejoicing in greater vagaries than ever, his real self being surmounted almost everywhere by his own inverted image cocked up at an absurd angle. Towards the Seven Islands it seems clear of ice to within a short distance of the land and we steam across. It proves, however, that a thick barrier does lie between us and Parry's Island and so we hook on to the lee side of a piece of grounded ice near Nelson's Island, and a very curious piece it is, too. It has a nearly rectangular surface 200 by 300 or 400 hundred yards and this surface, tho' sloping upward considerably towards the island, as it proves to be aground on a shelving bottom, is very even (& slippery from the late frost). It has indeed the appearance of being merely a piece of a level floe or field, but of a very great thick one, as on the side where we hook on, its vertical face projects 10–12 feet above the water and on the far side 14–30 feet. Imagine then our astonishment when on sounding we find it to be aground in 18½ fathoms of water so that its total thickness is over 120 feet! A mere "floe piecie," say the connoisseurs. "*Why David Gray's father* [Peterhead whaling captain John Gray (?-1856)] *sails once along the edge of such a floe for 12 h, believe it who can.*" [David Gray, Jr. (1829–1896), so named after his famous whaling grandfather David Gray (1776–1833) was, like many whaling captains, a keen naturalist. He was a close acquaintance of Leigh Smith]. There is promise of a splendid set of photos from here, but the sun is too low for them now and, from old experience, we little expect to find ourselves in such a berth when we turn out. The drift ice is indeed coming down with the tide in heavy bodies, and at a rate that makes us feel, as it rushes against the weather side of our protector, what a very bad chance we should have if at the turn of the tide, it should attack what is now the lee side, always supposing that we were foolish enough to wait for it.

August 29, Friday.

Shift of wind and turn of tide, sent us again on our travels early this morning and we hooked on after a time to a drifting floe. As after some time no change for the better turns up, we determine to run for one of the bays or indentations of the coast just to the southward of the North Cape. We have a fair smooth passage and by midday come to anchor in a shallow bay close to the N. Cape. Eaton goes botanising, the owner and the skipper ascend the hill for a lookout and I take an exploring walk up the valley, stretching from the head of the bay, and running towards the sea beyond, as the range of hills of which the North Cape is the extremity are of no great width and but a neck of land separates me from the shores of the bay beyond, that between Cape Platen and the North Cape. There is besides a doubtful question of geography here: Parry in his account of his Spitzbergen expedition states that a sound runs right thro' from what he names Beverly Bay [Beverlysundet] into this other large bay beyond, this tho' I understand was not actually explored, and it has since remained a moot point. The opinion of the Swedes on the subject was divided. After leaving the bay, I take up a low flat valley and, in the bottom of this, lie some lagoons, one brackish. The valley is full of driftwood, and its soil consists chiefly of rounded granite boulders, and there are many whalebones about. After a time, I meet a high barrier of these rounded boulders and the tail end of a gently sloping glacier coming from the hills on my right, and as when I surmount this the same even slope leads to the sea; on the other side it is not hard to guess that this was a sound and that, too, within the recollection of this driftwood, some of it sound enough yet. The whole neck is not more than three miles across. Then I find myself on the shore near the wider [?] and southernmost of the two Castren Islands. I keep along the shore a bit, but soon find walking on the glacier which comes close down to it preferable, as tho' very slippery it is not steep and the frost has given it a very firm surface. I keep on this as far as the

Figs. 9.39 and 9.39.1 Tiny but high Nelsonøya (middle) and the south of Parryøya, depicted by Chermside from the *Diana* across the heavy drift ice from west to southwest during the first attempt to reach Sjuøyane in second part of June 1873. The contemporary image (from 2 July 2016) by Andreas Umbreit shows much of the same scenery from a more southwesterly direction

entrance of a sound with a very narrow mouth that I come across, and then have to climb along the very rugged uneven blocks of granite, of which the sides of the mountains that rise from its shores consist. In places, too, cliffs occur which have to be circumvented, but even this seems preferable to the narrow strip of shingle at the water's edge, blocked every few yards by rocks and lumps of ice. This extremity of the Nord Ost Land is of the same formation as the Seven Islands, that is, the rocks are "Granite Gneiss." Among them one finds also argillaceous slate with a rather crystalline fracture, hornblend and talc, and at times quartz and mica with apparently no felspar, then again, some lumps seem to be almost entirely of that substance, that is, porphyry or at any rate porphyrous granite. There are among the granite numerous veins, some white, some pink, marbled looking veins, of various thickness running from *above down* [and] composed mostly of rather coarse-grained masses of pegmatite. Some of the rocks, indeed, had a horizontal layer or crown of this, like the coating of sugar &c on a wedding cake. The components of this rock: quartz, mica, and feldspar were indeed mixed up in such various proportions that within a short distance one met the most different looking specimens of rock: white, grey, black and white and cream-coloured, and red, the latter sometimes coarse and sometimes very fine grain. The red colour, indeed, we thought was not owing to any oxide of iron or manganese, but to the pale pretty red of the feldspar. The quartz was also fine and very translucent, and some of it seemed to contain a good deal of iron pyrites. The mica in many specimens seemed to be distributed in such an irregular fashion as to show the rock to be granite, the lamellae of it being, I believe, parallel in gneiss. This mica often occurred in thick lumps of endless laminae or plates pressed together; these laminae seeming of rather unusual size, sometimes from one to two by two to three inches, and as transparent as talc, very easily separated too, and when separated tough and pliant. Being driven up continually to avoid the cliffs, I climbed the highest hill above the bay or sound I was exploring and found it to be about 900′ above the sea. Gryteberget, the hill across the sound, I made but 1,500 instead of 1,700 as given in the charts [it is in fact 1,690 feet high (515 m)]. I saw from this hill enough to make me continue my explorations and finding climbing among the rocks on the hillside monotonous I descended as there was now fast ice below. The rocks on the hillside were very thick or rather the beds, and intersected by numerous very deep cracks, immense masses being often thus cut out ready as it were for quarrying. The vegetation was unusually scanty and few mosses and *tripe de roche* [rock tripe] and that was all. The floe was sound and had frozen, but it was like walking in a labyrinth, as the numerous pools were frozen over, but did not bear safely. Continually keeping on to my right I come in sight of the sea, and after an hour or so find myself at the edge of the floe in Beverly Bay, and for the future we at any rate are convinced that this is an island. [In fact, Chermside would confirm Parry's tentative observation that the indeterminate peninsula was an island, with Nordkapp as its northern extremity. In his account of the Swedish expedition, Nordenskiöld published an updated map of the area and named the newly recognised island after Chermside. It is now known as Chermsideøya.] Here I met a whaleboat, the owner and Captain having come out to explore. This has luckily been done, as the fast ice would have put an end to the boat expedition, altho' they might have left her and gone on. As there is no hunting to be done here, we return to the ship. I again go on shore after a feed, as it is now fine enough for photography, tho' I fear too dark, but we leave this tomorrow. So, I take a dry plate or two to try and get a view of the North Cape. Then I set to work to climb the hill, as my main object in coming is to get some angles. This is soon done, as the cliff is only 450 feet above the sea level. Its sides are steep, but in some places, there is a good deal of vegetation. At its foot or rather where it rises from the high shelving beach of rounded boulders are some rocks bearing apparently a very well-marked water line, and the smoothed polished marks of glacial abrasion. The latter are doubtful, as the weathering effect of combined humidity and frost on the granite is here very evident, some of the boulders high up on the hill side being from these effects almost as rounded and bearing the same marks of attrition, as those but lately washed by the sea. The extremity of the line of hills is only an isolated projection of rock, separated from the rest by a bridge or rather saddle a few feet across and with a sharp edge like the ridge of a house. The sun was now close to the horizon but hidden by clouds and, I fancy, must have sunk for a few minutes. There was a strong keen air up here and a beautiful lookout and the charm of such a view, tho' less novel, is to me as powerful as ever. Barren as is the land here; it supports a few reindeer, tho' we did not kill any, as they range wide, the trouble of the quest, and the distance they have to be carried make them hardly worth the trouble. The lookout across the ice is by no means encouraging, but the islands seem accessible and to them we are bound for a lookout. After 1½ hours work on the hill, I returned.

August 30, Saturday.

Early this morning, as we weighed anchor, there was a very beautiful sight, the refraction of a distant and pointed small berg, gave to it and its inverted image the appearance of a lighthouse, as from some polished surface on it, the sun's rays were reflected back in all their brightness. We steamed across thro' very scattered ice towards the islands, but the thickness of the ice round it, made our grounded piece near Nelson's Island, an unsafe berth. There were some seals on the ice and they lay seemingly well enough. We steamed close to one in the ship but failed to hit him in the head so lost him. After hooking on to the fast floe between Parry's and Phipps' islands, the dinghy went after another, who from the warmth of the sun was so sleepy that, in spite of the noise made in crashing thro' the new ice, he let the boat get close to him and was easily polished off. An enormous brute, he was the largest we have got yet, a bearded seal and almost as big as a sea-horse that we got a little later in the day. This brute came up close under the stern, apparently smelling at the bearskin trailing from it in the water, this seems a by no means uncommon occurrence, and the seahorse possibly seeing one of his natural enemies in the water, hopes to take him at a disadvantage. A ground seal even if he gets hold of him can shake him off or drown him in the water, at least so the Norwegians say. This seahorse was wounded with a rifle from the ship, and a boat then lowered. After missing with the harpoon gun, they got fast with a hand harpoon and after the usual salvo, he was despatched. After taking a photo or two and getting lunch we all except P, start for Phipps island, distant some three miles across the floe. This island consists of a lot of isolated hills connected by low sea beaches and we have no doubt the cluster of islets here represented in the old Dutch charts, was at that time correct. Eaton and I branch off southwards, as should there be fast ice in the sound between Phipps and Martin islands, I wish to explore the latter, both for surveying and also to search for the Swede's boat. The beach separating the sounds between Parry and Phipps, and Phipps and Martins islands, is not more than half a mile across and never more than 50′ above the sea level. It consists of rounded granite boulders and there are immense quantities of driftwood and some whalebones on it. We found also an oar of the ordinary description, with a flat iron plate not much decayed at the end of the handle. The sound was full or nearly so of drift ice not very closely packed, and two or three miles of this work would have brought us across, but we did not think it right to try as if before our return it opened much with wind or tide, they would have a great deal of bother in bringing us off. We, however, determined to follow it along on the side we were on, as we could even in that way, examine closely enough more than half the shore of the island. When we arrived at the eastern opening of the sound, there was no longer any beach. We had in fact for some time been climbing along the face of the cliffs and we now determine, instead of climbing round the promontory and returning by the gorge between the two hills beyond, to try and climb over. An hour's climb brought us to the top and a curious top it was; this special hill being a sort of isolated ridge about a mile long and at the top open only a few feet wide. The gully up which we came was one of the few places where it was accessible and up some of the clefts of this for it was a crack rather than a gully, we had to at times shiggle up like a chimney sweep, our backs against one side and our knees against the other. I don't think I should have tried it without the assistance of Eaton, a very good cragsman, and an Alpenstock. The top was where we reached it first 1,100 feet, and on the eastern pinnacle 1,220 above the sea level. We found some argillaceous slate on the top and a good deal of talc. Barren as the hill was, we had seen many traces of reindeer and the owner's party, who had ascended the hill 1,400′ the other side of the gorge and were now descending, caught sight of two. They look very jolly beasts now as their coats are redder and thicker and their legs white, very solid chubby little deer they are. They were very tame and came towards them, and were easily killed, a difference from the Wüde Bay deer. It is, however, very satisfactory having killed them on these islands, the most northern known land in

Figs. 9.40 and 9.40.1 Small but high Nelsonøya (middle) and the south of Parryøya, depicted by Chermside from *Diana* across the heavy drift ice from the West during the first attempt to reach Sjuøyane in second part of June 1873. The contemporary image (from 2 July 2016) by Andreas Umbreit shows much of the same scenery from a more southwesterly direction, but closer

the world. Our descent was to me terribly steep, and the lower half one single snow slope with rocks here and there so that as it was frightfully hard and slippery, the *glissard* [*glissade*: a motion of sliding or gliding down a steep slope with support of an ice axe] was the only method of descent, and would have been charming, had not the steering made it rather uncomfortably nervous work. In the bottoms were some lagoons, and here we meet with some very beautiful and curious ice. We have indeed frequently met with "penknife ice" of which Parry's description is so good that I may as well quote it: "*[A great deal of the ice over which we passed to-day] presented a very curious appearance and structure, being composed, on its upper surface, of numberless irregular needle-like crystals, placed vertically, and nearly close together; [their length varying, in different pieces of ice, from five to ten inches, and their breadth in the middle about] half an inch, but pointed at both ends. The upper surface of ice having this structure sometimes looks like greenish velvet, a vertical section of it which frequently occurs at the margin of floes, resembles, [while it remains compact] the most beautified satin-spar, and asbestos, when falling to pieces.*" [From: William Edward Parry, *Narrative of an Attempt to*

Reach the North Pole. London: John Murray, 1828, p. 61.] But this should be called "razor ice," at least, it was a strip some half mile long and 200 or 300 yards broad, and instead of penknife crystals, these were in thin curved plates, curved both in surface and edge, each plate from 2 to 12″ deep and of about the same width and all set vertically.

After getting all the angles &c that I required, we struck for home, this time taking across another and narrower gap between two of the hills. This was almost completely filled by a gently sloping glacier, and *glissards* on this were charming from the absence of stones. The owner's party took back thro' the next gap northward, the point at which we had made the island first this morning. The northern side of this is bounded by the highest and most continuous group of hills in the island, the whole island being six or seven miles in length. The part of the hills that rise above the gap in which lies one of these low granite beaches is very curious: half a crater, it looks like forming a very perfect amphitheatre, bar the terraces. We made home across the floe, the pools of which were hard frozen. Some of the crew came to the island to assist with the deer and the remainder were occupied in sliding. It was indeed for a great rarity up here, a splendid opportunity for skating, tho' usually up here the snow falls so soon as to make it impossible. That of the snow which had been thawed by today's sun, was now freezing into needles and penknife ice in places, and I think Parry's explanation of the drops of rain, [is] at best only one of the causes. I fancy any time that the snow has thawed soft enough, when the frost comes on, it is frozen into needles, and the ice being more dense and compact. The interstices are found in support of this. Where pools are formed on the flow and then dried up one almost always finds ice somewhat of this description in the bottom of the pool. Where also masses of ice are submerged under others, the snow does not as a rule freeze into solid ice, but crystals of a similar perhaps rather coarser form. The rainfall is indeed so uncertain and its distribution so general over one floe that this phenomenon can, I think, hardly be accounted for in this way. We find on our return that a successful bear hunt has taken place. It was descried from the masthead, lying asleep near a hummock beside the "kreng" of a seal it had killed, and off which it had dined. Three guns started to cut it off from Parry's Island and drive it towards the floe edge. Potter who had been away in the dinghy was recalled and, with a hand who had seen the bear, made for some hummocks near the floe edge. Four shots are fired but the bear is only slightly wounded. He, however, comes close to P's well-chosen ambush and one well-placed bullet finishes him. We certainly have lost all faith in the ferocity and terrors of the Polar Bear, the "lion of the north" and all that twaddle not only from our own slight experience but from all we hear. The owner indeed has killed a good many. All that the Norwegians say confirm his and our experience, and numberless as are the bears the various members of the crew have seen killed, and as great is their capacity for yarns, none of desperate encounters with bears is forthcoming. The bear is indeed a large, powerful beast, and has the better of you in powers of going on land, ice, or water, but I fancy it would be a very great stress of hunger that drove him to attack a man. His chief object, like that of almost all wild beasts, is to get away from you, and even when wounded seems still to try and fly and if only let alone will, I fancy, prove harmless. They do indeed follow your footsteps, but one gentleman who I know, that met one, unarmed, succeeded easily enough in driving it off with stones. A shotgun, I feel sure, would set it flying, and when you read as we did two days ago in an account of an Arctic expedition, of a bear with eight bullets in him, surrounded by 20 men mostly armed with rifles, and a pack of Esquimaux dogs to distract his attention, and are told that "*it was a desperate moment for all,*" it seems simply ridiculous. In fact, as an object of terror he ranks, I think, lower than a hungry wolf, and a single wolf can be an object of no great terror to an armed man. *Àpropos* of wolves, I cannot help thinking Arctic wolves inhabit these islands. The Swedes thought they saw one here, so did we before. Well, today we found endless fox tracks in the snow and sand, for both sand and clay found in considerable quantities from the decomposition of the granite, and also both parties found tracks considerably larger and if those of a fox, that fox must be the grandfather of all foxes. The central and nail-less toe, moreover, was of a different shape. Two of the freshest and clearest prints that we measured were five inches long by three inches broad. The owner moreover saw three beasts once cross a plain near Cape Torrell, dark and shaggy, but not large enough for deer. [There are no wolves on Svalbard; in fact, only three species of land mammal survive there: Svalbard reindeer, Arctic fox, and the sibling vole.] The sun tonight set splendidly and for the first time for months we saw him, dip for a few minutes and then reappear, and a wild looking sunrise it is.

August 31, Sunday.

This morning it is blowing a hurricane from the SE, and then is a very heavy blinding snowstorm going on. This is very annoying as I have but two points to visit to complete a surveying sketch and triangulation of the islands, and one of the points is on Parry's Island, from which we are not more than 1½ miles distant. The lookout yesterday N and E was very discouraging and there seems only this bight to the NW to be explored. We have had some difficulty in holding on this morning even with steam and three warps out, and our "krengs" that we hoped would cause the brunies to rally round them, have gone floating away. The new ice, which yesterday covered the water about here, has utterly smashed up. After taking a photo, we steamed out thro' a good deal of ice, and 'ere long it came on to a blow, and we came into a heavy sea. At nightfall, this rather increased. Before the gale came on again, we met a heavy long smooth sea from the W. The islands and everything look vastly different today from what they did yesterday, and the floe is so much broken up in places that you can in one place almost row ashore. The floe is some inches deep in snow, also the decks, and skating would be pretty well over. Besides, it is thawing and sludgy. The whole scene indeed is changed, and the dark mountain tops of yesterday are all snow-clad tops slopes, with here and there bare patches on the steeper slopes. Indeed, the steepness of these slopes prevents them, I expect, from ever being very deeply covered and enables the poor deer to get at the moss on them. How on such a barren place, deer can be so fat and jolly, one can't imagine. They range, I suppose, very wide. There is, indeed, on these islands one very "muckle stag" [a stag of extraordinary size and cunning], from his footprints, that is. We none of us saw him. We make the ice about Lat 80°35′ and keep along the bight reaching Lat 80°40′. A point then stretching SW obliges us to fletch out, and we then heave to, and with the main yard aback proceed to sound and fetch up water from the bottom, rather risky for the thermometers &c as it is now blowing very strong. We got a heavy knock today that slightly stove in part of our bulwarks and smashed one of the stanchions. As it has now come on thick, there is not much to be done. The wind is now from the NE and, getting as much in the lee of the ice as possible, we first dodge, and then lie to with the mainyard aback.

September 1, Monday.

Towards midday it gets calm, and clearer and we get a meridian attitude. Our highest point in exploring these bights to the N was about 10.30 a.m. this morning: 80° 54′. As we come out of the bight, we are still in sight of the islands, their tops showing clearly but a great deal of haze and refraction about their lower parts. We keep on exploring the ice edge, and tho' sometimes among drift ice as in going through point ends &c, come always to a hard, well-defined pack edge. There are not many birds about, tho' there are a fair number of Dovekies in winter plumage but they, I fancy, keep always near the ice edge wherever it may be. We have been of late very successful in catching the small fish among the ice, upon which so many of these birds' prey. They are a sort of whiting (*Merlangus polaris*) [a form of cod] and live in holes and crevices in the ice. At the broken edge of a floe, you can see them going in and out and at the least alarm taking refuge in their holes. Where a hole occurs in the floe, they come up into the pools above. There are two kinds, apparently, a dark and a

Figs. 9.41 and 9.41.1 Here, Chermside took a photo of southernmost Phippsøya with Hytteberget (left) and Utkiksnosa (middle) through Straumporten and onto southernmost Martin's Island (misspelled by Chermside for what is actually Martens Island (Martensøya)) in the background, far right and looming through between the mountains in the middle. The picture was taken from the *Diana* across the hindering dense drift ice from the west during the first attempt to reach Sjuøyane in second part of June 1873. A contemporary image (from 2 July 2016) by Andreas Umbreit shows parts of this scenery from a somewhat different position

white kind, very much alike in everything but colour. There was a fine sunset tonight and the sun was altogether gone for some time, tho' it was never nearly dark, and just as he rose again it came on to a blow. We took another sounding tonight.

September 2, Tuesday.

The gale continues all the morning and after a time we stand off the ice edge a bit, as a good many detached streams come drifting away from it. Towards midday, it gets calmer and we stand away southward to make either Moffen Island or Wüde Bay, and we then get up steam. Grey Hook, the western extremity of Wüde Bay, is in sight this afternoon but we find we can neither fetch that nor Moffen Island on this wind. It is therefore determined to make for Magdalena Bay, where we should find a bottle giving tidings of the *Sampson*. Thence we may either go to the Foreland or the Ice Fiord if there are many bottlenose whales about, or try up north again to the westward, as there may be some chance in longitude further W. Of course, we no longer expect to get any great distance N and, from what we have seen of the northern pack, no distance at all.

But up there we might find fish and if by any chance we find a bight of water or slack ice, a day's steaming will take one a long way. We get up steam and amongst other things nearly run on top of a seahorse, cruising about by himself in the open sea. We only get a distant rifle shot at him and miss. When in sight of the Norway Islands, it comes on to blow, and a fog comes on as well. Harder it blows and harder, and we have to claw off from the land and face the gale in the open sea. We hove to tonight in sight of the ice and sounded during a lull, or rather took a temperature for we did not dare to let out much line. We saw an Arctic Puffin here today; rather unusual I fancy. It is our regular old form, and after the first few smashes or upsets when leaving the islands, everything has been snug enough. Tonight, as I turn in, there is the old diversion: seas down the companion.

September 3, Wednesday.

Nothing to be done but wait and endure the gale. The wind is very cold, but the surface water very warm. It does look as if the end of the Gulf Stream found its way up here. It is rather poor fun: a heavy gale when you are merely lying to, especially if it is blowing from the direction you least wish, and it is now from the N, the direction most unfavourable to us. When you are tearing along the way you would go, you put up with it well enough. Well it's no use growling and I think, as we stand about the decks in big coats, whitened all over with snow, we are happy enough for all our Arctic appearance, and a pipe certainly is a great solace. Too rough for sounding tonight and not much good to anyone, if we did, as we have but hazy notions where we are and where we may have driven to. We heave to for a great part of the day, but at last have to sail a bit off the ice as that is only a few miles from us and streams of it come drifting down on us. At times, it is blowing with hurricane force as if it could blow no harder. We have only two absurd little patches of canvas on the ship, a close-reefed main topsail and close-reefed foresail, and yet she lies over a great deal. The sea is, however, by no means, as heavy as one would expect, and the terrors of the Average Arctic Gale, as far as our experience goes, and we have had an unusual bout of them, are greatly exaggerated. In fact, under most circumstances these regions are about the best place, bar a harbour, where a ship can experience one. One knows, of course, those pictures of storm-tossed ships on their beam ends among crashing bergs and huge masses of ice and the reality is somewhat different. If the ice be as thick and frequent as generally shown in those pictures, as far as we have seen, it acts like oil on the waters, in fact so deadens the swell, that if a ship, in a gale have a pack on its lee, if it can do so the safest thing for it to do, is to take the pack. There is risk, of course, as the pack may be tight so that it cannot get in far, and the swell sometimes penetrates a long way. Once in, as we were in the straits, it is very comfortable. Even among heavy drift ice, the swell is so much deadened as often to give one a good berth. If the wind blows any way but this one can almost always get under the lee of some projecting point and there ride out the gale in comparative safety and comfort, and when it blows dead off the ice, altho' the detached streams have to be avoided, the small sea that is raised in a heavy gale is a great blessing. There, too, in almost all gales during the season that ships are up here, one has the immense advantage of daylight. The only very dangerous time, I think, would be among wide scattered drift ice, too frequent to be easily steered, not close enough to keep the sea down.

September 4, Thursday.

Blowing hard still, but the barometer is up, and it is clearer. As there is a good deal of ice round us, we stand towards the land. The gale gradually diminishes but a strong breeze holds and we bowl along gaily, 'til we sight the land, about midday. It is a splendid sailing day, tho' cold, and very jolly on deck, tho' there is not much room among the boats which the late gale obliged us to take in board. We get up steam and do from seven to eight knots per hour and make out the Norway Islands and Hakluyts Headland. We were lucky in not having drifted further and can just lay a course for Magdalena Bay. Then the wind gets more aft and we tear along better than ever. The whole of the western land, so abundant in jagged acuminated peaks, is now powdered with snow, and we see that it is fine weather there, from the sunlight on the distant hills. By 6 o'clock, we are steaming into Magdalena Bay, and in another hour are at anchor round a small projecting peninsula near its extremity on the southern side. The small cove formed here is very sheltered and forms a very snug landlocked anchorage. It is called English Cove, and I suppose received its name centuries ago from the English whalers, one of whose bays this was. The bay is now quite free from ice, fast ice that is, tho' there are floating about many hard, transparent lumps of washed ice, fragments of the numerous glaciers. The scenery like all I have seen on the western coast is very striking, high hills of a dark sepia, often almost purple, rising from its shores often almost precipitously, and are composed of a sort of transition rock, and the ravines and gorges between them are filled by glaciers. Doubtless the glaciers have carved them out, but the hills here are so much steep and jagged than those of the North East land, that besides seeming to resist better the levelling action of the glaciers, the effects of weathering on the rocks are not nearly so apparent. Their summit often consists of a series of jagged spikes, or *épines* as they call them in the Vosges, separated by cliffs. Lit by the evening sunlight, those inaccessible needles standing clearly out against the skyline form a fitting tiara to the amphitheatre of hills that encircle the bay, the end isolated as is almost always the case by the icy cliffs of a glacier, the remains of that, that scooped out the bay. Heavy snow is over everything, but from the force of the wind it was in drifts on the mountains, many dark portions of the scars being this bare so that the whole has the effect of a drawing in sepia and white, except indeed the edges of the glaciers showing in striking enough blue and green tints now, as owing to the season fractures are continually being made. Several of these glaciers are hanging, being placed in small folds or hollows of the mountains following in their outline steep and almost precipitous. The largest and most noticeable of these is just above the cove in which we are anchored and looks almost as if in the act of falling, suspended as it seems from 200 to 300 feet above the sea, as unlike the others the edges of these glaciers do not touch the sea. This is the first really safe anchorage we have been in since we left Diana Island in Wüde Bay, that is safe from a heavy sea, and what is worse ice, tho' even here the fragments of the glaciers annoy us, and we have ere the night is out to payout cable several times, under the attacks of some of the larger ones, as these get on a great deal of way, owing to the strong sharp squalls that soon [portend] the fury of another gale. We call at the *Poste Restante* [remainder post, or general delivery], a mere rock washed by the tide, but there are no tidings forthcoming for if there were any, the bottle has been washed away. This mistake was owing to the fact that these rocks appear as islands in the special survey of this bay made by the H.M.S. *Trent* in 1823, and whatever they may have been then they are completely washed over by heavy seas at high tides now. The central portion of the projecting peninsula, which encloses one side of the cove, is high ground, and as on its summit appeared a sort of cavern with a projecting point. We landed and made our way to it, thro' snow in some places already four feet deep. We found however nothing. This is a sort of cemetery and contains a very large number of graves, with the sort of rude wooden coffins [that] we have often met elsewhere. The water of this bay is of a thick rather turbid green and proves saltier than that of the bays on the N coast. Large masses of the main glacier at the end of the bay are continually falling into the water with a tremendous roar, and even at the distance we are from it (1½ to two miles) cause now and then quite a sea. The other glaciers indulge in this amusement to a much smaller extent. Later on in the evening, we see a fox on the peninsula and landing on the neck about 400 yards across cut him off. We then advance and after an exciting hunt shoot him (*Proh pudor* [shame] *vulpiade!!!*). Unfortunately, his coat is spoilt by the bullet, as he was too cunning to give us a shot with small shot. The skin was a sort of light purple on the back and white elsewhere.

September 5, Friday.

A nasty thick morning, blowing and snowing hard, and clouds on the hills that look bad for our chances of photography. At last, however, as we get to work in despair, it clears and for two or three hours. It is lovely and we get a fair series. Then in the evening it comes on to blow, where from, I should not be rash enough to say, as down these deep mountain-embayed fiords it blows any and every way. We get another fox today, a black one. All day long there is at intervals a sound as of distant thunder, and this is due to two causes, one the fragments of glaciers, the other the avalanches of snow and rocks that keep up a perpetual, and at times very distant rumbling among the hills. Well as these rocks seem to weather, they must indeed have a hard time of it, now in the fall; the melted snows of the summer still fill their crevices, and as this water freezes masses are broken off, often hundreds of tons at once, and hurled down the sides into the valleys beneath. Late in the evening, I take the dinghy and go to explore a valley running inland from the left of the main glacier. A very hard and wet pull brings me to the head of the berg and the ice is there so thick that I have some difficulty in getting thro', and a stiffer haul still to get the boat out of the reach of the waves singlehanded. Whilst standing within 50 yards of it, I had the pleasure of watching the immense masses falling from the glaciers from time to time. I now struck up the gorge formed by the intersection of the glacier with the hill on my left. The gale had already had a great effect on the snow, the land being in some places nearly bare and the drifts in others of great depth. The walking was thus excessively laborious, and as bad as any we have ever come across, for although the snow was very soft and thus far higher than the clinging heavy sludge we used to meet, and out of which one could hardly pull one's legs, the extremely uneven distribution of depth caused one to put forward one's foot each time with extreme caution. Nor is the bed of a torrent full of boulders a lively road, when these are so carefully concealed that one's legs got every now and then caught between rocks as in a trap so that my shins soon felt as if I had been playing a football match. There was still a good deal of water in the stream. This issued from under the glacier at the point I lost it by a curious dark tunnel about 20′ wide by four feet [in] height [and] perfectly rectangular. I then took to the hillside. This was by no means bad reindeer ground, some patches being very fertile, chiefly however in moss, about which they don't seem much to care. This bay, however, was once a frequented resort of theirs, but they have apparently, from Norwegian and other accounts, been scared away, or deserted it for some other reason. I saw no traces of any, but foxes both white and black were numerous enough. I did not, however, fire at any, as my bullets spoil their skins. They were out hunting and I disturbed several with their prey—the little purple sandpiper—but what they live on, when these go southward, I can't imagine. I also saw the traces of some very small animal, possible an ermine, the existence of which in Spitzbergen is doubtful [the ermine, or stoat, is found throughout the circumpolar north—with the exceptions of Iceland and Svalbard]. I was now in a long flat valley curving North and West. We had seen indications of this valley and wished to know whether it ran 'round into the South Gat. The scene was now magnificent, dark hills rising on either side to a great height and looking all the more beautiful for their winter coat. Behind me the glacier and beyond that the dark green foaming waters of the fiord. Above and beyond these the hills again, their snowy peaks lit by the crimson tinge of the stormy sunset. I could see some way inland, up the vista of a slow winding glacier, and could see endless peak upon peak and mountain upon mountain of either rock or ice. As I faced up the valley, I met the squalls in all their fury, and the thick clouds of powdered snow, they drove before them were almost building. The mountain tops, indeed, as they were caught by the whirlwind seemed smoking like volcanoes. The valley was flat but very rocky. It was barer than the hillside, and I think of limestone formation. After toiling about two miles up this in as many hours, I found that its northern end was closed by a glacier, and as I should have had to go a long way further I determined to return, for it was now very dark and the hill tops wreathed in clouds, snow was falling, and I thought a heavy storm might if it obliterated any footmarks be dangerous.

Returning was far easier work, dark as it was, but the wind was at my back, and my own footsteps showed me what depth of snow to expect. On reaching the shore, I had some difficulty in getting the dinghy launched thro' the surf, and got a ducking, and the boat nearly stove by the guiding ice. I also nearly succeeded in running on to a sunk rock soon after. When clear of the lee of the glacier, I experienced a cross wind and the squalls that occurred every two or three minutes were very severe. In fact, had it not been so dark I should have hoisted a signal of distress. However, with care I got on board all right after a long bout of it, plus a good deal of sea water as I couldn't do much bailing and pulling at once. The sun was now showing patches of light here and there, it had risen 'ere I turned back, but its rays were hidden by the snowstorm and the hills. Soon after I got on board it came on worse than ever if possible. We had now both our remaining anchors out, and a lot of cable to each. We had, however, dragged, and the ship's stern was but 30 yards from the shore. There was no sea here like in the main fiord that I had crossed, but the wind as powerful as ever. During one squall, we heard a distinct snap and thought the cable had parted. Steam was up and we got up both our anchors with some difficulty, the fluke of the large one was snapped clean off, and the stock of the kedge was broken right in two so that, of our three anchors, one is lost and the other two smashed. A glance at the landlocked cove we are in, in the chart, will show that for such accidents to happen, my statement of the excessive severity of the squalls is no exaggeration. After some time, we come to anchor again (the anchors being tinkered and adjusted as well as circumstances will allow) in rather a different position right under the hanging glacier, and orders are given for steam to be kept up.

September 6, Saturday.

We have again dragged, but slightly however. It is snowing at times and thawing and the squalls are fierce almost beyond belief. Never in my life, except in September 1864, do I remember to have seen it blow harder. Here is the ship in a snugger berth than in nine out of ten harbours, and at times she heels right over down to the sail. Unluckily if our cables part now, even steam could not prevent her going on the rocks of which I have spoken (the *Poste Restante*), as at the entrance to the [ILLEGIBLE] fiord, and over these the sea is breaking wildly. It is, indeed, proposed to run her ashore, as there is not sea enough in the cove to do her much harm. In the afternoon, there is a lull and we go away in the dinghy after some foxes that had been seen on the peninsula. I say a lull, but nevertheless to make the 150 yards to the shore took us an immense time and then we made it at a bad point for cutting the foxes off. *Tallyho* gone away—there goes one across the neck, *tallyho* again as we view another. I run hard to cut him off, but fail, getting however a long shot (we have all brought shotguns); *forrard* on, as tho' not crippled he is pretty well peppered. Two guns and a beater take the hill on the peninsula for the others, and P and I go after the wounded one. Another long shot is fired and the chase maintained by view on footmarks in the snow. The hounds in the shape of Rover are well on him too, we soon hoop him to earth, and Rover is now the terrier. Unfortunately, he funks, until the foxie being fast to the toe of a boot is being drawn out, when he makes a very brave onslaught. "Gone away" again and this time a white fox breaks across the neck, without giving a shot, as our forces are in no regular order by this time. One fox, a black one, still remains and gives a most exciting hunt, to the guns on the peninsula, who go rushing about and tumbling in the deep snow in the most amusing manner. Presently the Captain gets a shot and misses. The poor bewildered beast makes straight at him and nothing but the difficulties of the safety spring and the Captain's inability to get down that left "do-head" can give him a chance. Three times he tries to discharge his piece and fails. The poor beast now closely hemmed in makes a brave rush to break the line and the beater actually hits him with his cap. He is now crossing the neck and has still to run the gauntlet of Potter and myself. He

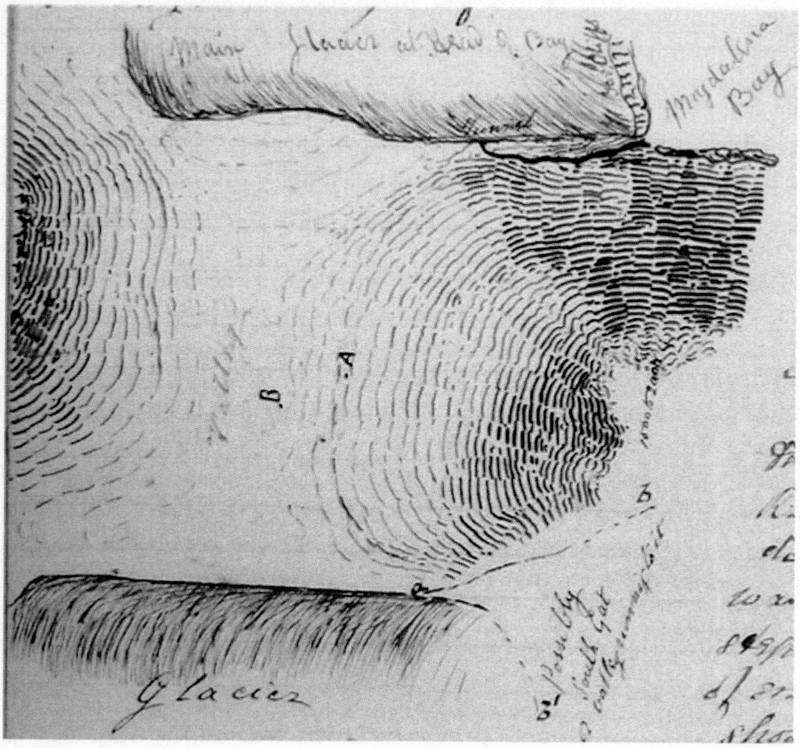

Sketch 9.8 South Gat. I first reached the point A and thence took down the hill to B and finding myself unable from either to see beyond the point, gave it up. From the configuration of the ground, the valley is probably closed by the glacier in the line a-b, but it is possible it might run as a-b' and the valley lead thence into the South Gat

is very cunning and gives but a long shot, enough however to stop his going. The dog, however, funks and tho' he has the legs of the wounded fox allows him to escape among the rocks on the hillside, spoiling unluckily our chance of finishing him, tho' he deserved his share of a cartridge. The result of our *chasse* was a very pretty nearly white fox. The squalls soon came on again and heavy rain fell during a great part of the day. This soon melted much of the snow and we then saw what a number of graves there were on the top of the peninsula. They were almost all rude wooden coffins filled in with stones and numbered, I should think, nearly a hundred. There is a famous "Rotchie" hill in this bay and one can imagine the din in summer. The *Sampson* when here this year in June killed endless numbers of them. There was then fast ice in the bay and there were a good many seahorse in it, of which she killed five. The birds have now nearly all gone except the dovekies who are assuming their winter plumage, and a few broods of eiders. We go in pursuit of these in the evening, but they swim away from us, making more way in the sea that is running off the point than we can. We again land and soon view a fox and, after a long hunt, find his whereabouts by tracking him. Then we get a view and, getting 'round him, he is driven and gives an easy shot. His skin is one of these blue white ones and perhaps the handsomest we have got. Tonight, the weather is still rainy and very coarse, but shows signs of mending.

September 7, Sunday.

It is a fine bright morning and the "early bird" is up and doing, in the shape of a "smackie" bound S that has passed the bay. We go on shore to leave a bottle and, about 9 a.m., anchor is up and we are underway. When we get outside it is lovely weather, strong puffs of a shifty wind blowing at times. It is determined now to make S for Ice Fiord. We have indeed just before the late gale been right up to the pack edge, further N and further West than at the time we suspected. Nor 'til our eight hours steam and sail to approach the Hakluyt's Headland did we know it so that the idea or hope of finding a slack pack rather to the W and N of Spitzbergen seems at an end. The late gales would, in all probability, have done little towards shifting the ice materially. Indeed, from the sky, it is doubtful whether it is not nearer the headland than before, and open water to the N can do but little good to us now, except to enable us to attain a high latitude. It is to the E or NE from the Seven Islands that we wish to penetrate. Open water does often exist here, and was in sight on August 30th from Phipps island, an impracticable pack however preventing us from reaching it. Our attempt, indeed, ended in the short besetment a few days previously that I have described. This open water is possibly formed annually under the lee or west coast of unknown land, in the same way, as that on the west coast of Spitzbergen, which is supposed to be due to ocean currents, and the direction that the centrifugal force of the globe impels them to take, I think, this may be taken rather as a fact than a supposition. Should other land and similar currents exist N and E of Spitzbergen, it by no means follows, as is assumed by English Arctic explorers, that this is an impracticable, or at any rate not the best route for Arctic exploration. Now, however, from the lateness of the season and appearance of the ice, little change can be expected to the E, nor unless the northern pack had a good offing of 40 to 50 miles from the land, would a fair chance of retreat be open. It is by no means certain that we shall not have another look at this, but it is considered advisable first to keep rendezvous with our consort in Ice Sound. So, we steam and sail south, along the coast and past the wondrous Seven Ice Mountains. This was the first part of Spitzbergen that we saw, and yet it seems striking as ever. It is very characteristic of the country and perhaps the finest part of the coast to visit for any one merely wishing to get an idea of the country, and by far the safest. That is equally true of course of all the western coast as far as ice is concerned, but here you can approach it without danger and coast its shores, whereas south of the Foreland, the sunken rocks and reefs are numerous and intricate, and the entrances to the harbours difficult of navigation. It is a soft warm sunny day and very clear and of ice nothing but here and there at very wide intervals a solitary floating block, just to show one is in Arctic regions. Were it not for the very heavy swell from the W, it would be quite a "ladies" yachting day and I know several ladies quite good enough sailors to enjoy the scene in all its glory. Presently we are abreast of Kings Bay, and we can see the distant "Three Crowns" (alas, we never got a photo, we didn't then understand the working of chemicals in these regions) rearing up their snow-crowned summits. Then we come off Prince Charlie's Foreland, a long narrow mountainous island about 50 miles long. We determine to take the outside passage, as from the narrow width of the safe channel inside, more time would be taken for the passage. A ridge of mountains runs along this for about 30 miles and is then succeeded by lowland connecting them with three more mountains at its lower extremity. Extremely deceptive this lowland is, as the end of the range above it seems the end of the island and a bay running in here, one might in less clear weather easily run a ship right on to it. The bay moreover contains several rocky islets, but the water is deep close in shore. The mountain range in its centre rises into three lofty peaks said to be 6,000 to 7,000 ft high, but I think not really above 4,500 [Chermside is here correct: the highest, Monacofjellet, is 3,456′ (1,081 m)], tho' from the distance of the observations I took any slight error would greatly influence the height. This is said to be the only regular mountain range in Spitzbergen by an eminent Swedish geologist, but this we think must arise from an erroneous understanding on his part, of the usual acceptation of the word "range," and possibly to his not having personally visited the neighbourhood of Kings Bay. We passed the smackie (inshore of us) 'ere we reached the foreland. All day we continued steaming down this, the wind being paltry, but a very long heavy sea running. This perhaps accounted for our not seeing bottlenose whales as we expected. There was little or no snow on any but the highest hills of the foreland. These are very dark and barren, but there are a great many long wide valleys running up and plains in places between the hills and the sea. These are very green and extremely fertile looking, but their beautiful colour is due to luxuriant moss, for which reindeer don't seem to care, as the owner on landing last year saw none or traces of any and only got a fox. It is, however, excellent ground for geese, who frequent these mossy places in great numbers, but are as difficult to shoot here as elsewhere, having as a rule, a sentry posted on higher ground than the rest keeping a look out. In fact, the only way we have found at all successful is to put them up at once, one gun then to lie down on or near the ground they have been feeding on, the other if they have settled near to again flush them and then lie down on that ground, one or other gun in this way usually gets a shot, and with No. 1 they are not hard birds to bring down at a long shot. Their feeding grounds may be known by numerous little holes bored in the ground by their bills, which at first, we thought due to grubs or worms, or possibly lemmings. They breed far inland and when flying about these grounds in early summer can be tolerably easily killed. They are then, however, hardly worth the killing and carrying. It is now dark enough at night, but tonight very clear and there is a beautiful sunset, and ere we turn in an equally fine sunrise to the NE over the dark hills and glaciers of the interior.

September 8, Monday.

We are by 8 a.m. off Dodmansoren (Dead Man's Ears) [Lexfjellet], the northern headland of the large bay called Ice Fiord or Ice Sound. Like several of the bays in this land, it is at no great distance from its entrance divided into two more great bays. Some of these again divided. These fiords show the beds of ancient glaciers and are usually of very even width. This sound runs far into the heart of the country, its northern arm Dickson Bay [Dicksonfjorden] being separated by only ten miles of mountain land from West Fiord [Vestfjorden] in Wüde Bay. The other arm ending in Sassen Bay [Sassenfjorden], also runs 50 or 60 miles into the country, and is only distant 15 or 20 miles from the shores of Stor Fiord [Storfjorden], a long valley, according to Norwegian accounts, running right across to Agardh's Bay [Agardhbukta]. We have not made as good progress as we expected, as owing to a strong northerly current, our patent log differs by five knots this morning, in the course along the Foreland with that we have made along the land. There is but little snow on any but the distant hills and all the land we see as we enter the bay is very mossy and fertile. The hills, indeed, are softer, rounder, and mostly of insignificant height. Not a piece of ice to be seen, so were it not for a distant glacier or two, there is but little of an Arctic scene to strike one. It appears, indeed, to us to strike one as far more homely and less bleak and savage than the dark mountains and rocks of Shetland. Green Harbour [Grønfjorden], the first bay on the southern side, is our destination and we soon make out the *Sampson* lying snugly at anchor about halfway down it. We steam up to her and before we can do likewise have to beg an anchor, and this on board we anchor close to her on the western side of the bay. Hearing from the *Sampson* that cod are plentiful here we at once proceed to rig lines. Those that I made first were on the same principle as those of the Norwegian fishermen, with two hooks and a crossbar, however, instead of a single hook, as I did not expect to fish in very deep water. The gear consists of strong line (of this we laid in a stock at Shetland) about as thick as an ordinary sash cord. At the end of this an iron crossbar about one foot long, with a sinker attached, as when the fish are being taken quickly it is of great importance to get the line to sink quick. To each eye of the crossbar is attached a snood, or short length of finer line, with a large cod hook a foot or so from the bar. On this usually works a small lead fish, and as a substitute we used bits of old lead pipe. These can move up and down on the hook and line. It is well to scrape and brighten them continually. Overboard it goes and I commence jigging in the same way as at the herring fishery, tug-tug and in less than a minute I haul up two small cod. Another line is soon rigged and by lunchtime we have a dozen or two. There are some salmon lakes near here, and at their outlet, during the owners stay in the fiord last year, the Norwegians fished with great success, with only a few fathoms of net. This outlet lies beyond the western promontory of the bay, right out of the harbour, but it is determined to explore one of the valleys running inland from the harbour, as an easier way of reaching it may exist. The owner, P, and I land for this purpose, and we soon see the aptness of the name Green Harbour, the terraced slopes rising from the shore, being deep in luxuriant moss, and these fertile slopes embrace the shores of the harbour on both sides. The hills around the harbour are curious, as everything is in straight lines, nature by no means rejoicing in her usual curves. These hills are mere prisms with crests almost as level and regular as geometrical solids. They look indeed like huge earthworks, one edge across the valley rises above it, its end being as it where the profile of a properly finished traverse of a fortification and it runs some miles with a regular top and almost equally regular section. How it finishes at the other end I cannot say, but as seen from this side it appears thus (see Sketches 9.9 and 9.10).

To the left of the valley lay another mass like a small prism on top of a larger crest lines of each, so lovely and regular as to appear unnatural. Further inland, mountains of the same ugly geometrical forms were surmounted with pyramids; some of course like those of the three crowns, rather irregular, but two especially noticeable, of almost regular tetrahedral form. We soon left the first valley as it evidently did not lead thro' to another and took across a spur to the next. There was here on the far or southern side an exposed face of high cliff, which in a geological light is extremely interesting. The strata have here been tremendously distorted. At first, they are inclined to the vertical at an angle of 30°. They then bend as shown in the section. These strata here disappear into the ground. Finding that the valley we are in soon narrows and is hemmed in beyond by mountains, we take over a spur of its southern bounding hill and get into another running E and W. The soil of the lower part of this valley is very deep dark-coloured and soft. The gullies of the water-courses often having channels over 50 feet deep in it, the crossing of which is rather laborious, in two or three miles we come upon a lake. Frozen over in the late frost with ice about half an inch thick, this has been broken up by the gales and is piled up on the northern shore in layers of flat plates often two or three feet deep. Beyond this is a second collar separating this from a valley beyond, and as once in Wüde Bay we find here a lake with two outlets, balanced as it were on the ridge between the watercourses of two valleys, and it is hard to tell by which outlet it would discharge, were its surface a few feet higher. We cross into the valley below which are a chain of four lakes, and we walk seawards along the largest some four miles in length. There are curious hills on both sides of this valley, some apparently on the eastern side, composed of sandy shale, their sides steep and slipper and the surface so smooth as to be hardly broken by a boulder here and there. We here lose traces of fossils, of which hitherto we have seen an endless number, tho' from want of a hammer we did but little collecting. Beyond the shaly hills on the same side comes the rugged grooving of strata exposed and showing a very wavy line, on the precipitous face of the cliffs; these faces having a very uneven surface, some parts of the scarped parts of the hills projecting a good deal on the far or western side beyond the lake, the steep slopes run upwards being ribbed with ridge and furrow almost as regular as those of a ploughed field, tho' the furrows of course converge. A deeply excavated winding gully leads from the lake to the sea and it is at the mouth of this that the Norwegians have had several successful catches of salmon. We take homewards along the rising land above the sea, passing under some steep cliffs, evidently a malley head. The rocks of these cliffs are highly coloured with mosses and lichens, giving them a picturesque appearance. There is but little snow about them, but along the slopes of it there are a bright red colour (like cochineal) being covered with the minute plants (diatomasae) that grow on them in the fall of the year. This indeed has been the case wherever we have been for the last month. On the point or extreme promontory of the harbour we see the ruined skeleton of a house. Near here, we come across a fine white fox. He is also aware of us. Presently we make out two men, and these afterwards proved to be Eaton and one of the

Sketches 9.9 and 9.10 Gronfjorden

engineers. At this point, about six miles from the ship, they found the ruins of a stone house and near it on the beach the remains of a seal club, of a pattern not now in use and leaning against the house a pair of snow shoes, which they brought along with them. On an open bit of plain near here they also found human remains, unburied, lying perfectly exposed, possibly one of the Russian hunters who used to winter here years ago. Bits of his clothes remained and by him lay his old-fashioned powder horn. Among the remains, they found the rosary that I suppose lay on his breast. It is of metal and contains, in all, five enamelled plates of saints &c, and is in very good preservation. With it they brought away some of his beads. The few deer that have been here have been chased away by the numerous visitors to the fiord and we saw nothing but a few geese, and our walk became very tedious as we were nearly famished, having walked over 16 miles of rough ground ere we reached the ship. A boat has been away fishing as the best ground lies about ½ of a mile off, opposite a run of water and we have by now half a cask of cod. Some of us again go fishing tonight and have capital sport.

September 9, Tuesday.

It is a cold thick morning and no photography can be done, and we must lie here today as it is too thick to venture outside. Two smacks who we saw standing out from Coal Harbour [Colesbukta] last night have put back. The owner and I spend the morning on board the *Sampson* and I may as well here record her doings. She remained in Loome Bay for two or three days after our departure, as some seahorse appeared, and of these she captured three, completely cutting out the smackies who got none. She then left the straits but, like ourselves, found the ice very thick. During the heavy gale, she was driven into Treurenberg Bay, and this was so full of ice that she had a very rough time of it, and was nearly driven ashore. By stealing alongside of the land, and in the narrow strip of water between it and the ice, she managed to get 'round Verlegen Hook. Off this, she discovered sunken rocks in a very shoal water. Of these, she took soundings and got clear without any damage. Her progress hence along the N coast was very slow. Owing to the ice, Wüde Bay was not open, the mouth being completely jammed with it. She crept close along shore after passing the mouth of Liefde

Bay, which she was afraid to enter, lest it might prove a complete trap. When at last she got to the Norway Islands the ice led her away from the land a bit and she rounded Hakluyt's Headland, and sailed thence into Magdalena Bay, here she found no remains of ice, and no seahorse, but got two ground seals, one by blinding it with a shotgun, and then getting in a harpoon. Up the valley that I explored the other day, they trapped two foxes (black) and there were also the tracks of what Captain Walker asserts to be a wolf, but they could not trap it. I saw the footmarks on the night I was there, as fresh as those of the foxes I disturbed feeding. From Magdalena Bay, the *Sampson* has brought a fine collection of shells &c, as she has been persistently dredging and travelling since we parted. Thence she sailed for a cruise off the Foreland hoping to fall in with the bottlenose. Of these, she saw none, and found a continual heavy sea running. Off the Foreland, she spoke a schooner, the captain and crew of these were Norwegian, but she had on board (so they say) a Spanish party of exploration. They informed the latter of the state of the ice and they have never I fancy been along the N Coast, but it is possible that it is they who robbed our Post Office, as one was left by the *Sampson* not on the tide-washed rock, but in a position where had it remained there still, we should have found it. From the Foreland, she had sailed down to her present harbour. Neither here or elsewhere in the Fiord had she seen bottlenoses, tho' there were lots of herring whales, and altho' she chased these almost daily could not get a shot. With the cod, she had been luckier, having taken in all nearly three tons. She has on shore a small tent for smoke drying, which we inspected. She has killed but few deer as from the numerous hunting parties at this part of the sound they are very scarce. One that they got has to be carried for eight hours. It may (from accounts) be said that the deer, tho' numerous in the land about the shores of the fiord, have almost always to be carried long distances as they are chiefly to be met with in the inland valleys. Sassen Fiord, the extremity of the southern arm of the whole sound, is a very good place, and it is probably to this that an English yacht who has been uprooted as seen in these waters was bound. If so, she is doubtless Captain Wade Browne's craft, as the owner recommended him to try Sassen and Advent Bays [Adventfjorden] in this fiord, as places likely to give good sport, and that can be visited in an unprotected vessel without the slightest danger of ice. The *Sampson* besides the cod fishery has been very busy in dredging and trawling and has got several curious species of fish by the latter process, some like one, that we got in Magdalena Bay, more like a "Poggie" or "Bullhead" than anything else, another a lump of fish. This cod fishery is a great success. It has always been said that the fish in the Spitzbergen waters were very scarce, and here are cod in abundance, almost sufficient for a company to start a fishery. Several expeditions have been made to Coal Bay by the *Sampson's* people to hunt deer &c. On one journey up the valley opposite our anchorage they came across a very fine specimen of the snowy owl, but having only rifles, and from the common occurrence of this bird in the straits, took no trouble to secure the specimen. There have been several smacks in the neighbourhood or boats crews left by them to fish nets, boats, tents, and provisions being left for them. These are now all gone with the exception of one crew between this and Coal Harbour. The smack these men belong to has not yet put in an appearance, but she is doubtless the one we passed off the Foreland. These men are in a very bad way for want of provisions stores &c and have done no good in the fishery (white whale). But for their speedy prospect of relief, the owner was for sending off to them today. Another smack was anchored off the outlet of water from the lakes and caught abundance of salmon, with a stop-net stretched for 24 h across the mouth of the river, on to which they say the salmon wanting to home call in here, so that they have frequently heard of us, mostly yarns. The Swedes, too, put in here, and left a letter asking us if we went there again to bring their boat the *Annie* from the NE corner of Phipps Island. This must be, I think, a mistake, as we explored that [place]. They mean, I expect, Martin's Island. It is decided to give up trying for salmon, as the place I have spoken of is a ten-mile pull, and as we should have to be away some time the men would be very tired, and we hope to be under way again tomorrow. So, it is hardly worth it, as we should be unable to leave the net down for 24 h in the Norwegian fashion and the lake itself is a long carry for a net and not very favourable for dragging without a boat. We could of course drag at the mouth of the river, but even for that we don't know the right times of tide, and as a rule everything depends on local knowledge, these sort of proceedings not usually succeeding at first. The weather besides is so unsettled that the boat might be easily weatherbound and a good deal of delay caused. The Norwegian boats crew of whom I spoke as living across the bay, indeed, got their boat so damaged in a gale the other day that it is useless. There are several runs of water nearer but as a successful haul, straight away out of hand, is so problematical; it is not considered worth wetting the net, as it is a great trouble drying it. We continue the fishery for cod, all the afternoon and evening with great success, getting fish from one to ten lbs in weight, often getting them as fast as we can haul. We have now several lines rigged on the *Sampson's* model, three large hooks bound together like a grapnel, and lead, something in the shape of a fish run over their shanks. This one scrapes bright with a knife now and again. A heavy sinker and over it goes. We row or drift slowly until we come to a good spot, and then when the fish begin to take anchor, usually in 20 to 30 fathoms of water. The line is perpetually jerked up and down, a little hauled in and let go, the fish attracted by the lead swim round it, and then a sharp pull of the grapnel strikes in one or more hooks. Then you have to haul up and tremendous work it is perpetually jigging the line and then hauling in 30 fathoms with a heavy sinker and an unwilling fish at that end. Sometimes, if only lightly hooked, he gets off on the way up and then all your labour is gone. With the grapnel hooks the fish may be hooked anywhere, hardly ever in the mouth and when hooked near the tail pull very hard. It is a very deadly way of fishing and is prohibited on some parts of the Norwegian coast. With the hooks the way I rigged them, the fish are usually caught by the mouth, and strangely enough though it may be mere chance when these are luckiest the grapnels do nothing and vice versa. I should think we caught nearly a couple of hundred today, some very fine ones, tho' it is exceptional to get them over 10 lbs. We turn in expecting to be roused by the sound of the propeller and the cheery song of the men at the capstan.

September 10, Wednesday.

Heavy snow and sleet, a gale, and a fog, so we can't budge this morning. Weatherbound, in fact, a very unusual thing for us, as the little *Diana* is not as a rule very particular what weather she goes out in. But navigating this part of the coast in thick weather is out of the question owing to the rocks so with steam up we wait for a change. It comes, in the shape of a heavy rain, everything else as before. We have room for more water, but the weather is too bad for the boats to work. Some trawling is done today and we get a few of the smaller fish of various sorts and great numbers of large shrimps. These have a hard shell-like crabs and are rather larger than prawns, they prove very good eating. We also get a fine hermit crab, the first we have found here. The whole of the land around this large fiord is geologically a subject of great interest. Here in this harbour, within a mile or two of the ship, you come across three formations. The land rising from the shore being tertiary, thin veins of dark and variously-coloured shale running at intervals towards the hills, these beds being nearly vertical. Above this, you come to Jurassic formation, and above this again to mountain limestone of which the hills above the lake are formed. In the valley beyond, you meet with various sandy and argillaceous shales and slates, and we also found numerous blocks of sandstone. Fossils, chiefly of marine plants and animals, are very numerous. There is, however, at the northwest promontory of the bay a curious vein ending in an isolated rock rising like a wall from the sea. This vein of slaty laminated shale is very thin and appears projecting from the ground at an angle, and in this, tho' rare, are to be found fossil forms. The Swedes, indeed, who have spent much time in this fiord, have found here other vegetable land fossils, one a sycamore leaf. On the other side of the bay, at the point below the hills, the formation is tertiary and near the point is a seam of coal five feet thick. Of this, two boatloads have been brought away. Although the outcrop it burns well —tho' it contains a good deal of sulphur—it leaves no cinder only a white ash. This coal is lignite and is found in many places along these shores. I forgot to say that crowning the limestone hills, we came across our old friend of the mountain, Hyperite. In the afternoon, I went fishing in the dinghy and had a great take, and later on the casey was away but was not so successful tho' she got about 40. We also caught a dozen or so from the ship tonight. A plant of the _____ order has been found here by Eaton; it is, I believe, new to Spitzbergen. We, at any rate, have not met with it before. Rather before midnight a smack arrives, the one we expected, and anchors close to the *Sampson*. She has had a hard time of it in a southwester outside, but the wind is now shifting to the N.W. She is informed of the whereabouts of her boat crew and intends fetching them tomorrow. She is from Smeerenburg and like nearly all the rest has had a bad voyage, only getting seven whitefish. This has been a disastrous season for the Norse fishers; Mack who got the buried casks, the steamer *Sofia*, and one other, being apparently the only three exceptions. Next year they will, I fancy, like him make a spring voyage for seals up the western ice, and come on here, as their boats are left lying here as elsewhere.

September 11, Thursday.

A clear morning and a hard frost, the decks and everything covered with frozen snow, and the hills of late so dark one rolling sheet of white. We get a hasty photo, and by 11 a.m., we are under way, the smackie also following as she is going to pick up her crew. The *Sampson* is now all snug for the rough weather on the journey home, her main topmast housed, and her topgallant yard stowed. She is to visit Bel Sound [Bellsund] and dredge &c, if practicable, and then to bear up, taking soundings and temperatures whenever practicable. We are now bound S for a try round the South Cape [Sørkapp] and to the eastward, as there is a possibility of reaching Kong Karls Land or doing something in that direction, tho' I believe as a rule, the block of ice in those waters is much more rarely clear than on the other side. When we get outside about 12.30 p.m., it is blowing fresh from the N.W. and we lie a course for the South Cape. There is rather a heavy sea, but a beam wind is this ship's best sailing point, and with all available canvas bent, and steam knocked off, we get along faster than I have ever seen her before. She is, however, very tender and, by the p.m., rolls and strains so occasionally that we have to shorten sail a bit. It is trying to be a fine day but every now and then we get enveloped in the blinding darkness of heavy snow showers. These would pass and then we got a fine view of the land until another came. All the land is as before, high and jagged and above Bel Sound, there are some very prominent peaks. This, like Ice Sound, is divided near its entrance into two bays. Across the entrance of the northern one stretches Axel Island [Akseløya], which bids fair soon to close it, the smackies even failing to negotiate the narrow passage except in boats. We saw a finner today. The sea continues quite free of ice and will do so, I expect, except at the mouths of the bays, until February, unless the winds drive in the pack on the land which, I fancy, is of very rare occurrence.

September 12, Friday.

Early this morning we are off Horn Sound [Hornsund]; see the high peak that rises above it to the S. We now fall in with a very strong current and for nearly two hours, the bearings of the land hardly alter at all. At last, however, we seem to be making way and by noon are abreast of the hills rising above the South Cape of Spitzbergen, the cape itself not being very clearly visible as it is the extremity of a low flat island lying off the mainland. We give it a pretty wide berth as it is, I believe, foul ground. It is a lovely day, fairer winds and weather we could not have if we chose them. The former as we alter our course eastward is shifting and is soon blowing from the S.W. We keep on all day sailing across the mouth of Stor Fiord [Storfjorden], up the western side of which we get a view past the Hedgehog's mountain [Hedgehogfjellet], and so far it looks unpromisingly barren, no plains or fertile valleys, piles of rock for hills and mountains the rest glacier. We see far more loons and rotges than we have of late in the water here, and still not a morsel of ice. The stream of ice that pours from here in the spring does not seem continuous in the summer and fall, like that of the western ice. Early in the year, it stretches away from some point between Nova Zemlya and the North Cape of Norway, embracing Bear or Cherry Island [Bjornøya], and running on W until it unites with the western pack. After the breach between these two packs effected by the warm stream running N from the coast of Norway, it still continues to form a discharging stream of ice and is called the S.E. pack. Now, however, tho' the pack usually remains in a pretty well-defined condition, the stream of ice no longer pours south and this, combined with the reputed lightness of the ice, of which we shall soon judge for ourselves, seems to argue land to the E. We see a bottlenose whale this evening, but he goes down at once, and does not as is often the case come and play around the ships bows. Sunset at a quarter to eight, the nights seem so long and dark, after the liberal allowance of daylight in an Arctic summer.

September 13, Saturday.

Early this morning the breeze fell, and the sea got very calm. By 8 a.m., we are steaming along between Hope Island [Hopen] and Stans Foreland [Edgeøya], the SE point of Spitzbergen, and called generally by the English, Edge Island, owing to some disputed claim of discovery. Between us and it lie a dangerous and very numerous groups of small islands called the Thousand Islands [Tusenøyane], and it was either near this eastern promontory of Stor Fiord or further up it at Tymens Straits [Freemansundet] that the *Diana* was driven ashore by the ice in 1871. The high dark ridge of Hope or Sea Horse Island lies to the S of us, and we keep NE across calm open water for some hours. About 3 p.m., we come into water filled with numerous grounded icebergs, or as the charts say frozen islands. In another hour or so, besides these being very frequent, the whole of the sea was covered with a thin coating of ice, "pancake" ice as it is called from its appearance. This is the new ice of the season. Owing to the motion of the water, it is not one level sheet, but a number of various shaped flat cakes. These overlap one another as if a table were covered with leaves of paper, completely coating the surface of the sea. At first, this new ice is but thin and for some hours we steam gaily through it. Besides the bergs, we have now come across older ice either streams or the pack, but by keeping E a little we keep clear of it, until it is dark, when we take the pack, and knock off for the night.

September 14, Sunday.

During the night it has frozen pretty sharp. The bay ice cements the pack together pretty closely. We are still just in sight of Stans Foreland and the ice there rests against the land some way further south than here, and seems also closer pack. To the SE, then, there is also pack, but possibly only a point end. Kong Karls Land is not discernible with any certainty from aloft, but we are in sight of the Ryk Ys or Witch Islands [Ryke Yseøyane]. Our only chance now seems eastward, so forcing our way out, we try in that direction. Two or three-hour steaming serves to show this is of no use as the ice continually runs us southward and we double back, and again press on in the old direction. Last night was one of the most glorious we have seen in these regions, calm and clear and a beautiful setting sun over Stans Foreland and the long icy ridge, called King John's Glacier [Kong Johans Bre], and behind the ship as she sped across the calm frozen sea, lay a long glittering track of water just her own width, stretching away like a dazzling mirror as far as the eye could see. Today, the track closes less than ¼ of a mile behind the ship and by 10.30, the bay ice is so thick that she can hardly push through it. We recognise after a consultation that we are baffled again and by a more hopeless enemy, an unusually early frosty season. This is very annoying because totally unexpected, and a calamity, preventing our seizing almost the only opportunity of penetrating the ice, we have seen. This is by no means, barring the bergs, very heavy pack, and what is more is not at this point closely packed. From us towards our goal extends a pack, that though impracticable for a sailing vessel is not so far for a steamer like this. As far as can be seen from the masthead progress, possibly very slow, is possible, and beyond this there is indication of water; it may be only a lane or pool, but still water. And now our advance is arrested by this new ice, so thick has it become that it is a question if further progress, through the closer parts of the pack be possible, and even if such be the case, common prudence forbids the attempt. Should the present frost and fine weather brought in at the recent change of the moon, and therefore likely to continue, hold for two or three days more, retreat would be impossible, and it is questionable, whether with 15 miles or more of ice between us and the open sea any gale later in the season would liberate us. The ice is here from two to three inches thick. It would indeed hardly bear me, but some of the men were on it last night. A snowstorm now would so thicken it, as it freezes into a sort of compound mass, that we should have hard work even from here, and from the interior of a pack cemented by young ice, should find it an impossibility. After getting our meridian observation, we are again under way and steer for Hope Island. After leaving the edge of the frozen sea, we come across at intervals for miles, smooth oily-looking unruffled patches of water and this proves to be water on which the ice crystals are just forming. Before leaving the ice, we shot a large ground seal from the ship. He lay well owing to the strong sun, but we unluckily lost him. He was one of the largest sized seals and close on the water's edge as they always are. It is useless, of course, and this is where tyros [novices] usually fail and many old hands too, from the tempting marks their huge bodies offer, to shoot them anywhere but in the head, which is very small compared to their size, but the skull of which unlike that of the walrus is very thin. Well, this one kept putting his head up and down as they do, sleeping about ten seconds at a time and then looking round, and the first two bullets both hit him in the head; of course, instant death, the beast rolling over on his back at the second, hanging just over the water. Unluckily by now a third bullet is fired into his body and its consequence a spasmodic muscular twitch causes him to slide slowly off, and 'ere a harpoon can be put in, we see our prize slowly sinking thro' the clear water, far out of reach. This spasmodic twitching is very noticeable in seals and seahorse and causes the sailors to say when they are flensing them even an hour or two after death that they are still alive. We take several soundings but 90 fathoms are the ones we have met with at greatest depth in these seas. It is dark 'ere we reach Hope Island, and after some time we come to anchor in 14 fathoms, in a very open roadstead, the bottom being merely flat rock, on which the anchor drags about without any hold. We are at anchor on the E side near the north-eastern end, in a sea, free from ice, bergs excepted. These are not very numerous and mostly aground.

September 15, Monday.

We all rowed ashore this morning, a distance of about two miles. There was a strip of beach beneath the cliffs on which the boat was hauled up. We landed near a gully about four miles from the N.E. point of the island. But little is known of this island, and until in 1871 its position was approximately fixed by Lieutenant Weyprecht of the Austrian Navy, its longitude was incorrect and it's latitude over 40 miles to the bad. It is described in the chart as quite barren and has a completely fancy shape. The sea beach on which we landed is only a narrow strip of shingle 50 to 100 yards wide, and above it rises the steep walls of the ridge that forms the island. This beach has been apparently pushed up by the ice, and does not bear the same traces of upheaval, that so many we have met with do. It is covered with immense quantities of driftwood and we also came across several very large jawbones of whales. On searching the collar at the dip between the two hills, we found the island here about 1,200 yards across. This first dip in the ridge is distant three miles from its north-eastern extremity. We here came across numerous bear tracks and the skeleton of one luckless brunie but none of us saw one during the day. On either side of this collar or saddle, itself only 50′ feet above the sea, the ground rises and we here separated on our respective pursuits. I took northwards for surveying purposes, and soon gained the long and rolling sort of table land. This varies from 500 to 800 feet above the sea, rising in some points to nearly 1,000 feet above it. The sides of the ridge towards the sea on both sides are very steep, with rugged grooved gullies seamed and eaten into the cliffs by the streams of melted snow, here and there. There is a narrow strip of shingle below, but often the rugged scaurs rise precipitously from the water, the whole escarpment being in few places accessible from below. The ground on the top is level and now hard frozen, and at the northern part of the island is very mossy, and evidently from the countless old nests, eggs, and feathers a great breeding place, for loons, dovekies, geese, burgies, and even ducks, tho' now all but the burgies, kitties, and mallies are gone, except indeed very large flocks of snow bunting assembling possibly for an intended migration. As I walk from point to point of observation, I am aware of numerous white foxes, plainly discernible against the dark moss, at long distances on this smooth unbroken ground. "Bonnie beasties" they are, skipping about like great Angora cats, and quite as clever as them in catching the poor snow buntings, but very annoying they are when they catch sight of one. For it is not lively whilst taking observations, to have four or more foxes sitting at a very respectful distance on their haunches, keeping up a most discordant concert by howling and yelling at one with all their might. I did for one though, after finishing at one point. I dropped on the ground, and crawled ignominiously to a brow that concealed me, peep over, there he is sure enough about 200 yds off, howling at nothing, and I stealthily bring my rifle up for a steady aim. Not good enough for he sees me and is off. Presently, as I am setting up my stand at another point, yap-yap from the edge of the cliff, I run to the edge, and am greeted by a defiant yap some way on and also down the cliff. I run along, look over and see nothing, foxie having meanwhile run just the opposite way, on a lower ledge. Unfortunately for him he gives a triumphant bark at this and goes on but, just as he has to jump a gap, I view him and a snapshot sends him over the cliffs. By getting along the face I find he is on a strip of shingle not in the water, but tho' a bonnie white fox, the skin after one of my "express" bullets is not worth a five- or six-mile tramp so I leave him to the burgies. The end of the island is crescent-shaped and only 500 yards broad, cliffs of a dark slate-coloured shale 250 feet above the sea forming the horns and having the W one at least a sharp razor edge, with not even footing on the top. Below is a beach some 250 yds wide, a mausoleum for seahorse, but none have been recently killed there, as a whaleboat that visits it from the ship reports. Even on this high land, we find brunie tracks. We, between us, explored the island for about ten miles of its length as far as the second gully or collar, beyond which rises a curious sugarloaf hill of which we took a photograph. The island is about 20 miles long and at its SW extremity rises to a height of 1,400-odd feet. The main ridge is divided into five by these collars, always low and if no great width, and in no part is it over 2,000 yards wide. The cliffs are all of stratified rock, the strata dipping NE at angles varying from 8° to 4½° and the hills are composed of shale, we rather think of tertiary formation, tho' we found no fossils, or only very questionable ones. Not a trace of deer any of us see. On the beach, we found endless dried sponges, cup or funnel shaped, the cup being hollow and a stem projecting downwards from the apex of the inverted cone and forming the stem or root that attaches the sponge to the stone or shells. The water around the island, especially on the western side, was shallow and of uneven depth and numerous sunk rocks occur. The bottom is merely a surface of rock, sloping ledges of which at times almost reach the surface, so that in sounding you get an abrupt change from one to 13 fathoms. Plants are scanty on the islands, but besides the moss already mentioned, which is not so abundant in the centre of the island, there is a great deal of Iceland moss. It was very clear at times in some directions in a rather peculiar way: from the northern extremity which is nearest to it, Stans Foreland was but dimly visible, while at the same time from the centre of the island, it was clearly visible and the distant South Cape as well, the ice to the E however being better discernible from the former position, its whole extent indeed being clearly mapped out. It extends from the E coast of Stans Foreland eastward, embracing the Ryk Ys islands. The hue forms a curve running NE from its starting point, and it is the northernmost bights or recesses of this we have been exploring, and thence it sweeps round S so that in a direction nearly due S from the island we can see a projecting point end of it. This accounts for the smooth water and the formation of the young ice, as we have been working in a large sheltered ice bay, instead of in open sea, as we imagined. The southerly trend of the pack is very annoying as our next move was to have been eastward, just to inspect the lie of this pack of which but little is known. However, it seems now of no use, and it is also decided not to enter Stor Fiord for fresh meat, as it is a very dangerous rocky coast there and with this young ice forming the game is hardly worth the candle. We got back to dinner and a boat was afterwards sent away for a load of driftwood. No results were got by dredging, except a fine live specimen of a sponge. The weather is lovely and calm, which is very lucky, as we have no hold with our anchor, which is merely dragging about on the shelves of rock. There was a beautiful sunset tonight.

September 16, Tuesday.

"Come brail up your main topgallants, my jolly boys.
For the wintry star I see.
And it's time for to leave these barren shores.
And for England to bear away, brave boys.
And for England &c".
(Old whaling song).

It is alas our last morning in Spitzbergen. We had intended exploring the pack edge for a day or two towards Nova Zemlya, but its southerly position as seen yesterday from Hope Island, and the prevalence of young ice seem to make it only a waste of time. All hands called at 2 a.m. in the first shimmering twilight, we get up the anchor and steam along the island, running a base for the rough survey of it as we go. There is a fine sunrise and we determine the variation of the compass by it. We steam gaily along the shores of the island in calm water, its cliffs with the strata so plainly showing bearing the same character throughout. It is about 20 miles long, and at its southern extremity the hills rise to a height of 1,400 feet ending in a bold bluff cliff. This is astern of us by the time of the second edition of breakfast and we take thence a departure and, at the same time, our last look at the distant shores of Spitzbergen. In a few hours, we have a light, favourable breeze and come into a rolling sea, which we must expect all our homeward passage. The wind is easterly and as we have "bore up" we steer S by W so as to pass near Bear or Cherie Island, which lies midway between Spitzbergen and Norway [and] a great resort for seahorse in the days of yore and from that cause productive of much profit to the old English "Muscovy Company." The seahorses have long since deserted it, except perhaps in winter, but an adventurous skipper of Tromso [Tromsø], not many years since, persuaded a crew to winter there in hopes of great *chasses*, but unluckily not a vestige of a seahorse did they see to relieve the dreariness of their sojourn. It and a rock off it called Gull Island [Måkeholmen] are a great place for sea fowl, and it seems to promise well in minerals, as lead ore and lead glance have been found there and also specimens of Virgin silver. At its northern end are three mountain peaks and then lowland connects these with the southern portion, from which rises a mountain called Mount Misery [Miseryfjellet].

Our "wee foxie" is adrift. He is certainly the most untameable little buck I ever saw, and only comes out of his cask when all is quiet. He has for some time been rather subdued, as a few days ago, when disporting himself on the taffrail rail, one of the mizzen guy ropes caught him and swept him overboard, and a boat had to be lowered for his rescue. Gradually, however, he recovered his boldness. He is not, I think, afraid of Rover, and barks and growls at him whenever he comes near him. But for the "Burgies" he is no match. They will stand near him, opening their immense maws to their fullest extent, keeping them open close to him, until he moves, and then pecking him. The other day he came down the companion and took refuge underneath my berth and from his vantage point among my clean clothes snapped and snarled and resented very determinedly all efforts to draw him, and succeeded in biting me, 'ere I effected it successfully. Now he had taken up his abode among the casks in the hold and defies all attempts to get him out and, if he lives, we may expect not to see or at any rate catch him before we reach Dundee. Under steam and canvas, we make from six to seven knots all day. Everyone is very busy on deck, stowing gear &c, but below, to our regret, though we expected it, we are at a standstill, as one can just about manage to write at times and no more, so that our packing, arranging specimens, observations &c has all to be left to its own devices. We already feel a difference in the temperature now we are so far from the ice, and the water is also much warmer on the surface. We take a good many soundings today and find this a very shallow sea. At night there is a fine aurora in the south and west, but we expect to see better ones as we get farther south. We are now making for the Greenwich meridian in about latitude 70° or 69° as once there, wind from the E, the N or the W, would all help us to make a passage. I could not have believed that after being so long afloat, we should have been so loathe to quit Spitzbergen. I suppose it is because we have been so unlucky in our main objects.

September 17, Wednesday.

It is a foggy morning, with a strong wind, and there has been a good deal of rain in the night. We are, at 8.30 a.m., pretty close to Cherie Island. There is a great deal of sea on, but the wind falls in the afternoon and we commence to steam. The deck is in awful confusion, as they are taking in the whaleboats, washing the lines, and stowing the gear. We try a deep-sea sounding, but the sea is too heavy to let out much line.

September 18, Thursday.

Confusion still reigns on deck. There is no place that one doesn't find oneself in the way. Foxie seems still to the good, as he has taken in a loon to his retreat. We saw a good many finners today. One came near the ship; their "blast" is very strong and high. Sea still very rough and, in spite of the boats being in, we roll a great deal. We took a sounding tonight; no bottom at 560 fathoms.

September 19, Friday.

'Til noon there is but little wind, tho' there has been plenty here of late from the state of the sea. We see a good many whales today, finners, herring whales, and bottlenoses; one of the latter was quite close to the ship for some time, and we had a very good view of him. The *Sampson* had great sport last year in killing one of them. A deep-sea sounding at 550 fathoms showed a temperature below the surface of 54½°, the surface water being only 44°. There was a display of the aurora tonight, but it was too cloudy to see it well. We also saw a good many falling stars.

September 20, Saturday.

A strong variable wind and heavy rain. Nothing to be done below but read. We see lots of these finners, and a great many skuas, some of each of the three kinds found in these regions. A sounding at noon gives a higher temperature below than at the surface of the water, as we have always found the case here, but we do not again touch such a warm current as yesterday. Tonight, it began to blow and the wind began to head us. The sea for the last few days has been running higher than we have seen it since we left England in May, and now we are getting the benefit of a lumpy cross sea, and as the sailors say, she won't try "to stand up to it." We saw some swordfish today. One that was close to the vessel seemed about 12 or 14′ long.

September 21, Sunday.

Only a light wind and steaming again. We came across both finners and herring whales, and see a large patch of blood, over which numerous mallies are squabbling, supposed to be the result of a conflict between a finner and a swordfish or thrasher. It comes on to blow 'ere night and we shorten sail. The night is clear, and the aurora very bright. It is too rough to sound today. We have, indeed, been very unlucky in this respect, as in spite of our having sufficient apparatus on board for sounding in over 1,000 fathoms, plenty of deep-sea thermometers, vessels for bringing up water from the bottom &c, we have the whole summer through when in the seas where there is sufficient depth or soundings are likely to afford results of interest and value, almost invariably met exceptionally rough weather, so that not only is the gear much endangered, but it becomes almost impossible to judge when one gets to the bottom. The wind has now begun to head us and we make but slow progress. Our fair wind has gone and we get a head sea in addition so that the decks are full of water, and there is 'ere morning the old business of a sea down the companion, and on the stewards arrival I am greeted by the sight of my boots being emptied like water cans, and find a lake all over the floor of my cabin.

September 22, Monday.

Early this morning the wind headed us and we stowed all but the fore and aft sails and commenced steaming. There is a heavy cross-sea and, the ship being no sea boat, rolled tremendously. Experienced as we ought to have become 'ere this in foul weather, a few smashes occurred, but the meals were, on the whole, triumphantly negotiated, and did immense credit to the cook and steward. It was clear tonight and we witnessed such a display of the aurora, as I have never seen before. In the W there were continually playing streams of shooting lights and ever and anon these would assume an intense brilliancy and be woven as it were into a sheet of flame. Then this would grow and unroll southward and upward and spread like a curtain of gleaming dancing quivering flame, until reaching the zenith. The whole southern and western part of the dome of the heavens were a mass of light. Then it would fall until the curtain reached the eastern horizon. Then as it paled away, instead of one sheet, you saw endless bright and glancing streamers waning now increasing in brilliancy shoot in all directions, sometimes all converging at others diverging and so on with bands and horizontal wreaths and curved bars of light all over the heavens lighting up the long dark sea, and the mass of white crested surges. The light was of an unusually bright yellow, the usual displays that I have seen showing more red. As is usually the case, the more vivid displays only a last few minutes. The strong gale that is blowing seems to produce more than an ordinary vibration in the ever-moving streams of light. The steam is found of little use against the heavy head and cross-seas we are continually meeting, as the screw is jerked continually in and out of the water. We are now coming into the latitude of shipping and carry our lights at night, as we might at any time come across an Archangel ship.

We did indeed see our first ship today.

September 23, Tuesday.

Still blowing hard from the west, and we meet a very heavy cross-sea. We commenced to steam this afternoon, but the ship is so tender and jumps so that we had to give it up in a few hours as the screw was so much out of water as in a more unpleasant noise or annoying sensation, I don't know. Ever since we bore up, the sea at night has been very phosphorescent. Tonight, it is certainly a grand sight: a ship ploughing along a foaming sea plunging her bows into a wave of fire at every jump and leaving a long bright trail behind. A dark cloudy night and nothing but the long black waves in sight, but as the fierce gale drives their crests away in spray, each drop is a little globe of fire, so that for a long distance one can see the same light playing wildly over the troubled waters and around the ship itself it is, of course, most intense, leaping up in shower after shower as the cross seas catch her. Saw one or two finners today and some swordfish. We had again to let the steam run down as the knocking about was so heavy. I started counting the revolutions of the engines and once in 50 revolutions, 30 were made out of the water. Saw a bark today bound SE.

September 24, Wednesday.

A great deal calmer as we are now under the lee of Shetland. It is a very warm soft day, and we begin to enjoy the mild weather, though of late we as well as many of the crew have suffered a great deal from the heat (never yet up to 60°) and the sort of muggy heaviness of the atmosphere after the pure Arctic air. Our barometers, all four of them, have behaved disgracefully, standing at an unparalleled height (as regards this voyage) for some day's past, in spite of all the foul weather. We start stowing and packing things today. The night is clear and starlight and we linger long on deck, enjoying the calm soft night, and still more the relief of being able to sit and stand and walk on deck without holding on to something, after our long buffeting. The phosphorescence is very clear tonight, and one can plainly see the revolving blades of the fan and the rudder at night. We see a seal today, our last I suppose. We are now, too, among different birds, having seen solan geese two days ago, the common or English skua, and other birds not met with in the North. We saw several ships today, and one smack passed close to us bowling along gaily.

September 25, Thursday.

A stiff southerly breeze springs up and we find ourselves in a heavy sea again, right ahead, so we make very slow progress. At noon, we are about 65 miles from Kinnaird Head. Luckily the wind goes down towards evening and we make better way. Tremendous scrubbings and paintings go on, but thank goodness, there is no more paint to pay things over with, but they manage to grease all the unpainted woodwork. Our chronometers seem to have gone well, as our position by observations with them corresponds very well with the soundings we get.

September 26, Friday.

Before I turned in last night, we sighted Kinnaird Head, or rather the lights. Early in the morning, we first sighted the land near Buchan Ness. It has, however, all day been rather hazy. We come pretty close in shore, and very tame the coast looks: no snow, no sharp hills, and fields, houses, and trees look very strange. We see, of course, endless shipping, and pass by Aberdeen, Gourdon, Montrose, Stonehaven and many other places of interest; one a fine old castle. The tide helps us down and at 7 p.m., we sighted the Tay lights. Then we had a flare up for a pilot and about 9 p.m. entered the river [Tay], a strong ebb tide running against us that made it a very "dour" steam. Stopped to be searched at Broughty Ferry, and then steamed on and came to anchor [at Dundee, alongside the quay on the north side].

End of Volume 3 and diary.

Appendix 1: Spitzbergen/Svalbard Place Names, 1873 and Current

Chermside, 1873	*Modern*
Advent Bay	Adventfjorden
Agardh's Bay	Agardhbukta
Aldert Dirkses Lakes	Dirksbukta
Amsterdam Island	Amsterdamøya
Angelins Mountain	Angelinberget
Augusta Bay	Augustabukta
Auk Head	Alkefjellet
Axel Island	Akseløya
Barents Land	Barentsøya
Bastion Islands	Bastionøyane
Bear/Cherry/Cherie Island	Bjornøya
Bel Sound	Bellsund
Beverly Bay	Beverlysundet
Biscayers Hook	Biskayerhuken
Black Mountain	Svartberget
Blomstrand Harbour	Blomstrandhamna
Brandy Bay	Brennevinsfjorden
Cape Atker	?, likely Kapp Payer
Cape Barth	Kapp Barth
Cape Fanshawe	Kapp Fanshawe
Cape Foster	Fosterneset
Cape Selander	Selanderneset
Cape Linderman	Lindemanodden
Cape Smyth/Leigh Smith	Kapp Leigh Smith
Cape Mohn	Kapp Mohn
Cape Payer	Kapp Payer
Cape Petermann	Kapp Petermann
Cape Platen	Kapp Platen
Cape Selander	Selanderneset
Cape Torrell	Torellneset

(continued)

Chermside, 1873	*Modern*
Cape Wrede	Kapp Wrede
Carl Island	Carlsenøya
Castren/'s Island	Castrénøyane
Coal Harbour	Colesbukta
Cloven Cliff	Klovningen
Cross Bay	Krossfjorden
Danes Gat	Danskegattet
Danes Hand	Danskeneset
Dane's Island	Danskøya
Diana Island	n/a
Dickson Bay	Dicksonfjorden
Dodmansoren (Dead Man's Ears) Lexfjellet	
Duym Point	Tommelpynten
East Fiord	Austfjorden
Edge Island	Edgeøya
English Cove/Bay	Engelskbukta
Eol's Cross	Eoluskrossen
Fair Foreland	Fuglehuken
Fair Haven	Fairhaven
Foot Island	Foottøya
Fosters Islands	Fosterøyane
Foul Bay	Fuglefjorden
Gillie's/Gillis Land	Kvitøya
Green Harbour	Grønfjorden
Grey/Gray Hook	Gråhuken
Gryteberget	Gryteberget
Gull Island	Måkeholmen
Hakluyts Headland	Hakluythovden
Hamburger Bay	Hamburgbukta
Hecla Cove	Heclahamna

(continued)

P. J. Capelotti (ed.), *The Coldest Coast*, Historical Geography and Geosciences,
https://doi.org/10.1007/978-3-030-67880-7

Chermside, 1873	*Modern*
Hedgehog's mountain	Hedgehogfjellet
Helis Sound	Heleysundet
Hinlopen Straits	Hinlopenstretet
Hope Island/Sea Horse Island	Hopen
Horn Sound	Hornsund
Hyperite Island	Gyldénøyane
Ice Cape	Isrundingen
Ice Fiord/Ice Sound	Isfjorden
Karl Alexander Island	Karl Alexanderøya
King Charles Land	Kong Karls Land
King Fredericks Island	?, likely Langeøya
King John's Glacier	Kong Johans Bre
King's Bay	Kongsfjorden
Kobbe's Bay	Kobbefjorden
Liefde, Liefde Bay	Liefdefjorden
Little Table Island	Vesle Tavleøya
Loom Head	Lomfjordhalvøya
Loome Bay	Lomfjorden
Lordragers Fiord	Zordragerfjorden
Low Island	Lågøya
Magdalena Bay	Magdalenefjorden
Martin's Island	Martinsøya
Middle Hook	Kaldneset
Moffen Island	Moffen
Moss Island	Moseøya
Mossel Bay	Mosselbukta
Mount Middendorff	Middendorffberget
Mount Misery	Miseryfjellet
Murchison Bay	Murchisonfjorden
Nelson's Island	Nelsonøya
Nord Ost/East Land	Nordaustlandet
North Cape	Norkapp
Norway Islands, Outer and Inner	Yttre og Indre Norskøyane
Parry's Island	Parryøya
Phipps/Phipp's Island	Phippsøya

(continued)

Chermside, 1873	*Modern*
Prince Charles/Charlie's Foreland	Prins Karls Forland
Quoad Hook	Kvadehuken
Red Bay	Raudfjorden
Red Beach	Raudstranda
Ross's Island	Rossøya
Ryk Ys/Witch Islands	Ryke Yseøyane
Russian Gat	[?] possibly Linnévatnet
Sassen Bay/Fiord	Sassenfjorden
Seven Ice Mountains	Dei Sju Isfjella
Seven Islands	Sjuøyane
Shoal Point	Langgrunnodden
Smeerenburg	Smeerenburgfjorden
South Cape	Sørkapp
South Gat	Sørgattet
Spitzbergen	Svalbard
Stans Foreland	Edgeøya
Stor Fiord	Storfjorden
Swedish Foreland	Svenskøya
Table Island	Tavleøya
Three Crowns	Tre Kroner
Thousand Islands	Tusenøyane
Thumb Point	Tumlingodden
Treurenberg Bay	Sorgfjorden
Tymens Straits	Freemansundet
Ulve Bay	Ulvebukta
Unicorn Bay	Einhyrningbukta
Verlegen Hook	Verlegenhuken
Vibe Bay	Vibebukta
Vogel Sang	Fuglesangen
Wahlenberg Bay	Wahlenbergfjorden
Wahlberg's Island	Wahlbergøya
Walden Island	Waldenøya
Welcome Point	Velkomstpynten
West Fiord	Vestfjorden
Wood	Woodfjorden
Wüde Bay	Wijdefjorden